Behavioural Responses to a Changing World

Behavioural Responses to a Changing World

Mechanisms and Consequences

EDITED BY

Ulrika Candolin
Department of Biosciences, University of Helsinki, Finland
and
Bob B.M. Wong
School of Biological Sciences, Monash University, Australia

Behavioural Responses to a Changing World. First Edition. Edited by Ulrika Candolin and Bob B.M. Wong.
© 2012 Oxford University Press. Published 2012 by Oxford University Press.

OXFORD
UNIVERSITY PRESS

Great Clarendon Street, Oxford OX2 6DP,
United Kingdom

Oxford University Press is a department of the University of Oxford.
It furthers the University's objective of excellence in research, scholarship,
and education by publishing worldwide. Oxford is a registered trade mark of
Oxford University Press in the UK and in certain other countries

© Oxford University Press 2012

The moral rights of the authors have been asserted

First Edition published in 2012

Impression: 1

All rights reserved. No part of this publication may be reproduced, stored in
a retrieval system, or transmitted, in any form or by any means, without the
prior permission in writing of Oxford University Press, or as expressly permitted
by law, by licence or under terms agreed with the appropriate reprographics
rights organization. Enquiries concerning reproduction outside the scope of the
above should be sent to the Rights Department, Oxford University Press, at the
address above

You must not circulate this work in any other form
and you must impose this same condition on any acquirer

British Library Cataloguing in Publication Data

Data available

Library of Congress Cataloging in Publication Data

Library of Congress Control Number: 2012933275

ISBN 978–0–19–960256–8 (hbk.)
    978–0–19–960257–5 (pbk.)

Printed and bound by
CPI Group (UK) Ltd, Croydon, CR0 4YY

Links to third party websites are provided by Oxford in good faith and for
information only. Oxford disclaims any responsibility for the materials contained
in any third party website referenced in this work.

UC dedicates this book to her daughters, Erika and Heidi, whom she loves more than words can say.

BBMW dedicates this book to his parents, Steven and Lesley, who supported his interests and nurtured his dreams.

Contents

Foreword xiii
Nicholas B. Davies

Preface xv
Acknowledgements xvii
List of Contributors xviii

Part I Mechanisms

1. **Understanding behavioural responses and their consequences** 3
   Andrés López-Sepulcre and Hanna Kokko

   1.1 Introduction 3
   1.2 What causes changes in the average behavioural phenotype of populations? 4
      1.2.1 Covariance between trait and fitness: viability and fertility selection 6
      1.2.2 Between and within individual variation 7
   1.3 When does behaviour change adaptively? 8
   1.4 Demography as a cause and consequence of behavioural adaptation 10
      1.4.1 Does adaptation always enhance persistence? No 11
   1.5 Conclusions: beyond changes in the population mean of a behaviour 12

2. **Environmental disturbance and animal communication** 16
   Gil G. Rosenthal and Devi Stuart-Fox

   2.1 Introduction 16
   2.2 Signal production 18
      2.2.1 Acoustic signals 18
      2.2.2 Visual signals 19
      2.2.3 Chemical signals 19
      2.2.4 Signals acquired from the human environment 20
      2.2.5 Matching signals to altered habitats 20
   2.3 Signal transmission 21
      2.3.1 Acoustic signals 21
      2.3.2 Visual signals 22
      2.3.3 Chemical signals 24
   2.4 Signal detection 25
   2.5 Population-level and evolutionary effects on signals 26
   2.6 Conclusions 27

3. The endocrine system: can homeostasis be maintained in a changing world? 32
Katherine L. Buchanan and Jesko Partecke

    3.1 Introduction 32
    3.2 The endocrine system 32
    3.3 Environmental disruption of the endocrine response 33
    3.4 Photoperiodism and climate change 35
    3.5 Urbanization and its ecological effects 36
    3.6 What do we know about the effects of urbanization on hormonal responses? 36
    3.7 Chemical pollution and endocrine disruption 37
    3.8 Conclusions 41

4. Experience and learning in changing environments 46
Culum Brown

    4.1 Introduction 46
    4.2 Learning and its role in the development of behaviour 46
    4.3 Social learning 48
    4.4 Interaction between innate and learnt responses 49
    4.5 Costs associated with learning 50
    4.6 Learning and evolution 51
    4.7 Learned responses to human induced environmental variation 52
        4.7.1 Learned responses to urbanization 54
        4.7.2 Learned responses to climate change 55
    4.8 Conclusions 57

Part II Responses

5. Dispersal 63
Alexis S. Chaine and Jean Clobert

    5.1 Introduction 63
    5.2 Dispersal: a balance of costs and benefits 64
    5.3 Dispersal is a plastic behaviour 65
    5.4 Acquisition of information 67
    5.5 Dispersal in a changing landscape 69
        5.5.1 Habitat quality 69
        5.5.2 Habitat fragmentation 71
        5.5.3 Dispersal as a mechanism for invasion and range shifts 74
        5.5.4 Ecological traps 74
    5.6 Conclusions 75

6. Migration 80
Phillip Gienapp

    6.1 What is migration? 80
    6.2 Environmental change and migration 81
    6.3 Migration time and fitness 82
        6.3.1 Migration time and fitness in birds 82
        6.3.2 Migration time and fitness in salmon 83

		6.4	Effects of climate change on migration time	84
			6.4.1 Birds	84
			6.4.2 Salmon	86
		6.5	Climate change and migration—consequences for populations	87
		6.6	Conclusions	89

7. Foraging   93
Ronald C. Ydenberg and Herbert H.T. Prins

	7.1	Introduction	93
	7.2	Effects of changes in food on foraging behaviour	94
	7.3	Effects of changes in predation danger on foraging behaviour	97
	7.4	Consequences for populations	99
	7.5	Consequences for communities and biodiversity	101
	7.6	Behaviour as a diagnosis tool	102
	7.7	Conclusion	103

8. Reproductive behaviour   106
Anders Pape Møller

	8.1	Introduction	106
	8.2	Domestication and its effects on reproductive behaviour	107
		8.2.1 Domestication and reproductive behaviour	107
		8.2.2 Domestication and changes in mate choice and mating behaviour	109
		8.2.3 Domestication and changes in parental effort and parental care	109
	8.3	Urbanization and its effects on reproductive behaviour	109
		8.3.1 Urbanization and reproductive behaviour	109
		8.3.2 Changes in fear responses due to urbanization	110
		8.3.3 Urbanization and changes in timing and duration of reproductive seasons	111
		8.3.4 Urbanization and changes in life history strategies	111
	8.4	Global change and its effects on reproductive behaviour	112
		8.4.1 Global change and behaviour	112
		8.4.2 Changes in singing behaviour in response to climate change	112
		8.4.3 Changes in intensity of sexual selection and climate	113
		8.4.4 Changes in infanticidal behaviour and climate change	114
		8.4.5 Changes in human harvesting and composition of animal populations	115
	8.5	Synthesis	115
	8.6	Future prospects for research	115

9. Social behaviour   119
Daniel T. Blumstein

	9.1	Introduction	119
	9.2	What environmental factors might influence sociality and how do humans impact them?	120
	9.3	Adaptive social behaviour has demographic consequences	122

	9.4 Individual based models link environmental drivers with demographic outcomes	124
	9.5 Possible consequences in the Anthropocene	125
	9.6 Prospectus	126

10. Species interactions 129
Shelley E.R. Hoover and Jason M. Tylianakis

10.1	Introduction	129
	10.1.1 General mechanisms of impact	130
	10.1.2 Range shifts	131
	10.1.3 Temporal shifts	131
	10.1.4 Ontogenetic changes	132
	10.1.5 Altered behaviour	132
10.2	Effects of GEC on different types of behavioural interactions	132
	10.2.1 Mutualisms	132
	10.2.2 Competition	133
	10.2.3 Parasitism/pathogens	134
	10.2.4 Consumer–resource interactions (predation and herbivory)	134
10.3	Consequences of network architecture for the effects of GEC on species interactions	136
10.4	Interactive effects of multiple drivers on species interactions	138
10.5	Conclusions	139

Part III Implications

11. Behavioural plasticity and environmental change 145
Josh Van Buskirk

11.1	Introduction	145
	11.1.1 The special role of behavioural plasticity	146
	11.1.2 Potential fitness effects of behavioural plasticity	147
11.2	Assessing the fitness consequences of behavioural plasticity	149
	11.2.1 Optimal plasticity	149
	11.2.2 Beneficial plasticity	150
	11.2.3 Maladaptive plasticity	152
11.3	Outlook	154

12. Population consequences of individual variation in behaviour 159
Fanie Pelletier and Dany Garant

12.1	Introduction	159
12.2	Should we expect a link between behaviour and population dynamics?	161
12.3	Whose behaviour might be more likely to affect population dynamics?	162
12.4	From the population to the individual level	163
12.5	From the individual to the population level	166
12.6	Is there potential for feedback between behaviour and population dynamics?	168
12.7	Concluding remarks and future directions	171

13. Ecosystem consequences of behavioural plasticity and contemporary evolution 175
Eric P. Palkovacs and Christopher M. Dalton

   13.1 Introduction 175
   13.2 Behavioural effects on ecosystems 176
      13.2.1 Consumption 176
      13.2.2 Nutrient cycling 178
   13.3 Rapid behavioural trait change 179
      13.3.1 Behavioural plasticity 179
      13.3.2 Contemporary evolution 180
   13.4 Reaction norms and ecosystem effects 183
   13.5 Conclusions 186

14. The role of behavioural variation in the invasion of new areas 190
Ben L. Phillips and Andrew V. Suarez

   14.1 Introduction 190
   14.2 Behaviours influencing the process of spread 191
      14.2.1 The mechanics of spread 191
      14.2.2 Dispersal behaviour during spread 192
      14.2.3 Behaviour and population growth during spread 192
   14.3 The effect of behavioural variation on spread 193
      14.3.1 Plastic responses 193
      14.3.2 Evolved responses 194
   14.4 Behavioural variation and the impacts of invasive species on natives 195
   14.5 Conclusion and future directions 196

15. Sexual selection in changing environments: consequences for individuals and populations 201
Ulrika Candolin and Bob Wong

   15.1 The importance of sexual selection 201
      15.1.1 Population-level consequences 202
   15.2 Sexual selection and the environment 202
   15.3 Consequences of environmental change 204
      15.3.1 Resource allocation and trade-offs 204
      15.3.2 Interactions among sexually selected traits 205
      15.3.3 Honesty of behavioural displays 206
      15.3.4 Impacts on population dynamics and selection processes 207
   15.4 How can animals respond? 208
      15.4.1 Phenotypic adjustment of behaviour 208
      15.4.2 Genetic changes 210
      15.4.3 Population responses 210
   15.5 What next? 212
      15.5.1 Taking account of the complexity of environmental change 212
      15.5.2 Multiple signals and multiple sensory modalities 212
      15.5.3 Is population rescue possible? 213

16. Evolutionary rescue under environmental change? 216
Rowan D.H. Barrett and Andrew P. Hendry

16.1 Introduction 216
16.2 Key questions 217
    16.2.1 How important is genetic (as opposed to plastic) change? 217
    16.2.2 Will plasticity evolve? 219
    16.2.3 Is evolution fast enough? 220
    16.2.4 Standing genetic variation versus new mutations? 222
    16.2.5 How many genes and of what effect? 223
16.3 Constraints on evolutionary responses to environmental change 225
    16.3.1 Limited genetic variation 225
    16.3.2 Trait correlations 226
    16.3.3 Ultimate constraints 227
16.4 Conclusions 228

17. Ecotourism, wildlife management, and behavioural biologists: changing minds for conservation 234
Richard Buchholz and Edward M. Hanlon

17.1 Introduction 234
17.2 Anthropogenic behavioural disturbance of wildlife 235
17.3 Is behavioural change bad? 236
17.4 What is conservation behaviour and how can it help? 237
17.5 Recent literature in recreational disturbance of wildlife 238
    17.5.1 Conservation behaviour and the wildlife disturbance literature 239
    17.5.2 Methodological problems in the wildlife disturbance literature 239
    17.5.3 Wildlife disturbance science and conserving biodiversity 244
17.6 Conclusions 245

Index 251

Foreword

Nicholas B. Davies

Our generation will surely be the last to take the natural world for granted. When I was a young boy, I thought there would always be skylarks singing and cuckoos calling to greet the spring, and on hot summer days swifts would forever scythe the skies in search of their insect prey. But populations of these and many other familiar species have declined alarmingly during my lifetime. What can behavioural ecologists do to help?

This inspiring book makes a powerful case that we can contribute through a better understanding of how organisms adapt their behaviour to a rapidly changing world. A change in behaviour is often the first response. Sometimes an individual's flexible behaviour (for example: earlier breeding, hiding from predators) is sufficient to adapt it to new conditions. Or a behavioural change will expose the organism to new selection pressures and hence promote the genetic changes necessary for adaptation. So behaviour can often play the lead in evolution.

Organisms have evolved in response to environmental change ever since life began, with changes on the grand scale of shifting continents and ice ages. For thousands of years they have been faced with human-induced changes, too, as our ancestors cut trees, burnt grasslands, or flooded and drained the land. Indeed, skylarks were likely beneficiaries of some of these changes. But the current scale and pace of change is unprecedented, involving: climate change, habitat destruction and fragmentation, ever more intensive farming and fishing, urbanization, and a new biotic environment of invasive species, pathogens and parasites. Can organisms adapt fast enough to avoid extinction?

Many chapters describe examples of rapid behavioural changes to suit the new conditions. Birds, frogs, and whales have adapted their vocal signals for more efficient transmission in a noisy human environment on land and in the seas. Some bird populations, like the blackcap, have changed their migration routes in response to milder temperate winters; their shorter journey to western Europe rather than tropical Africa not only saves on the stress of a long journey but also enables them to arrive on their breeding grounds sooner, to claim the best breeding territories. Furthermore, different arrival times on the breeding grounds also leads to assortative mating by wintering area, and hence restricted gene flow, which has likely contributed to the rapid evolution of the new migration behaviour.

However, in other cases species are suffering in a new world. Changes in water chemistry are impairing the responses of fish to alarm and sex pheromones. Increased water turbidity is obscuring fish visual signals and leading to increased hybridization. Some migrants now arrive too late on their breeding grounds to catch the earlier springs, perhaps because there are no corresponding environmental changes in winter quarters to cue an earlier departure. In Greenland, the migration of caribou has not advanced sufficiently to match the earlier spring plant growth on the calving grounds, so reproductive success has declined. Some species are caught in 'ecological traps', because the stimuli used to guide their behaviour are no longer reliable cues to habitat quality; so some birds are settling in non-native vegetation with poor food and insects are ovipositing on tarmac roads, fooled by the reflective surface.

The book shows that changes in behaviour can also lead to unexpected broad scale community changes. With the return of wolves, elk have become more cautious in feeding close to cover, so thickets are regenerating and affecting populations of other species, too. Range shifts in response to climate change are exposing communities to new predators (shell-breaking crabs in Antarctica, for example) and hence influencing food webs. Our agricultural monocultures are affecting bee behaviour and hence pollination in the wider countryside.

In many ways, this volume is a cry for help. Few studies have identified whether the changes in behaviour reflect genetic change or phenotypic plasticity. If the latter, then is the behavioural repertoire sufficient to adapt to the full range of environmental change? Some studies have shown that species with more flexible behaviour, and with larger brains, are better able to cope with novel environments. Nevertheless, long-term studies are revealing some surprising differences even between populations of the same species. Some populations of great tits, for example, have advanced their breeding entirely through phenotypic plasticity, to keep track of earlier spring food supplies, while others have not adapted and their populations are declining. Why do populations differ in their responses? The book points to the need for new theory to identify whether evolutionary change can be sufficiently rapid for populations to avoid extinction. In some cases, strong selection on individuals to maximize their fitness in changing environments can lead to a lower carrying capacity for the habitat, and hence drive a population closer to extinction.

The book also raises many new questions. As the environment changes, will individuals simply disperse to search for their old habitats or will they stay and adapt? What cues will they use to determine whether they leave or stay? Answering these questions will influence whether we try to conserve species by land-sharing, namely getting biodiversity into our fields, or land sparing, namely keeping biodiversity and our crops separate, with corridors to aid dispersal. The plea in the closing chapters is for behavioural ecologists to join more in conservation efforts to help save our natural world in the face of change, so there will continue to be skylarks and swifts in our skies, both for their own sakes and to inspire our future generations.

Preface

Humans have left an indelible mark on the planet. From the Arctic tundra to the desert outback of central Australia, the reach of human activities has touched even the most remote places on Earth. The changes entrained by such activities are having a profound impact on the natural world. For animals, survival in rapidly changing environments comes down to three options: disperse, adjust through phenotypic plasticity, or adapt through genetic changes. Although environmental changes have been taking place long before the arrival of humans, changes linked to anthropogenic activities are resulting in conditions that many species have never before encountered. Worsening the situation, evolutionary processes are seldom able to keep pace with the sudden ecological changes that humans are causing. Instead, the survival of populations—and ultimately, species—hinges on the plasticity of traits that have evolved under past conditions. The faster the changes are, and the more the conditions differ from those experienced during a species' evolutionary past, the greater the risk of population decline and, in the worst case scenario, extinction. Here, behavioural responses can play an important role in helping individuals to rapidly adjust to new conditions, and to survive and reproduce in the altered environment.

Behavioural adjustments often represent the first response to changing conditions. A bird, for instance, may adjust its vigilance in response to the presence of humans, or a butterfly may have to move to a different patch in search of host plants for laying its eggs. With such responses, animals attempt to increase their probability of survival and reproduction in the changing environment. In addition to these direct responses, environmental changes can also affect behaviours by interfering, for example, with the sensory systems or physiological processes needed to mount an appropriate response. The behavioural alterations that follow (if any) can be adaptive or maladaptive, depending on how they influence fitness. If the responses of individuals alter key demographic parameters (e.g. birth, death, or migration rates of the population) then the dynamics of the population will also change—sometimes for the better; other times not. Changes in the demography of one species can, in turn, influence others and, eventually, the whole community to which it belongs. This can result in further changes to the environment through feedback loops that can, ultimately, impact the entire ecosystem.

Behavioural responses can also have important evolutionary consequences. Responses that help counter drastic population declines can give the population additional time for accruing genetic changes (evolutionary rescue). This is particularly crucial when the behavioural response does not fully rescue the population, and genetic adaptation is required for persistence in the longer term. On the other hand, changed conditions that differ drastically from those experienced by populations during the course of their evolutionary history can constitute major obstacles to persistence that are unlikely to be surmounted by behavioural responses alone. They can also trigger maladaptive behavioural decisions. These so called 'evolutionary traps' can be quite common under human-altered conditions, potentially driving populations into decline.

This book aims to provide insights into the behavioural responses of animals to human-induced environmental change and how—by impacting on ecological and evolutionary processes—such responses can affect the fate of individuals, populations, and ecosystems.

The book is organized into three interrelated parts. Part 1 focuses on the mechanisms underlying behaviour. It discusses how behavioural responses are dependent on the environment, and provide an important context for understanding how anthropogenic changes can modify the way in which animals respond. This section begins with a theoretical framework for understanding how environmental change can affect behaviour at the population level. It then considers the impact of environmental change on animal communication and the endocrine system, as well as the role of experience and learning as potential mechanisms for coping with human-altered conditions. Part 2 explores behavioural patterns and processes under anthropogenic change, including dispersal, migration, foraging, reproduction, social behaviour, and species interactions. Part 3 considers the implications of behavioural responses for populations, ecosystems, and biodiversity. This section begins by exploring the potential role of behavioural plasticity in changing environments and then discusses the impacts of altered behaviours on population dynamics and ecosystem function, as well as the effects of invasive species. Evolutionary implications are further explored in the context of sexual and natural selection, and the potential for plastic and evolutionary responses to rescue populations from decline. The section concludes with a discussion of the importance of behavioural research in conservation science and the role that behavioural scientists can play in providing insights into the impact of anthropogenic activities.

Acknowledgements

This book began as an idea conceived in Ithaca during the International Society for Behavioral Ecology Congress. We mulled over it during the poster sessions, discussed it on our coffee breaks, and by the end of the conference dinner had decided that this was a book that needed to be written.

Since that fateful meeting, we have had the opportunity to work with an amazing bunch of people to bring our idea into fruition.

We begin by acknowledging the efforts of our contributors for their enthusiasm, dedication, and hard work.

A special thank you to the many reviewers who provided insightful, timely, and constructive feedback on chapter drafts.

We thank the team at OUP, particularly Ian Sherman for persuading us that we were up to the task and Helen Eaton for making sure we could deliver.

We are grateful to the staff at the Tvärminne Zoological Station for feeding us during our stay and for providing a tranquil environment for putting together a book. BBMW wishes to thank Topi, Lasse, and Pirjo Lehtonen for their hospitality in Helsinki, Turku and Tampere.

Lastly, thank you to our families and friends for your constant encouragement, love, understanding, and support. We could not have survived this without you.

Ulrika Candolin
Helsinki, Finland
Bob B.M. Wong
Melbourne, Australia

List of Contributors

Rowan D.H. Barrett, Department of Organismic and Evolutionary Biology, Harvard University, 26 Oxford Street, Cambridge, MA 02138, USA
rbarrett@fas.harvard.edu

Daniel T. Blumstein, Department of Ecology and Evolutionary Biology, University of California, 621 Young Drive South, Los Angeles, CA 90095-1606, USA
marmots@ucla.edu

Culum Brown, Department of Biological Sciences, Macquarie University, Sydney 2109, Australia
culum.brown@mq.edu.au

Katherine L. Buchanan, School of Life and Environmental Sciences, Deakin University, Geelong 3217 Victoria, Australia
kate.buchanan@deakin.edu.au

Richard Buchholz, Department of Biology, University of Mississippi, PO Box 1848, University, MS 38677-1848, USA
byrb@olemiss.edu

Ulrika Candolin, Department of Biosciences, PO Box 65, University of Helsinki, 00014 Helsinki, Finland
ulrika.candolin@helsinki.fi

Alexis S. Chaine, Station d'Ecologie Expérimentale du CNRS USR 2936, Moulis, 09200, France
alexis.chaine@ecoex-moulis.cnrs.fr

Jean Clobert, Station d'Ecologie Expérimentale du CNRS USR 2936, Moulis, 09200, France
jean.clobert@ecoex-moulis.cnrs.fr

Christopher M. Dalton, Department of Ecology and Evolutionary Biology, Cornell University, Ithaca, NY 14853, USA
cmd273@cornell.edu

Nicholas B. Davies, Department of Zoology, University of Cambridge, Downing Street, Cambridge CB2 3EJ, UK
nbd1000@cam.ac.uk

Dany Garant, Département de Biologie, Université de Sherbrooke, Sherbrooke, QC, J1K 2R1, Canada
dany.garant@usherbrooke.ca

Phillip Gienapp, Ecological Genetics Research Unit, Department of Biosciences, University of Helsinki, 00014 Helsinki, Finland and Institute for Evolution and Ecology, University of Tübingen, Auf der Morgenstelle 28, 72076 Tübingen, Germany
phillip.gienapp@zool.uni-tuebingen.de

Edward M. Hanlon, Department of Biology, University of Mississippi, PO Box 1848, University, MS 38677-1848, USA
emhanlon@olemiss.edu

Andrew P. Hendry, Department of Biology, McGill University, 859 Sherbrooke St. W., Montreal, Quebec, H3A 2K6 Canada
andrew.hendry@mcgill.ca

Shelley E.R. Hoover, School of Biological Sciences, University of Canterbury, Private Bag 4800, Christchurch 8140, New Zealand and Centre for High-Throughput Biology, University of British Columbia, 2125 East Mall, Vancouver, BC V6T 1Z4, Canada, and Agriculture and Agri-Food Canada, Box 29, Beaverlodge, AB, T0H 0C0, Canada
hoover.shelley@gmail.com

Hanna Kokko, Department of Ecology, Evolution and Genetics, Australian National University, Australia
hanna.kokko@anu.edu.au

Andrés López-Sepulcre, Laboratoire d'Ecologie et Evolution, CNRS UMR 7625, École Normale Supérieure, 46 rue d'Ulm, 75005 Paris, France, and Department of Ecology and Evolutionary Biology, University of Arizona, Tucson, USA
alopez@biologie.ens.fr

Anders Pape Møller, Laboratoire d'Ecologie, Systématique et Evolution, CNRS UMR 8079, Université Paris-Sud, Bâtiment 362, 91405 Orsay Cedex, France
anders.moller@u-psud.fr

Eric P. Palkovacs, Division of Marine Science and Conservation, Nicholas School of the Environment, Duke University, Beaufort, NC 28516, USA
eric.palkovacs@duke.edu

Jesko Partecke, Max Planck Institute for Ornithology, Schlossallee 2, 78315 Radolfzell, Germany
partecke@orn.mpg.de

Fanie Pelletier, Département de Biologie, Université de Sherbrooke, Sherbrooke, QC, J1K2R1, Canada
Fanie.pelletier@usherbrooke.ca

Ben L. Phillips, School of Marine and Tropical Biology, James Cook University, Townsville 4814, Australia.
ben.phillips1@jcu.edu.au

Herbert H.T. Prins, Resource Ecology Group, Wageningen University, Droevendaalsesteeg 3a, 6708 PB Wageningen, The Netherlands
herbert.prins@wur.nl

Gil G. Rosenthal, Department of Biology, Texas A&M University, 3258 TAMU, College Station, TX 77843, USA and Centro de Investigaciones Científicas de las Huastecas "Aguazarca", 392, Colonia Aguazarca, Calnali, Hidalgo 43230, México
grosenthal@bio.tamu.edu

Devi Stuart-Fox, Department of Zoology, University of Melbourne, Victoria 3010, Australia
devis@unimelb.edu.au

Andrew V. Suarez, School of Integrative Biology, University of Illinois, Urbana, IL 61801, USA
avsuarez@life.illinois.edu

Jason M. Tylianakis, School of Biological Sciences, University of Canterbury, Private Bag 4800, Christchurch 8140, New Zealand
jason.tylianakis@canterbury.ac.nz

Josh Van Buskirk, Institute of Evolutionary Biology and Environmental Studies, University of Zürich, 8057 Zürich, Switzerland
josh.vanbuskirk@ieu.uzh.ch

Bob B.M. Wong, School of Biological Sciences, Monash University, Victoria 3800, Australia
bob.wong@sci.monash.edu.au

Ronald C. Ydenberg, Resource Ecology Group, Wageningen University, Droevendaalsesteeg 3a, 6708 PB Wageningen, The Netherlands
ydenberg@sfu.ca

PART I
Mechanisms

CHAPTER 1

Understanding behavioural responses and their consequences

Andrés López-Sepulcre and Hanna Kokko

⊃ Overview

How do populations respond to environmental change? We aim to provide a conceptual overview using the Price equation, which decomposes the mean change exhibited by a population into four components: viability selection, within-individual changes over their lifetime, fecundity selection, and parent–offspring differences. Mechanisms such as phenotypic plasticity, learning, genetic adaptation, maternal effects and cultural evolution can all be understood via their influences on these components. However, we also highlight the fact that population size effects should often be considered more explicitly than this breakdown of components achieves. For example, phenotypic plasticity may help or hinder adaptive evolution, and adaptation does not necessarily lead to a better maintenance of large population size.

1.1 Introduction

Since the very inception of evolutionary theory, animal behaviour has been seen as a trait upon which selection can act. Darwin's theory of sexual selection, exposed in the *Descent of Man* (Darwin 1871), sparked some of the earliest research on the adaptive value of behaviours (e.g. Noble and Bradley 1933), yet the modern synthesis of the 1940s didn't pay much attention to traits that one would nowadays call behaviours (Birkhead and Monaghan 2010; Kokko and Jennions 2010). It took close to a century before the adaptive framework began to dominate the study of behaviour—thanks to the work of Niko Tinbergen and Konrad Lorenz (Tinbergen 1963). This solidified the link between evolutionary biology and the behavioural sciences that Charles Darwin had suggested. Like organs, behaviours represent adaptations to the environment. Tinbergen, for instance, demonstrated that the sticklebacks' fierce reaction against the colour red represents an adaptation to exclude attractive sexual competitors (Tinbergen 1963), while Lorenz studied the impulse of goslings to follow the first object they see after hatching, which ensures they remain safe with their mother and learn how to be adult geese. These landmark studies sparked decades of search for the adaptive function of different behaviours (Owens 2006), which later translated into the behavioural ecologists' modern obsession with fitness consequences of behaviour, ideally in the real ecological context. Behavioural ecology was born (see Birkhead and Monaghan 2010, Kokko and Jennions 2010).

There is, however, some irony in the images that these early studies of adaptation convey. There are famed pictures of geese, which happened to be imprinted on Konrad Lorenz's boots on hatching, courting the ethologist as if they were conspecific adults, and stories of Niko Tinbergen's sticklebacks wasting energy on aggressive displays towards the reflection of red cars which would pass by the window next to their tank. It is hard to see the adaptive value of those behaviours. Naturally, we all know that humans are rarely present when goslings hatch,

Behavioural Responses to a Changing World. First Edition. Edited by Ulrika Candolin and Bob B.M. Wong.
© 2012 Oxford University Press. Published 2012 by Oxford University Press.

and red car reflections do not represent a frequent sight for most sticklebacks. Perfectly adaptive behaviours can become maladaptive when taken out of context, and we can only expect organisms to adapt to what has been relevant for a substantial part of their evolutionary history. But history changes. In an era of massive human-induced environmental change, goose anthropophilia and stickleback paranoia are the least of our conservationist worries. While behavioural ecologists argue about the 'optimality' of behaviour (Fox and Westneat 2010; Gardner 2010; Kokko and Jennions 2010), entire species are disappearing as they fail to adapt to rapid changes in their environments. The catastrophic population consequences of island birds' inability to escape introduced predators represent a clear example (Blackburn et al. 2004). The list of catastrophic behavioural maladaptation is long. A fatal attraction to lighthouses can claim thousands of seabird lives per night (Jones and Francis 2003), human use of tactical sonars or seismic surveys appear to cause whales to strand on beaches (Weilgart 2007), and dragonflies lay eggs on the tarmac which, under their polarized vision, looks just like the best of ponds (Horváth et al. 1998).

Other organisms seem to adapt to change much better, and this might allow them to mitigate any negative population consequences, sometimes to the extent that the change proves beneficial. Trinidadian guppies *Poecilia reticulata* can evolve their escape ability upwards within a few years of changing their predatory environment (O'Steen et al. 2002). Quolls *Dasyurus hallucatus* in Australia have increased their survival by learning to avoid eating introduced toxic cane toads *Rhinella marina* (O'Donnell et al. 2010). Torresian crows *Corvus orru* have gone beyond learning and survival, and have spread—through cultural transmission—their ability to feed on cane toads by turning them on their bellies and eating their non-toxic innards (Donato and Potts 2004).

Behavioural ecologists often argue about the likely population consequences of behavioural change (or the lack of thereof) when environments are changing. If bird migration timing becomes mismatched regarding weather and the consequent peaks of food availability, will the consequences be dramatic or mild (Jones and Cresswell 2010, Chapter 6)? Will mutualistic relationships between corals and their photosynthesizing symbionts break down or will coral hosts be able to switch to zooxanthellae partners that tolerate thermal stress better (Kiers et al. 2010)? Given such a diversity of examples, how and when do we expect the behaviour of populations to adapt to a novel environment?

The intention of this chapter is to lay the conceptual framework necessary to understand how changes in behaviour occur at the population level and mention the tools we have in hand to predict it. Throughout the chapter, we make a deliberate effort to understand behaviour as a phenotypic trait that can have a genetic basis while also depending on the environment. Fitness-related behaviours will, by definition, have consequences on birth and death rates, which means they will have an impact on population dynamics. The importance of this can be illustrated by a study on Seychelles magpie robins *Copsycus sechellarum*, where competition for territories and mates was shown to strongly influence the demography and extinction risk of this endangered species (López-Sepulcre et al. 2009).

These links between fitness and demography will allow us to refer to tools of analysis common in evolutionary and population biology. Behaviours are not always like any other trait, however; they possess a high degree of plasticity derived from a diversity of mechanisms (learning, conditioning, genetically determined reaction norms, etc.). Consequently, they have the potential to change at a much faster rate than many other traits of an organism, which causes a perhaps richer set of potential evolutionary trajectories than offered by many other suites of traits.

1.2 What causes changes in the average behavioural phenotype of populations?

The behaviours that a population exhibits can be described as a set of phenotypes. In evolutionary biology, our understanding of phenotypic change can be captured by the Price equation (Price 1970), a simple representation of the necessary and sufficient conditions for inter-generational phenotypic change. Although the Price equation has no predictive power beyond one generation nor, in its simplest form, does it explicitly account for all possible mechanisms of change, it is nevertheless a good starting point to structure one's thoughts on the mechanisms responsible for phenotypic change. One of its most common formulations reads,

$$\Delta \bar{z} = cov(w_i, z_i) + E[w_i \Delta z_{i,j}]$$

In prose, the equation states that the change in the average phenotype z in a population between one generation and the next (Δz) is the sum of two quantities: the covariance between an individual's trait z_i and its fitness w_i as quantified in the parental generation (first term on the right hand side), plus the expected (mean) trait difference between parents i and offspring j (second term on the right hand side). The former change captures selection, and the latter describes the bias in the transmission of the trait from parent to offspring (i.e. how consistently different the offspring are from their parents). Note than in the latter term, the average difference between parent and offspring $\Delta z_{i,j}$ is weighed by the fitness of the parent (i.e. that difference will be represented more often in parents which sired more offspring).

For example, in a migratory bird, the very earliest arriving birds might enjoy better breeding success than they would have before the onset of climate change. If we choose to measure the arrival date as an integer of days after January 1, such that early arrivals are expressed as low values of z_i, we expect a negative covariance between z_i and fitness. This, by itself, tends to make Δz negative, predicting that birds of future generations will arrive earlier. However, it may also happen that the arrival time of an offspring has little to do with the arrival time of the parent, for example, because arrival timing is influenced by weather, rather than a genetic disposition to arrive early, and offspring experience a colder year than their parents (which makes them arrive later and thus have higher z_j). In this case, the early parents with an unusually low z_i, whose fitness w_i is high in the second term of the equation, will tend to have offspring who arrive later, hence $\Delta z_{i,j}$ is positively biased. For those parents with later arriving times, their offspring may arrive earlier than them, creating for those parents a negative $\Delta z_{i,j}$. Is $E(w_i \Delta z_{i,j})$ then zero, given that the population features both positive and negative biases $\Delta z_{i,j}$? No: because the fitness w_i of the latter type of parents is low, they have less weight on the mean, and the net effect is a positively biased $E(w_i \Delta z_{i,j})$. In other words, offspring are arriving later than they would if the covariance between fitness and trait value was the only factor at play. This outcome means that the fitness advantage of parents with early times is diluted by a low fidelity of trait transmission, and thus the phenotype does not change in an adaptive manner. The net change in arrival timing is small or null when most variation is environmentally, rather than genetically, determined.

By describing inter-generational change, this formulation of the Price equation does not explicitly account for changes in the trait within generations. Since many of the examples of interest involve organisms with overlapping generations, we need to make this explicit and rewrite the equation:

$$\Delta \bar{z} = \overbrace{\underbrace{cov(s_i, z_i)}_{\text{viability selection}} + \underbrace{E[s_i \Delta z_{i,i}]}_{\text{individual change}}}^{\text{intra-generational change}}$$
$$+ \overbrace{\underbrace{cov(r_{i+}, z_i)}_{\text{fertility selection}} + \underbrace{E[r_{i+} \Delta z_{i+,j}]}_{\text{parent-offspring differences}}}^{\text{inter-generational change}}$$

The first two terms represent changes *within* a generation (intra-generational change), which are described as the covariance between survival s and the trait (viability selection, $cov(s_i z_i)$) plus the expected change between one time step and the next in the trait values of survivors (individual change $E(s_i \Delta z_{i,i})$). The second two terms determine the change *between* generations (inter-generational change), which consists of the covariance between the reproduction of surviving individuals r_{i+} and the trait (fertility selection, $cov(r_{i+} z_i)$) and the expected parent–offspring differences among reproducing individuals (parent–offspring differences, $E(r_{i+} \Delta z_{i,j})$). It is good to check that this equation reduces to the first one for non-overlapping generations: there is simply no survival (the first two terms of inter-generational change are zero) and fitness is determined entirely by reproduction ($r_{i+} = w_i$).

Our intention in this section is not to suggest that all research conducted in the field should use the four components of Equation 1.2. However, we do find it useful to let this breakdown of components help organise thoughts on whether phenotypes will change as environments change, because any combination of mechanisms that we claim to cause a population change in a given phenotype represent a combination of those terms. Each of those terms should be accounted for when arguing about changing populations (see Fig. 1.1), which means that focusing on one is only sufficient if the others can be convincingly argued to be negligible. Considering the Price equation thus ensures that our discussion on mechanisms of behavioural change is logically complete. We now discuss the different biological mechanisms governing each component of the equation.

1.2.1 Covariance between trait and fitness: viability and fertility selection

Unsurprisingly, given the attention that behavioural ecologists pay to adaptive functions of a trait, selection has been the main focus of both theoretical and empirical studies of behavioural ecology (Owens 2006). The first and third terms of Equation 1.2, viability and fecundity selection, capture this line of thought. In principle, documenting selection is easy: individuals with 'better' traits have improved survival or reproduce more effi-

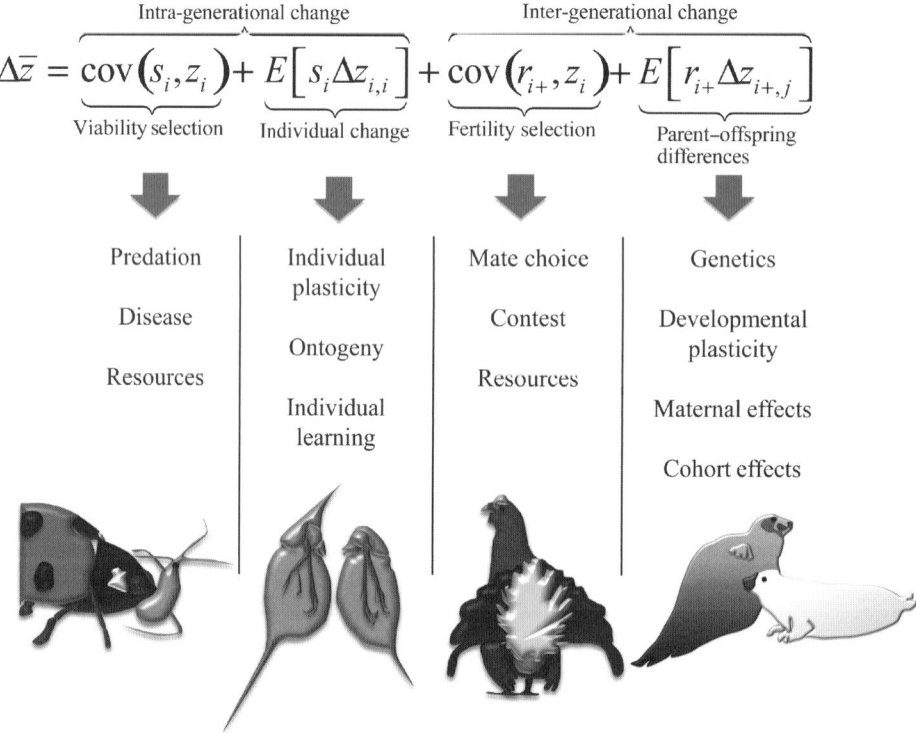

Figure 1.1 Decomposition of phenotypic change using the Price equation for overlapping generations, showing examples of mechanisms that affect each component.

ciently. It is rare, however, to see the full covariance between fitness and the phenotype calculated. Often only survival or reproduction is correlated with the trait. This can mislead, since survival and reproduction may trade off each other.

For example, in the presence of novel predators, some individuals might exhibit better antipredatory behaviour than others. All else being equal, one might expect these individuals to now have higher fitness than before the predator was introduced, but strictly speaking this argument implies that the survival advantage does not trade-off with reproduction. This would allow us to argue that the covariance between the trait of surviving individuals and reproduction is zero, but in reality, an individual constantly hiding in a refuge may not have high mating success (yet see Jennions et al. 2001 for the lack of evidence for such a trade-off). What is true is that, all else being equal, the arrival of a new predator probably does shift fitness in the direction that favours timidity more now than it did before. Parameterizing the Price equation to answer such short-term questions is, as such, not difficult, given that it merely involves the calculation of means and covariances. The difficulty lies in obtaining such comprehensive data. It is here —where one appreciates the enormous value of intensive individual-based field studies—that traits, survival, reproduction and parentage relationships are tracked (Clutton-Brock and Sheldon 2010).

It is important to highlight here the conditional nature of the reproductive term in Equation 1.2. The fertility selection component in this formulation is only the covariance between the trait of surviving individuals and their reproductive output: it does not include those individuals that do not survive (hence the notation $i+$). Survival obviously indirectly affects reproduction because dead individuals do not reproduce (in fact, when considering evolutionary change, survival is only relevant as a means to reproduction), but to avoid double accounting, this is assigned to the survival component.

1.2.2 Between and within individual variation

While behavioural ecology devotes much attention to the study of selection, the remaining two terms of Equation 1.2 are less often explored. The second term describes the degree of trait variation within individuals over their life time, while the last term describes differences between parents and offspring analogously to the second term in Equation 2.1. It is possible, but in reality not very likely, that both values equal zero. This requires that individuals do not change the behavioural trait in their entire lifetimes (e.g. they always display to mates with the same intensity), and that the trait is perfectly inherited such that parents and offspring are identical. If these conditions are not met, we must consider alternative routes to phenotypic change.

Let us first consider individual variation on the phenotype of individuals throughout their lifetimes (second term in Equation 1.2). Behavioural traits are often highly labile. If there is either adaptive or non-adaptive plasticity (reaction norms), we can expect environmentally driven changes in a population from one year to the next (Chapter 11). For example, if individuals tend to flee from predators, then years of predator abundance will see more movement than years when predators are scarce. Similarly, individual learning (see Chapter 4) or ontogenetic (developmental) change can produce differences in behaviour over time.

Such changes can be usefully classified according to their degree of reversibility. Some behaviours respond to immediate environmental conditions: fleeing from predators can function in this way. But antipredatory behaviour can also change with age or prior experience in a unidirectional way. Unidirectional changes are common among morphological traits such as the size of an organism that show either determinate or indeterminate growth. Ontogenetic changes in animal behaviour constitute one of the four big questions posed by Tinbergen (1963), and potential examples include the onset of breeding (which may be a plastic trait with respect to the social situation, Carrete et al. 2006), irreversible effects of early experience on future habitat choice (Stamps 2006), and the increase in helping behaviours at senescing ages (Richardson et al. 2007). Understanding the effect of directional developmental change on changes in the mean phenotype requires tracking the population's cohort or age structure (for a Price equation approach expanded to account for the component of age-structure, see Coulson and Tuljapurkar 2008).

Consider the most extreme case: completely reversible environmental plasticity where individual behaviour is instantly and completely determined by the environment, and the environment varies over time but is experienced in the same way by all individuals. In such a case, individual phenotypes vary over time but no individual differs from another (because they respond to environmental variations in an identical trait). Consequently, all selective components (covariances) disappear. The observable change in behaviour is now completely determined by plasticity tracking the environment (second term in Equation 1.2). Note that this scenario does not say anything about whether the behaviour is adaptive.

Finally, behavioural ecology must consider the last term in Equation 1.2, parent–offspring differences. There are several reasons why this component is important; above we already discussed parent–offspring differences as a source of variation that can counteract selection in the context of migratory bird arrival times. These differences are impacted by several different biological processes. Maternal effects can be important in determining offspring traits such as dispersal (Massot et al. 2002) and territorial behaviour (Stapley and Keogh 2005). Similarity between parents and offspring (i.e. a diminishing of parent–offspring differences) can also occur for habitat choice through imprinting on the natal habitat (Davis and Stamps 2004) or, more simply, by common environmental effects due to spatial heterogeneity and low dispersal. Social learning is another common route to similarity (as may happen with cultural evolution, e.g. Bentley et al. 2004; see also Chapter 4).

Of course, the most commonly studied mechanism of parent–offspring similarity in evolution is that which arises through genetic factors. For genetic factors to be of great importance, parent–offspring differences arising via other mechanisms should be small, leading to a dominating assumption in much of behavioural modelling that the mode of inheritance is not a constraint for evolution in the long term (Parker and Maynard Smith 1990). This is justified in models predicting broad patterns of evolutionary outcomes in the long term (e.g. to answer questions such as whether higher relatedness promotes the evolution of cooperation across species, Cornwallis et al. 2010), considering that parent–offspring similarity is high in the long run even though environmental fluctuations may swamp it when viewed over short periods of time.

However, this justification for adaptation breaks down when we are concerned with population persistence under acute environmental change. In this case, our interest shifts from studying the long-term ability of natural selection in solving problems to estimating the rate of evolutionary change over a much shorter timescale when the population is unlikely to be at evolutionary equilibrium. In such a context, the question is often: will the behavioural response be adaptive in the timescale of the environmental change, or will it be maladaptive (Gomulkiewicz and Houle 2009; Futuyma 2010)? For this, deviations from zero in the last term in the Price equation play a key role. Behaviours display heritabilities similar to other phenotypic traits (Stirling et al. 2002), but they are simultaneously susceptible to an unusually diverse plethora of mechanisms governing parent–offspring similarities, including learning, imprinting, and cultural transmission. The relative role in each factor diminishing or heightening parent–offspring differences could be fruitfully studied in a more integrative fashion.

1.3 When does behaviour change adaptively?

The breakdown of components presented above serves as a framework for understanding how different mechanisms affect a mean change in population behaviour, but it also helps us to reason whether such change will be adaptive. For a change to be adaptive it must go in the direction indicated by the selective components (i.e. the two added covariances in Equation 1.2). However, simultaneously, a population can also manage to persist in novel environments without adaptive change. For example, if all individuals are able to respond to, and flee, from a novel predator, phenotypic tracking without adaptive change can be sufficient to rescue a population.

In general, there is no guarantee that all components of the Price equation will point in the same direction (have the same sign). According to the idea of counter-gradient selection (Conover and Schultz 1995), selection and genetic change may oppose an environmental effect, leaving no net phenotypic change. A non-behavioural example of this is the orange hue of Trinidadian guppies, which shows marked differences across populations sharing the same diet, but due to variation in food availability, most differences vanish in the field (Grether et al. 2005). This example highlights that a lack of phenotypic differences does not exclude genetic population differences.

The genotype–phenotype map has an interesting relationship to selection. Selection can only act on variation that is expressed at the phenotypic level, but it cannot result in an evolutionary response unless the differences in phenotypes have a genetic basis. Hence, highly plastic traits can experience strong selection but, due to their low heritability, little evolutionary response.

Adaptive plasticity, however, has been argued to favour evolution in a different manner: it brings the phenotype closer to the adaptive peak, increasing population size and, consequently, the opportunity to display genetic variation closer to the optimum (Ghalambor et al. 2007; Fierst 2011, also see Chapter 11). In the context of learning, this is known as the Baldwin effect, whereby learning to cope with a new selective pressure allows the population and its genetic variation to persist, allowing future selection and evolution on the learned trait (Baldwin 1896; Chapter 4).

The amount of additive genetic variation present is indeed a strong predictor of how well populations manage to evolve to new adaptive peaks. Heritabilities of traits can only become high if there is ample additive genetic variation. This results in high parent–offspring similarity, and phenotypic changes under these conditions are more closely dictated by selection. Interestingly, life history theory predicts that heritabilities should only remain high for traits that are not strong predictors of fitness due to higher genetic variation (Bulmer 1989; Charmantier and Garant 2005). This prediction arises because selection in the past should have weeded out inferior genotypes if these recurrently, generation after generation, give rise to individuals displaying poor fitness; yet a review of empirical studies showed that the low heritability of fitness-related traits is due to higher environmental variation, rather then lower genetic variation (Houle 1992).

In the context of environmental change, lower heritabilities can be a worrying result: it means that those traits that matter most to how well individuals perform might be the first ones to run out of genetic variation should a changed trait value suddenly become optimal. Horizontal transmission of traits will also tend to increase parent–offspring differences and therefore slow down adaptive change (Helanterä and Uller 2010). For example, if individuals learn new foraging techniques from their peers, the underlying genetic differences in foraging ability will be smaller, and the response to selection will be weaker. Under such a scenario, genetic adaptation will not happen at the same pace as one would see without plasticity and learning.

We have, above, concentrated on the factors that facilitate or hamper genetic adaptation. In a conservation context, the hidden assumption is often that only genetic adaptation can rescue populations in the long term (see discussion on evolutionary rescue in Chapter 16). However, our earlier example of individuals fleeing from predators shows that behaviour can also change (more fleeing) without genetic evolution. Likewise, if individuals are not genetically adapted to, say, urban environments, but can compensate by learned behaviours passed on vertically and horizontally, the population is just as likely to persist. Indeed, plasticity, when not too excessive, has been shown to buffer populations against environmental variability in a way that can enhance population stability (Reed et al. 2010).

A different and graver concern is that not all change, whether genetic or non-genetic, occurs in an adaptive direction. Maladaptive behaviours might be horizontally transmitted. There is, for example, evidence that individuals may choose their habitat based on conspecific attraction. If some individuals make sub-optimal choices, these can become perpetuated (Stamps 2001; Nocera et al. 2006). Distinguishing between mechanisms of genetic and cultural inheritance in models of social evolution has been shown to yield markedly different evolutionary predictions (Lehmann et al. 2008). In the following section we will have a more detailed look at maladaptation and adaptation by making the population-level response explicit in terms of population size. This shows that while maladaptation is often bad news for the persistence of a population, fitness changes can become masked by density dependence. More paradoxically still, there are scenarios where even adaptive evolution can hamper population size.

1.4 Demography as a cause and consequence of behavioural adaptation

For all the beauty of the Price equation approach, it is not exempt from drawbacks. Population change can be decomposed in many ways, and unsurprisingly, the attention of a researcher subsequently becomes focused on what the method singles out. The equation remains true when, say, population size varies over time, but since the approach does not contain explicit terms for such changes, it does not encourage thinking about population sizes *per se* (Rice 2008). The reason to focus on this is that changes in behaviour are likely to affect demography and population size while, at the same time, population size can be an important factor influencing adaptation (Kokko and López-Sepulcre 2007).

A relatively popular avenue of theoretical work is to examine whether adaptation can proceed fast enough to avoid extinction when the environment changes to a state where the original phenotypes perform so poorly that extinction will follow unless there is evolutionary change (Chapter 16). Among such modellers, Gomulkiewicz and Holt (1995) were the first to point out that it is not sufficient to predict a deterministic trajectory where a population first declines after the environmental change occurs and then gradually bounces back (growth now occurs as the remaining individuals are those whose traits allow them to survive and reproduce in the novel environment). The reason why this is not guaranteed to avoid extinction, even if some individuals possess traits that allow them in principle to form a new source population, is that populations can be driven to very low sizes during the dangerous maladapted phase, and small populations are known to be vulnerable to extinction through demographic stochasticity (the factor considered by Gomulkiewicz and Holt 1995) as well as many other stochastic factors (Traill et al. 2010 and references therein). The milder the initial maladaptation, and the larger the initial population, the better the prospects of an evolutionary rescue (the continued persistence of the population that relies on adapting to the new environment, Gomulkiewicz and Holt 1995).

Recently, Chevin and Lande (2010) have investigated the general question of both plasticity and genetic evolution in density-regulated populations; they also assume that the degree of plasticity can either be constant or can, itself, evolve. Again, the environment is assumed to shift abruptly, and then the population either experiences evolutionary rescue or, failing to do so, goes extinct. They show what Gomulkiewicz and Holt (1995) in their Discussion already suspected: the density-independent scenario investigated by Gomulkiewicz and Holt (1995) is a best-case scenario, because density dependence tends to depress population growth. Chevin and Lande (2010) however also pointed out that if we include phenotypic plasticity, it keeps populations afloat much better than mere genetic evolution is able to.

This can be exemplified with a specific scenario. In the context of habitat choice, it has long been known that environmental change can produce 'ecological traps' (Schlaepfer et al. 2002). In a trap situation, individuals use out-dated cues of habitat quality and prefer habitats that have become worse than the (non-preferred) alternatives, or prefer to reside in places that have not previously existed (manmade habitats) and that pose unanticipated dangers (also see Chapter 5). The population dynamics of such cases are interesting. If the population remains large, then many individuals will be forced to breed in the non-preferred but safe B habitat. But if for any stochastic reason (say, a harsh winter) the population falls to low levels, then most survivors can follow their maladaptive preference for A and the entire population may become a sink—followed by extinction via an Allee effect (Kokko and Sutherland 2001). In this setting, there is clear scope for evolutionary rescue. Kokko and

Sutherland (2001) showed that either genetic adaptation or learning can rescue populations, but a very simple rule of phenotypic plasticity works much better than all alternatives. If individuals simply imprint on the type of habitat in which they were born, and preferentially breed in similar habitats, then populations almost immediately switch to near-optimal habitat use. This is because most individuals that are alive at any point in time must have been born in habitats that allow for successful breeding.

We do not know at present, however, how general such findings are. Theoretical effort on this important topic appears relatively scattered, with no systematic effort yet existing to work out precisely what kind of phenotypic change (or lack of change), or which kind of population regulation, should impair population persistence under environmental change. Simple rules of thumb may often work well: large population sizes not only buffer species against the demographic processes that cause vulnerability, but also create a more optimistic outlook for coping with new evolutionary challenges. Reflecting such principles, birds inhabiting large landmasses have been shown to have faster rates of molecular evolution than those confined to islands (Wright et al. 2009), and in the realm of microbial evolution there is even direct empirical support for the role of initial population size and genetic variation in promoting evolutionary rescue (Bell and Gonzalez 2009). The issue of evolutionary rescue is dealt with more extensively in Chapter 16.

1.4.1 Does adaptation always enhance persistence? No

There is one more point that is as intellectually exciting as it is worrying: adaptation itself might not always be in the best interest of a species when environments change. To understand this somewhat counterintuitive point, it is important to remind oneself that selection is much stronger at the level of the gene, or individual, than at higher levels. A population-detrimental behaviour can spread if there is no active policing against it, and if the behaviour causes a relative fitness advantage for its bearer (Rankin et al. 2007). Sexual conflict is a clear example (see Chapter 15), and one does not even have to think about the damaging consequences of extreme male behaviours (that sometimes kill females that males are attempting to fertilize, Reale et al. 1996, Shine et al. 2001; for evolutionary and population dynamic predictions see Rankin et al. 2011) to understand that adaptation can lead to a decline of the population. For example, male–male competition often favours large males, and this can place a large energetic burden on the females that raise male offspring as well as reducing equilibrium population sizes if these large males continually eat more food per capita than the females do. This reduction of what is available for the reproductive fraction of the population then effectively decreases the carrying capacity of the habitat (Kokko and Brooks 2003). Thus, as interesting as the ability of adaptation to rescue populations, is the possibility that strong individual-level selection drives the population closer to extinction.

Conspecific competition can make the consequences of adaptation surprising, and this is not confined to the realm of male–female interactions. Consider, once again, migrant birds, and now assume that there are two behavioural options that are genetically determined: a bird might migrate or it might attempt overwintering at the breeding grounds. The advantage of migrating is that this allows the bird to escape the harshest conditions in midwinter, and thus (despite the dangers of covering large distances) presumably migration improves survival. The advantage of year-round residency, however, is that the bird will gain prior access to the best territories. This means that the best strategy is not simply determined by what balances the survival prospects: a bird may benefit by using a strategy that yields lower survival (the overwintering attempt), because it is balanced by better breeding prospects if it survives. Depending on exactly how severe the winter is, this type of a situation can yield full migration, partial migration, or full year-round residency (Kokko 2011), even if migrants always survive better.

If populations adapt to climate change that makes winters milder and more survivable, one expects the evolution of more residency. This indeed happens in the model of Kokko (2011), and at first sight, one might imagine that such adaptation would improve population performance too. Yet the opposite is predicted to happen: populations can dramatically decline once birds maximize their fitness by abandoning migration. How can there be such a decline in a model that, for once, is focused on a case where climate change is assumed to have only favourable effects (wintering at the breeding grounds is assumed to become easier, while migration mortality is assumed to remain unchanged)? The reasoning goes as follows. The basic trade-off in the model is between survival over the winter (this is better for migrants) and good breeding prospects (this is better for residents). If the relative difference between surviving in different locations diminishes—which is plausible, given that climate change is expected to have its largest effects nearer the Arctic—then birds will increasingly choose the resident option *even though it is still the more dangerous one*. They are rewarded for it during the summer season, which makes this choice fully adaptive, but in the summer season the success of one individual comes at the expense of another since not everyone can occupy the best territories (Kokko 2011). One way to express this is that territorial competition is a zero-sum game, and when individuals evolve to invest more in such games (the expected outcome when the relative importance of survival diminishes relative to intraspecific competition for territories), the population as a whole is expected to perform worse than it did with less investment in outcompeting conspecifics (Rankin et al. 2007). Birds battling climate change might thus, in some cases, decline not because they fail to adapt, but because they do!

1.5 Conclusions: beyond changes in the population mean of a behaviour

We have here encouraged readers to consider that there is more to behavioural change than mere adaptation. The degree to which parents and offspring resemble each other varies because of a multitude of factors, and a change from one generation to the next does not always reflect adaptive evolution. Nor should we expect populations that adapt fastest to necessarily fare best on a changing planet: sometimes plastic responses to new situations are more adequate than adaptive evolution, and sometimes adaptation itself can lead to population decline.

While we have aimed to be comprehensive in decomposing population change to all relevant categories, it is important to repeat that any categorization tends to focus one's attention to specific factors at play, sometimes at the expense of other features that can prove important. Our examples are phrased in terms of a mathematical construction that describes the change in the population mean of a single behavioural trait. For long-term population and evolutionary dynamics, it is important to consider changes in the variance of the trait (see Chapter 12; for the relevant mathematical analysis see Coulson and Tuljapurkar 2008). Sex-specific behaviours are a clear example where a single mean does not capture what is going on: males can be larger or smaller, behaviourally dominant or subordinate, and more or less numerous than females, with obvious consequences for population dynamics and selection (Chapter 12); likewise the degree of variation among males can impact female behaviour (Chapter 15). Entire suites of behaviours may evolve in interdependent ways, a scenario that can be represented by multivariate extensions of the Price equation and which has spurred the study of behavioural syndromes (Dingemanse et al. 2010). Finally, a simple iteration of the Price equation does not allow us to visualize a likely important phenomenon: the evolution of plasticity and behavioural reaction norms. In essence, this would involve allowing the individual change term in Equation 1.2 to evolve: that is to be itself subject to selection, parent–offspring differences, and within individual variation. This important topic is discussed in Chapter 11.

Acknowledging these omissions should not take away from our main point that behavioural ecologists should be open-minded about the mechanisms underlying the change in a trait—or lack thereof. The following chapters present exciting examples where learning, plasticity, genetic inheritance and other modes of trait transmission determine what happens when populations experience novel conditions. Simultaneously, we strongly encourage explicit consideration of population size, density, and sex ratio in behavioural contexts. These are not mere outcomes that a conservationist is interested in (see Chapter 17), but also feed back to impact, and coevolve with, further behavioural change.

Acknowledgements

We would like to thank Swanne Gordon, Andrew Hendry, the editors and an anonymous referee for comments on the manuscript. During writing the Agence Nationale de la Recherche's project EvoRange (France) provided funding to ALS, and the Australian Research Council to HK. ALS thanks Regis Ferriére and the University of Arizona for providing working space.

References

Baldwin, J.M. (1896). A new factor in evolution. *American Naturalist*, 30, 441–51.

Bell, G. and Gonzalez, A. (2009). Evolutionary rescue can prevent extinction following environmental change. *Ecology Letters*, 12, 942–8.

Bentley, R.A., Hahn, W.M., and Shennan, S.J. (2004). Random drift and culture change. *Proceedings of the Royal Society of London B, Biological Sciences*, 271, 1443–50.

Birkhead, T.R. and Monaghan, P. (2010). Ingenious ideas: the history of behavioral ecology. In In Westneat D.F. and Fox C.W. (eds) *Evolutionary Behavioral Ecology*, pp 3–15. Oxford, Oxford University Press.

Blackburn, T.M., Cassey, P., Duncan, R.P., Evans, K.L., and Gaston, K.J. 2004. Avian extinction and mammalian introductions on oceanic islands. *Science*, 305, 1955–8.

Bulmer, M.G. (1989). Maintenance of genetic variability by mutation-selection balance: a child's guide through the jungle. *Genome*, 31, 761–7.

Carrete, M., Donázar, J.A., Margalida, A., and Bertran, J. (2006). Linking ecology, behaviour and conservation: does habitat saturation change the mating system of bearded vultures? *Biology Letters*, 2, 624–7.

Charmantier, A. and Garant, D. (2005). Environmental quality and evolutionary potential: lessons from wild populations. *Proceedings of the Royal Society of London B, Biological Sciences*, 272, 1415–25.

Chevin, L.-M. and Lande, R. (2010). When do adaptive plasticity and genetic evolution prevent extinction of a density-regulated population? *Evolution*, 64, 1143–50.

Clutton-Brock, T. and Sheldon, B.C. (2010). Individuals and populations: the role of long-term, individual-based studies of animals in ecology and evolutionary biology. *Trends in Ecology and Evolution*, 25, 562–73.

Conover, D.O. and Schultz, E.T. (1995). Phenotypic similarity and the evolutionary significance of countergradient variation. *Trends in Ecology and Evolution*, 10, 248–52.

Cornwallis, C.K., West, S.A., Davis, K.E., and Griffin, A.S. (2010). Promiscuity and the evolutionary transition to complex societies. *Nature*, 466, 969–72.

Coulson, T. and Tuljapurkar, S. (2008). The dynamics of a quantitative trait in an age-structured population living in a variable environment. *American Naturalist*, 172, 599–612.

Darwin, C. (1871). *The Descent of Man, and Selection in Relation to Sex*. London, John Murray.

Davis, J. M. and Stamps, J.A. (2004). The effect of natal experience on habitat preferences. *Trends in Ecology and Evolution*, 19, 411–16.

Dingemanse, N.J., Kazem, A.J.N., Reale, D., and Wright, J. (2010). Behavioural reaction norms: animal personality meets individual plasticity. *Trends in Ecology and Evolution*, 25, 81–9.

Donato, D.B. and Potts R.T. (2004). Culturally transmitted predation and consumption techniques by torresian crows *Corvus orru* on cane toads Bufo marinus. *Australian Field Ornithology*, 21, 125–6.

Fierst, J.L. (2011). A history of phenotypic plasticity accelerates adaptation to a new environment. *Journal of Evolutionary Biology*, 24, 1992–2001.

Fox, C.W. and Westneat, D.F. (2010). Adaptation. Pp 16–31 In Westneat, D.F. and Fox, C.W. (eds) *Evolutionary Behavioral Ecology*, pp. 16–31. Oxford, Oxford University Press.

Futuyma, D.J. (2010). Evolutionary constraint and ecological consequences. *Evolution*, 64, 1865–84.

Gardner, A. (2010). Adaptation as organism design. *Biology Letters*, 5, 861–4.

Ghalambor, C.K., Mckay, J.K., Carroll, S.P., and Reznick, D.N. (2007). Adaptive versus non adaptive phenotypic plasticity and the potential for contemporary adaptation in new environments. *Functional Ecology*, 21, 394–407.

Gomulkiewicz, R. and Holt, R.D. (1995). When does evolution by natural selection prevent extinction? *Evolution*, 49, 201–7.

Gomulkiewicz, R. and Houle, D. (2009). Demographic and genetic constraints on evolution. *American Naturalist*, 174, E218–29.

Grether, G., Cummings, M., and Hudon, J. (2005). Countergradient variation in the sexual coloration of guppies (Poecilia Reticulata*reticulata*): Drosopterin synthesis balances carotenoid availability. *Evolution*, 59, 175–88.

Helanterä, H. and Uller, T. (2010). The Price equation and Inheritance.extended inheritance *Philosophy and Theoretical Biology*, 2, e101.

Horváth, G., Bernáth, B., and Molnár, G. (1998). Dragonflies find crude oil visually more attractive than water: multiple-choice experiments on dragonfly polarotaxis. *Naturwissenschaften*, 85, 292–7.

Houle, D. (1992). Comparing evolvability and variability of quantitative traits. *Genetics*, 130, 195–204.

Jennions, M.D., Møller, A.P., and Petrie, M. (2001). Sexually-selected traits and adult survival: a meta-analysis. *Quarterly Review of Biology*, 76, 3–36.

Jones, J. and Francis, C.M. (2003). The effects of light characteristics on avian mortality at lighthouses. *Journal of Avian Biology*, 34, 328–33.

Jones, T. and Cresswell, W. (2010). The phenology mismatch hypothesis: are declines of migrant birds linked to uneven global climate change? *Journal of Animal Ecology*, 79, 98–108.

Kiers, E.T., Palmer, T.M., Ives, A.R., Bruno, J.F., and Bronstein, J.L. (2010). Mutualisms in a changing world: an evolutionary perspective. *Ecology Letters*, 13, 1459–74.

Kokko, H. (2011). Directions in modelling partial migration: how adaptation can cause a population decline and why the rules of territory acquisition matter. *Oikos*, 120, 1826–37.

Kokko, H. and Brooks, R. (2003). Sexy to die for? Sexual selection and the risk of extinction. *Annales Zoologici Fennici*, 40, 207–19.

Kokko, H. and Jennions, M.D. (2010). Behavioral ecology: the natural history of evolutionary biology. In Bell, M.A., Eanes, W.F., and Futuyma, D.F. (eds) *Evolution Since Darwin: The First 150 Years*, pp. 291–318. Sunderland, Sinauer.

Kokko, H. and López-Sepulcre, A. (2007). The ecogenetic link between demograpy and evolution: Can we bridge the gap between theory and data? *Ecology Letters*, 10, 773–82.

Kokko, H. and Sutherland, W.J. (2001). Ecological traps in changing environments: ecological and evolutionary consequences of a behaviourally mediated Allee effect. *Evolutionary Ecology Research*, 3, 537–51.

Lehmann, L., Feldman, M., and Foster, K. (2008). Cultural transmission can inhibit the evolution of altruistic helping. *American Naturalist* 172, 12–24.

López-Sepulcre, A., Norris, K., and Kokko, H. (2009). Reproductive conflict delays the recovery of an endangered social species. *Journal of Animal Ecology*, 78, 219–25.

Massot, M., Clobert, J., Lorenzon, P., and Rossi, J. (2002). Condition-dependent dispersal and ontogeny of the dispersal behaviour: an experimental approach. *Journal of Animal Ecology*, 71, 253–61.

Noble, G.K. and Bradley, H.T. (1933). The mating behaviour of lizards: its bearing on the theory of sexual selection. *Annals of the New York Academy of Sciences*, 35, 25–100.

Nocera, J.J., Forbes, G.J., and Giraldeau, L.-A. (2006). Inadvertent social information in breeding site selection of natal dispersing birds. *Proceedings of the Royal Society of London B, Biological Sciences*, 273, 349–55.

O'Donnell, S., Webb, J.K., and Shine, R. (2010) Conditioned taste aversion enhances the survival of and endangered predator inperilled by a toxic invader. *Journal of Applied Ecology*, 47, 558–65.

O'Steen, S., Cullum, A.J., and Bennett, A.F. (2002). Rapid evolution of escape ability in Trinidadian guppies (*Poecilia reticulata*). *Evolution*, 56, 776–84.

Owens, I. (2006). Where is behavioural ecology going? *Trends in Ecology and Evolution*, 21, 356–61.

Parker, G. and Maynard Smith, J. (1990). Optimality theory in evolutionary biology. *Nature*, 348, 27–33.

Price, G. R. (1970). Selection and covariance. *Nature*, 227, 520–1.

Rankin, D.J., Bargum, K., and Kokko, H. (2007). The tragedy of the commons in evolutionary biology. *Trends in Ecology and Evolution*, 22, 643–51.

Rankin, D.J., Dieckmann, U., and Kokko, H. (2011). Sexual conflict and the tragedy of the commons. *American Naturalist*, 177, 791.

Reale, D., Bousses, P., and Chapuis, J.L. (1996). Female-biased mortality induced by male sexual harassment in a feral sheep population. *Canadian Journal of Zoology*, 74, 1812–18.

Reed, T.E., Waples, R.S., Schindler, D.E., Hard, J.J., and Kinnison, M.T. (2010). Phenotypic plasticity and population viability: the importance of environmental predictability. *Proceedings of the Royal Society of London B, Biological Sciences*, 277, 3391–400.

Rice, S.H. (2008). A stochastic version of the Price equation reveals the interplay of deterministic and stochastic processes in evolution. *BMC Evolutionary Biology*, 8, 262.

Richardson, D., Burke, T., and Komdeur, J. (2007). Grandparent helpers: The adaptive significance of older, postdominant helpers in the Seychelles warbler. *Evolution*, 61, 2790–800.

Schlaepfer, M., Runge, M., and Sherman, P. (2002). Ecological and evolutionary traps. *Trends in Ecology and Evolution*, 17, 474–80.

Shine, R., LeMaster, M.P., Moore, I.T., Olsson, M.M., and Mason, R.T. (2001). Bumpus in the snake den: effects of sex, size, and body condition on mortality of red-sided garter snakes. *Evolution*, 55, 598–604.

Stamps, J. (2006). The silver spoon effect and habitat selection by natal dispersers. *Ecology Letters*, 9, 1179–85.

Stamps, J. A. (2001). Habitat selection by dispersers: integrating priximate and ultimate approaches. In Clobert, J., Danchin E., Dhondt A., and Nichols, J.eds *Dispersal*, pp. 230–42, New York, Oxford University Press.

Stapley, J. and Keogh, S.J. (2005). Behavioral syndromes influence mating systems: floater pairs of a lizard have heavier offspring. *Behavioral Ecology*, 16, 514–20.

Stirling, D.G., Réale, D., and Roff, D.A. (2002). Selection, structure and the heritability of behaviour. *Journal of Evolutionary Biology*, 15, 277–89.

Tinbergen, N. (1963). On aims and methods of ethology. *Zeitschrift für Tierpsychologie*, 20, 410–33.

Traill, L.W., Brook, B.W., Frankham, R.R., and Bradshaw, C.J.A. (2010). Pragmatic population viability targets in a rapidly changing world. *Biological Conservation*, 143, 28–34.

Weilgart, L.S. (2007). The impacts of anthropogenic ocean noise on cetaceans and implications for management. *Canadian Journal of Zoology*, 85, 1091–116.

Wright, S.D., Gillman, L.N., Ross, H.A., and Keeling, D.J. (2009). Slower tempo of microevolution in island birds: implications for conservation biology. *Evolution*, 63, 2275–87.

CHAPTER 2

Environmental disturbance and animal communication

Gil G. Rosenthal and Devi Stuart-Fox

> **⊃ Overview**
>
> Even seemingly benign anthropogenic influences can profoundly change animal communication. Human impacts have the potential to alter the dynamics of communication at every stage of the process, from the production of signals to their transmission and ultimately their evaluation by receivers. In many cases, disturbance reduces the efficacy of communication by weakening signal production, distorting or attenuating signals as they travel to the receiver, or hampering their perception. More insidiously, changes to the environment can modify the distribution of signals in the environment or the modalities that receivers use to evaluate signallers. Alterations to communication systems can have far-reaching evolutionary consequences, particularly given communication's role in maintaining reproductive isolation among species.

2.1 Introduction

Communication, which we define following Rendall et al. (2009) as an individual's use of signals to influence the behaviour of a receiver, is fundamental to the well-being of individuals and populations. Social signals are ubiquitously used in finding food, avoiding predators, resolving conflicts, and selecting mates. Interference with communication, therefore, can seriously alter survival patterns, change the magnitude and direction of natural and sexual selection, and impinge on basic evolutionary processes like reproductive isolation and hybridization. Even seemingly innocuous disturbances can impact communication in surprising ways, wreaking havoc on social systems and generating irreversible evolutionary consequences.

The ecological effects of anthropogenic disturbance on communication are often insidious; in many cases, subtle alterations to the environment are only brought to our attention because of changes in animal behaviour. The evolutionary effects of impairing communication can reach far beyond disturbed areas in space and time, particularly if reproductive barriers among species are breached (Servedio 2004).

At its simplest, communication can be abstracted as an interaction between a signaller and a receiver. First, the signaller produces a signal, either by directly generating energy or, notably in the case of visual patterns and colours, modifying the distribution of energy from an external source. Second, the signal is transmitted through the environment, where it inevitably deteriorates in magnitude and quality, and it is finally perceived and processed by the receiver, who may produce a behavioural response (Bradbury and Vehrencamp 1998).

In this chapter, we describe how human impacts can alter communication, which we divide into three phases: signal production, transmission, and reception (Fig. 2.1). Where possible, we also address the ecological and evolutionary consequences of altered communication. The vast majority of relevant studies have involved auditory, visual, and chemical communication; we discuss possible effects on other modalities at the end of the chapter.

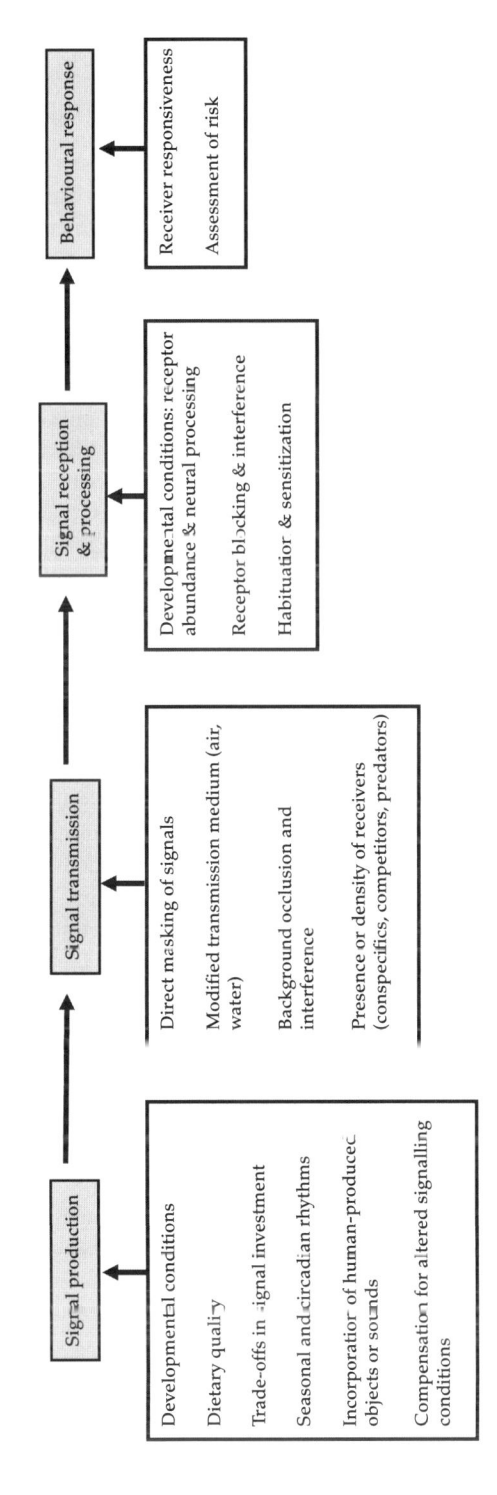

Figure 2.1 Phases of the signalling process (grey boxes) and major factors affecting each phase that have been shown to be modified by anthropogenic disturbances.

We focus on effects of anthropogenic disturbance on interactions among non-human animals, particularly conspecifics, and do not address the phenomenon of human–animal communication. In some cases, human impacts on animal communication are quite deliberate, as in the widespread use of pheromone traps for control of insect pests (Ridgway et al. 1990), and, more recently, invasive lampreys *Petromyzon marinus* (Li et al. 2007) or the use of simulated acoustic signals by hunters and birdwatchers. However, most such effects are unintended consequences of pollution, habitat degradation, or species introductions.

2.2 Signal production

One of the primary ways in which human alteration of the environment impacts signal production is via physiological effects on signal development and expression. In particular, contamination of the environment by metals and chemical pollutants can influence the development and production of signals via their impact on gene expression, endocrine function, and a range of other cellular processes. Perhaps the best documented effects of anthropogenic activities on the expression of animal signals involve endocrine disrupting chemicals (EDCs). EDCs encompass a wide variety of chemicals used in agriculture and industry, which, as their name suggests, affect the endocrine system. Their effects on behaviour and signalling are covered in Chapter 3. In this section, therefore, we focus on the potential effects of chemical and metal pollutants in general, as well as light pollution, on production of acoustic, visual, and olfactory signals. We also discuss situations where novel properties of the environment, such as artificial objects and mechanical sounds, may be incorporated into animal signalling repertoires. Finally, we note that signals are generally matched to local environmental conditions. Changes to signal expression often occur to enable effective signalling under altered environmental conditions, and can thus be a secondary consequence of environmental effects on signal transmission. We discuss these in the subsequent section on signal transmission, focusing here on changes to signal expression caused by physiological changes associated with various forms of pollution.

2.2.1 Acoustic signals

Pollutants such as chemicals and metals can affect many aspects of acoustic signal production. These include developmental processes, such as neural development important for song learning and memory (see also Chapter 4), resource allocation to signalling, and dietary quality affecting signal expression. Although empirical evidence is currently limited, a few recent studies have shown negative effects of pollutants on bird song. Great tits *Parus major* inhabiting sites with high levels of metal pollution have smaller song repertoires, and sing significantly less, than birds from less polluted sites (Gorissen et al. 2005). By contrast, European starling *Sturnus vulgaris* males produce longer and more complex songs when exposed to EDCs (Markman et al. 2008). However, EDC exposure also leads to immune suppression. By preferring males with more complex songs, females choose males in poorer health, suggesting possible population fitness consequences, especially in populations where males provide paternal care (Sandell et al. 1996).

As the previous example highlights, the effects of anthropogenic activities can be profoundly counterintuitive. Even well-intentioned interventions can have unintended consequences. For example, supplementing the diet of adult song sparrows *Melospiza melodia* results in increased clutch size, but male offspring have smaller song repertoires once mature (Zanette et al. 2009). Provisioning parents can thus make males less attractive to females, potentially facilitating heterospecific matings or inducing females to mate with males that produce less-fit offspring.

Apart from chemicals and metals, industrialization and urbanization generate noise and light pollution. We discuss noise in the section on transmission effects. Light pollution is likely to affect multiple aspects of communication, but its effects on the timing and expression of acoustic signals have been well-documented. Light pollution affects animal physiology and behaviour, thereby influencing signal production. This is unsurprising given that many animals show marked physiological changes in relation to both seasonal and circadian variation in natural light cycles (Navara and Nelson 2007). Bright city lights have induced widespread disruption of these natural light cycles.

Constant or changed exposure to light, even at levels comparable to the brightness of moonlight, can have major effects on circadian rhythms in a wide range of hormones, particularly melatonin (reviewed in Longcore and Rich 2004; Navara and Nelson 2007). Effects are exacerbated by the violet/blue wavelengths characteristic of artificial light. Melatonin has well documented effects on reproduction, protection against oxidative stress, and metabolism (Navara and Nelson 2007). Artificial light-induced reduction in melatonin has been implicated in the global increase in metabolic disorders and obesity in humans (e.g. Fonken et al. 2010). Clearly, light pollution has the potential not only to affect the timing of communicative behaviours (e.g. Kempenaers et al. 2010) but also the expression of sexual signals in animals. For example, frogs *Rana clamitans melanota* exposed to artificial light produce fewer advertisement calls (Baker and Richardson 2006). Despite recent recognition of the potential ecological consequences of light pollution (reviewed in Longcore and Rich 2004; Navara and Nelson 2007), we currently do not know how altered lighting regimes affect metabolism or trade-offs between signal investment and other physiological processes in natural populations.

2.2.2 Visual signals

An ever-growing number of studies have shown the harmful effects of human activities on visual signals, particularly secondary sexual ornaments. The development and expression of ornaments is often condition dependent and reflects an individual's level of, and ability to cope with, physiological stress (Buchanan 2000). Developmental stress from exposure to pollutants can influence a range of processes affecting the expression of signals. For instance, in the goodeid fish *Girardinichthys metallicus*, embryonic exposure to low concentrations of the organophosphorus insecticide, methylparathion, reduces male ornament size, colour, and courtship display rates (Arellano-Aguilar and Garcia 2008).

During adulthood, continued exposure to toxins can affect resource allocation, at the expense of signal expression. For example, animals experiencing higher levels of oxidative stress may allocate more antioxidants to reducing damaging effects of free radicals. This may compromise the expression of carotenoid-based visual signals, which is often correlated with levels of circulating antioxidants (Dauwe et al. 2006). For instance, yellow-legged gulls *Larus michahellis* fed a diet containing fuel oil from an oil spill had higher plasma levels of two types of antioxidant, vitamin E and carotenoids, and smaller red bill spots (Pérez et al. 2010a). Additionally, in free-living gulls exposed to an oil spill, the size of the red bill spot was positively correlated with body condition and negatively correlated with aspartate aminotransferase (AST), an enzyme indicative of liver damage in birds (Pérez et al. 2010b).

Several studies have similarly found that the intensity of yellow coloration on the breasts of great tits is negatively correlated with levels of metal pollution (Geens et al. 2009). In this species, however, there appears to be no relationship between total antioxidant capacity and carotenoid-based signal expression. Rather, Geens et al. (2009) propose that the differences in carotenoid coloration along the pollution gradient reflect pollution-induced differences in diet composition and quality. As this example highlights, various mechanisms could account for the relationship between pollution and signal expression. As is the case for acoustic signals, these mechanisms remain poorly understood in natural populations.

2.2.3 Chemical signals

Just as for visual signals, there is extensive evidence for disruption of chemical communication by a wide range of pollutants, in addition to endocrine disruptors. These pollutants can affect chemical information transfer, both within and between individuals, with potentially far-reaching consequences (Lurling and Scheffer 2007). However, it is often more difficult to distinguish which aspect of the signalling process (signal production, transmission, or reception) is being affected. Pollutants can alter chemical communication in three ways: (1) by affecting the quality and quantity of chemical signals (production); (2) by binding to the chemical signals themselves, reducing the quantity transmitted (transmission); or (3) by binding to receptors and influencing receptor function (reception).

The majority of studies on pollution and chemical signalling have focused on the effects of pollutants on receptor function (reviewed in Lurling and Scheffer 2007); however, a few have clearly shown that pollutants affect the production of chemical cues rather than their reception. For example, Ward et al. (2008) showed that exposure to the widely used surfactant 4-nonylphenol (4-NP) does not affect the ability of banded killifish *Fundulus diaphanus* to detect chemical cues but does affect their chemical signals. Similarly, exposure to nitrates, which are used extensively in agriculture, affects the properties of olfactory signals rather than their chemoreception in palmate newts *Lissotriton helveticus* (Secondi et al. 2009). Unexposed females preferred unexposed males over exposed males in olfactory—but not visual—mate choice tests.

Overall, a consistent conclusion of studies on the effects of pollutants on animal behaviour is that levels of exposure considered to be low (substantially lower than those causing mortality, mutation, or cancer) can have major effects on behaviour, including communication. In the most serious cases, this can cause 'behavioural castration' and population decline (Lurling and Scheffer 2007).

2.2.4 Signals acquired from the human environment

Animals can sometimes use features of human origin in their signal repertoires. For example, avian vocal mimics are frequently observed to incorporate mechanical sounds, like car alarms, into their acoustic repertoire (Clark 2001). Human activities can also affect the extended phenotype, such as external structures (e.g. nests and bowers) constructed by individuals to attract mates. For example, bowerbirds decorate their nests with both natural and artificial objects (e.g. coloured plastic and glass) that they collect from the environment (Marshall 1954). Colour preferences for bower decorations, both natural and artificial, have been extensively studied (e.g. Madden and Tanner 2003; Patricelli et al. 2003). Novel objects increase the range of potential signal innovations (see Endler et al. 2005) and their availability can affect both female preferences and male behaviour, such as rates of decoration stealing and destruction of the bowers of competitors (Hunter and Dwyer 1997). In black kites *Milvus migrans*, meanwhile, nest decorations including plastic and other scavenged items serve as territory-defence signals (Sergio et al. 2011).

Anthropogenic effects on nest design and building behaviour are not limited to birds. Three-spined stickleback *Gasterosteus aculeatus* males decorate their nests with coloured algae and, in a laboratory setting, will also use artificial objects such as colourful, shiny foil and 'spangles'. Females are more attracted to nests decorated with artificial objects than those without (Ostlund-Nilsson and Holmlund 2003), suggesting that availability of such objects in the wild could influence male nest decoration and female choice. In this species, changes to water flow regimes have also been shown to result in modified nest structure and building behaviour, with potential consequences for mate choice and sexual selection (Rushbrook et al. 2010). In general, human introductions of novel signal elements have great potential to affect receiver behaviour, as female preferences for novel male traits appear to be widespread in animals (Ryan 1998).

2.2.5 Matching signals to altered habitats

Signals are often matched to local environmental conditions (Endler 1992). Changes to the signalling environment (visual, olfactory, or acoustic) alter the efficacy of signals, thereby inducing changes to signal expression. For example, increased water turbidity not only affects behaviour and the transmission of visual signals, but also their expression. In palmate newts, water turbidity decreases the size of male secondary sexual traits, an effect that is not attributable to reduced foraging efficiency in turbid water (Secondi et al. 2007). One potential explanation for reduced investment in visual signals in turbid waters is that under conditions in which visual signal transmission is poor, males reallocate resources towards other activities such as mate searching or to other types of signals (e.g. olfactory). As discussed below, this can be accompanied by increased attention by receivers to more readily detectable cues (e.g. Heuschele et al. 2009).

As is the case for water turbidity, many human environmental disturbances primarily affect signal transmission, with changes to investment into signal expression being a secondary response. For example, anthropogenic noise pollution and urbanization drastically change acoustic signal transmission, which, in turn, alters signal expression (reviewed in Laiolo 2010; Slabbekoorn et al. 2010; Warren et al. 2006). As noise pollution tends to be both loud and low pitched, animals that signal in the presence of anthropogenic noise tend to increase amplitude (loudness) and/or increase frequency (pitch), so that they may be heard. Killer whales *Orcinus orca* increase the amplitude of their calls in relation to background noise levels decibel for decibel (Holt et al. 2009). Increases in amplitude are likely to require greater energetic investment, potentially influencing any relationship between the signal and other aspects of an individual's phenotype (i.e. influencing signal content as well as efficacy). As such, changes to signal expression tend to be a secondary consequence of changes to signal transmission, a topic that we will now discuss in greater detail.

2.3 Signal transmission

We define signal transmission effects as those which decrease the signal-to-noise ratio from the time a signal is emitted to the time it is transduced by a receiver. Human activities can alter signal transmission in a number of ways (Fig. 2.1). First, human activities can cause direct masking of animal signals, such as traffic noise masking bird or frog calls, or chemical pollutants interacting with pheromones. Second, they alter properties of the transmission medium, an example being changed light transmission through air or water due to pollution or eutrophication. Third, human activities modify physical structures that interfere with signal transmission. Both the construction of urban landscapes and drastic changes to natural landscapes (e.g. clear-felling, weed invasion) affect the acoustic, visual and chemical signalling environment. For example, urban environments tend to be characterized by large, flat, sound-reflective surfaces, which cause sounds to attenuate more slowly and to degrade due to reverberation. Lastly, human activities can alter levels of interference from other animals, including conspecifics, competitors, and predators, due to human-mediated changes in the population density and distribution of many species. In this section, we briefly review how human activities affect signal transmission for each signalling modality (acoustic, visual, chemical) and the consequences of such changes for animal communication.

2.3.1 Acoustic signals

By far the most attention on the effects of human activities on animal communication has focused on acoustic signals (reviewed in Laiolo 2010). Human-generated noise is widespread and often at levels substantially greater than those encountered in nature (Barber et al. 2010; Warren et al. 2006). Anthropogenic noise in both aquatic and terrestrial environments is characterized not only by high absolute levels, but also by a high degree of spatial and temporal heterogeneity in noise levels, and the prevalence of low frequency sounds (<1 kHz), such as traffic and boat noise (Slabbekoorn et al. 2010; Warren et al. 2006). However, noise pollution also includes higher frequency sounds such as those used to locate and measure objects underwater and to measure ocean temperatures (Slabbekoorn et al. 2010). Crucially, human generated sounds overlap in frequency with the hearing range of most animals as well as the frequencies of the calls of many species (Slabbekoorn et al. 2010), including low frequency specialists such as marine mammals (Clark et al. 2009). By masking acoustic signals, anthropogenic noise decreases the active space of individuals, that is, the distance from which a conspecific is able to detect an individual's call. Such a reduction in active space clearly has important implications for animal communication (reviewed in Barber et al. 2010).

In addition to producing noise, humans have altered the acoustic transmission properties of large areas, both through changes to vegetation structure and through urbanization. Natural vegetation structure is correlated with acoustic signal structure (Ryan and Brenowitz 1985), so altered vegetation should have marked effects on communication. Urban landscapes, moreover, are characterized by multiple, large, flat, often parallel surfaces that reflect sound. These have been termed 'urban canyons' because their acoustic properties resemble those of natural canyons (Warren et al. 2006). Such urban canyons create flutter echo, whereby sounds ricochet rapidly between parallel walls, causing slower attenuation (loss of amplitude) and signal degradation due to the multiple reflected sound waves arriving at different times (Warren et al. 2006). Thus, the structure of the urban environment is likely to exacerbate the masking effects of anthropogenic noise on animal signals.

The effects of noise pollution are likely to differ in aquatic and terrestrial environments because the sound transmission properties of air and water are very different (reviewed in Slabbekoorn et al. 2010). Due to the high molecular density of water, sound transmission in water is about five times faster—and therefore wavelengths are about five times longer—than in air. Sound also attenuates less and therefore travels much longer distances in water than in air. By contrast, light attenuates much more rapidly in water than air so many aquatic animals use sound rather than sight for navigation and use acoustic signals for long distance communication. Noise pollution may therefore affect different aspects of animal behaviour and lead to different responses in aquatic versus terrestrial environments.

Animals can respond to noise pollution in four main ways: (1) by changes to their spatial distribution or density to avoid localized areas with high noise levels (Bayne et al. 2008); (2) by changing the temporal distribution of calling behaviour (Fuller et al. 2007); (3) through an absolute reduction (or increase) in total calling effort (Sun and Narins 2005); or (4) by changing the structure of their calls. Changes to call structure include increased amplitude (e.g. Holt et al. 2009), changes to pitch (e.g. Parris et al. 2009; Verzijden et al. 2010), increased redundancy of call components (e.g. Brumm and Slater 2006) and use of narrower band widths (pure tones, see Slabbekoorn et al. 2002).

The changes exhibited by a species will depend on numerous factors, including the initial structure of the call. For example, Parris and Schneider (2009) showed that a bird species with a lower frequency call increased its call frequency in response to traffic noise whereas a species with a higher frequency call did not. Changes to call structure can have important implications for mate choice when there is a trade-off between signal efficacy and content. For example, in many species, frequency is negatively correlated with body size and larger, lower-frequency males are more attractive (Ryan and Keddy-Hector 1992). Larger individuals produce lower frequency sounds, yet high-pitched sounds are more audible in noisy environments. Thus, individuals face a conflict between attractiveness and audibility. Hu and Cardoso (2009) further suggested that bird species with naturally higher-frequency signals should fare better in urban habitats.

To date, the great majority of evidence for an effect of anthropogenic noise on animal communication derives from studies of birds (reviewed in Barber et al. 2010; Laiolo 2010). However, an increasing number of studies show similar patterns in amphibians (e.g. Cunnington and Fahrig 2010; Parris et al. 2009; Sun and Narins 2005). There is a growing awareness of the effects of anthropogenic noise on communication in aquatic environments (Clark et al. 2009; Slabbekoorn et al. 2010). For example, ship noise decreases the ability of toadfish *Halobatrachus didactylus* to detect mate attraction calls (Vasconcelos et al. 2007), and several studies have shown that cetacean communication is impacted by human activities (Foote et al. 2004; Miller et al. 2000). Human activities can even result in serious injury or death to echo-locating cetaceans (Jepson et al. 2003). The long range of sound in water suggests that anthropogenic noise could have a broad reach in aquatic environments.

2.3.2 Visual signals

Visual signals can be parsed into spectral, spatial, and temporal components (Rosenthal 2007), each of which can be susceptible to effects from disturbance. Transmission of visual signals depends on line-of-sight between the signaller and receiver, on spatiotemporal and spectral differences between signal and background, and on how the medium (air or water) attenuates and scatters light as a function of wavelength and distance. All of these have the potential to be drastically affected by disturbance.

Habitat degradation: changes in occlusion and background

Many disturbed habitats, from urban areas to cornfields to forest regrowth, contain higher densities of occluding structures. The most basic requirement of visual communication, namely that there be an unoccluded straight-line path between signaller and receiver, is thereby liable to fail. Conversely, disturbance like clear-cutting and coral-reef destruction can increase the line of sight, minimizing signal privacy and therefore rendering conspicuous visual signals more visible to predators.

Visual signals are frequently tuned to maximize contrast with the background. Habitat degradation, such as changes to vegetation structure, can alter the characteristics of the visual background, as can novel structures like buildings, roads, and underwater oil rigs. Novel backgrounds may not only differ in colour and spatial characteristics, but also in the temporal patterns exhibited in wind or current. In anoline lizards, for example, the characteristic 'head bob' display typically contains a high-frequency 'jerky' component that is distinct from the movement of background vegetation (Fleishman 1986). In two species of *Anolis* lizards, male signallers speed up body movements against a background of moving vegetation (Ord et al. 2007).

In some cases, the structure of complex visual signals may serve to minimize long-distance detection while conveying information at close range. This is notably the case in many tropical reef fishes, where high-contrast, high-spatial frequency patterns merge to closely resemble the background when viewed at a distance (Marshall 2000). In this case, both functions may be subverted since increased background contrast may be accompanied by reduced transmission efficiency through the medium.

The light environment

With the exception of a handful of organisms that generate their own light, visual signallers rely on modifying the distribution of light available in the environment (Endler 1978). Before it strikes a receiver, light is filtered through air or water as well as any overhead vegetation. The radiance (light intensity as a function of wavelength) of the light illuminating a signaller therefore varies as a function of vegetation type and vegetation density, and also varies as a function of space and time (e.g. leaves creating patchy cover and moving in the wind). Any alteration to habitat structure will thus change multiple properties of the incident light distribution. Across species of birds (Marchetti 1993) and fishes (Cummings 2007), visual signals coevolve with forest canopy structure. Any change in habitat structure should therefore compromise either signal privacy or signal efficacy.

In aquatic organisms, the filtering properties of water present another challenge to signal transmission, both in terms of the quality of light incident on the signaller, and on the way that signals are modified on their way to receivers. Pure water most efficiently transmits the middle (blue-green) wavelengths of the visual spectrum, while filtering out ultraviolet and, most severely, red light. Since phytoplankton absorb middle wavelengths for photosynthesis, eutrophication can flatten the filtering function of water and severely reduce the overall amount of available light. Humic substances, promoted by water runoff and organic decay can further reduce the amount of light available at shorter wavelengths (Fisher et al., 2006).

As is the case with canopy structure, signals coevolve with the transmission medium (Fuller et al., 2005). Changes to the bandwidth of available light can have severe consequences for communication. Two species of cichlid fish in Lake Victoria hybridize because females are unable to distinguish red males from blue males (Seehausen et al. 1997). In sticklebacks, the honesty of male-male agonistic signals is compromised in turbid water (Wong et al. 2007).

Although most work has focused on changes to the spectral environment, spatiotemporal cues are generally more salient to receivers (Rosenthal 2007).

These can also be affected by degradation. Overall light availability affects the detectability of fine spatial patterns. Many animals express spots, stripes, and complex textures which might not be discriminatable in turbid or low-light habitats. 'Veiling light' scattered by particles between the signal and the receiver can also interfere with signal detection, as well as disrupt the polarization signals used by many invertebrates (Shashar et al. 2004).

2.3.3 Chemical signals

To our knowledge, there are no studies demonstrating effects of airborne pollutants on chemosignal transmission in terrestrial organisms; however, the interaction between volatile hydrocarbons and atmospheric emissions is well known in pollution chemistry (Zhang et al. 2007). It is not unreasonable to speculate, therefore, that the active space of windborne insect pheromones might be reduced in polluted environments.

By contrast, numerous behavioural studies suggest that pheromonal communication can be compromised by chemical changes to the aquatic environment. In the water, chemical signals have the potential to interact with a host of dissolved and suspended substances. As will be discussed below, these substances can also affect receiver perception. These two effects are difficult to disentangle in studies of chemical communication. In visual and acoustic studies, physical measures of signal degradation (e.g. Nemeth and Brumm 2010; Seehausen et al. 1997) can decouple transmission effects from changes to receivers. Most studies of chemical signalling, however, rely on behavioural assays, making it difficult to disentangle effects on transmission and on reception. We briefly discuss case studies of chemical interference with aquatic communication, and then return to effects on receivers in the next section. Box 2.1 details the best-understood of these cases, the effects of pH on chemical communication.

Human impacts can markedly increase the salinity of freshwater and brackish environments (Beeton 2002). Salt concentrations can affect sensory responses to odorant cues (Velez et al. 2009) and salts could interact with the structure of chemosignals as well. In the Pacific blue-eye fish *Pseudomugil*

Box 2.1 pH and chemical communication in aquatic organisms

One of the best-studied vertebrate chemosignals is the alarm pheromone, or *Schreckstoff*. This substance, identified as hypoxanthine-3-*N*-oxide, is thought to be produced in specialized skin cells by all ostariophysan fishes, which include 64% of extant freshwater fishes (Helfman 2009). Upon injury, pheromone is released, causing a dramatic antipredator response in nearby fish. Because the signal consists of a single, identified molecule, it is possible to disentangle chemical effects on signals versus receivers; in the section on signal detection, we discuss how environmental impacts on chemoreception can affect perception of alarm pheromone.

Under weakly acidic conditions, the alarm pheromone undergoes a non-reversible covalent change that destroys its capacity to elicit a behavioural response (Brown et al. 2002). Since Brown et al.'s (2002) finding on two cyprinid species, a similar effect of reduced pH on alarm response has been found in multiple fish taxa (Leduc et al. 2004; Olivier et al. 2006). Though outside the realm of communication, learned recognition of predators is also impaired in acidic conditions (Smith et al., 2008; see also Chapter 4).

A widespread anthropogenic cause of acidification is acid rain. Leduc et al. (2009) showed that the responses to alarm pheromone of juvenile Atlantic salmon *Salmo salar* were impaired in small nursery streams following rainfall, in contrast to those of salmon in streams with higher capacity to neutralize acids. Sub-lethal changes to water chemistry can therefore have profound ecological impacts via their effects on communication.

Increases in pH, as commonly produced by eutrophication, may also impair chemical communication. In two freshwater snail species, response to predator odour cues is impaired in alkaline water (Turner and Chislock 2010). By contrast, Heuschele and Candolin (2007) showed that female three-spined sticklebacks increased their sensitivity to pheromone cues in alkaline water. The authors hypothesized that this increased sensitivity might compensate for the reduced salience of visual cues in eutrophic waters.

signifer, response to chemical shoaling cues is reduced by about 20% in saltwater relative to freshwater (Herbert-Read et al. 2010). This basic change in water chemistry can therefore have important impacts on communication.

Humans can also influence the concentration of dissolved organic chemicals in natural bodies of water. Humic acids (HA) are a ubiquitous byproduct of organic decay in freshwaters (but HA concentrations can be substantially increased by processes such as eutrophication, Thomas 1997). HA levels are also higher in disturbed environments associated with agricultural manure and other organic wastes (Kappler and Haderlein 2003).

Elevated levels of HA in aquatic habitats are of particular interest due to their apparent effect in impairing the recognition of conspecific sexual pheromones (Fisher et al. 2006; Hubbard et al. 2002). Mesquita et al. (2003) found that HA tend to dissolve relatively water-insoluble organic substances such as steroidal pheromones, reducing their biological availability to organisms that normally detect these chemicals as part of their communication. Small proteins can also be encapsulated by HA (Zang et al. 2000). Hubbard et al. (2002) proposed that the steroid portion of pheromones adsorbs onto the surface of HA microvesicles that form in water, rendering the pheromone effectively unavailable for olfactory detection. HA can also bind to odorant receptors, as discussed below.

In the swordtail fish *Xiphophorus birchmanni* females lose their preference for conspecific males over heterospecific males in water with elevated levels of HA (Fisher et al. 2006). This impairment of chemical communication may have major implications for the evolutionary fate of many such species because of the potential for interspecific hybridization. Similarly, zebrafish *Danio rerio* failed to distinguish conspecifics from goldfish *Carassius auratus* in HA-treated water (Fabian et al. 2007).

As will be discussed below, chemical disturbance can also have dramatic effects on receivers. With alarm pheromones as a noteworthy exception, very few chemosignals have been characterized for aquatic creatures, which as discussed at the beginning of this section are likely to suffer the greatest impact from disturbance of the chemosignalling environment. The interaction between aquatic pheromones and the aquatic environment is an open area for research.

2.4 Signal detection

Receiver response to a signal involves multiple steps. First, structures in the sensory periphery must transduce properties of the external environment (photons, pressure changes, volatile molecules) into neurochemical signals. These stimuli must be detected above background noise, and they must be perceived appropriately in order to produce a functional behavioural response. Each of these steps, from peripheral transduction to receiver behaviour, is susceptible to interference from anthropogenic effects.

Environmental disturbance can sometimes have a disabling impact on signal detection. Sensory noise can make detection difficult; this is discussed in detail in the section on signal transmission. In numerous forms, however, noise can have more lingering effects on sensation. Exposure to ambient acoustic noise during ontogeny retards development of the auditory cortex in rats (Chang and Merzenich 2003). The spectral quality of the light environment determines the relative abundance of different classes of cone photoreceptors; fish raised in monochromatic light are more sensitive to the wavelengths they have been exposed to during ontogeny (Fuller et al. 2005).

Since olfaction requires direct interactions between environmental chemicals and biological tissues, chemoreception appears to be particularly susceptible to disruption from environmental sources (reviewed in Lurling and Scheffer 2007). Exposure to low concentrations of cadmium impaired the alarm response of juvenile rainbow trout *Oncorhynchus mykiss* (Scott and Sloman 2004), and exposure to copper caused a reduction in the density of olfactory receptor cells and a loss of the alarm response in Colorado pikeminnows *Ptychocheilus lucius*; the authors argued that environmental concentrations of copper could occasionally cause such damage (Beyers and Farmer 2001). This damage could have long-lasting consequences: fathead minnows *Pimephales promelas*, exposed to copper as embryos, failed to respond to the alarm pheromone 90 days later (Carreau and Pyle 2005).

Comparative studies of communication mechanisms have emphasized sensation (but see e.g. Ryan et al. 2009). Very few studies have explicitly examined the effect of disturbance on perception and evaluation. However, there is evidence to show that the constraints imposed by environmental noise increase the difficulty of discriminating among signals; for example discriminating between conspecific and heterospecific signals (e.g. Fisher et al. 2006; Seehausen et al. 1997). Further, changes in the distribution of signals, discussed above, could yield habituation or sensitization to signals; changes in the encounter rate or detectability of particular stimuli can make receivers more or less likely to respond to them. Both habituation and sensitization have been documented in natural communication contexts (e.g. Reichert 2010) and these low-level learning mechanisms should be susceptible to disturbance-induced changes in signal distributions.

In many cases, responses to signals are learned (see Chapter 4), either by early exposure to signallers (Ten Cate et al. 2006) or by modelling the behaviour of other receivers (Dugatkin and Godin 1992). In either case, responses are dependent on the distribution of other individuals in the population. Changes in the distribution of either signal values, receiver responses, or encounter rates should therefore have substantial impacts on learned responses.

Behavioral responses to signals should also be susceptible to disturbance. For example, female mate choice is heavily dependent on predation risk; females tend to be less choosy in risky environments (Johnson and Basolo 2003). Introduced predators, habitat alteration, and trophic changes induced by disturbance could all change an individual's perception of risk and, therefore, its response to signals. These may have short- or long-term effects on receiver responses.

2.5 Population-level and evolutionary effects on signals

In this chapter we have focused on changes to signals at the individual level. However, anthropogenic changes can lead to population-level changes to signal distributions, that is, in the mean and variance of signalling traits. This can occur not only via changes to individual signal expression but also via population level processes such as selective mortality. The most obvious example of this is reduction in the average size of ornaments, such as antlers and horns, in species subjected to trophy hunting (Allendorf and Hard 2009).

Population level changes in animal signals can potentially result from rapid evolutionary responses to radically altered selective regimes, a topic covered in Chapter 16. Alternatively, changes can result simply from major reductions in population size and connectivity, due to massive habitat loss and fragmentation. This can affect phenotypic variation just as it affects neutral genetic variation. For example, inbreeding depression directly affects the expression of acoustic signals in field crickets *Teleogryllus commodus* (Drayton et al. 2007), and small population size causes a reduction in signalling effort in wolf spiders *Hygrolycosa rubrofasciata* (Ahtiainen et al. 2004) and birds (Laiolo and Tella 2008).

Habitat isolation and loss also affect signal expression indirectly due to both genetic and cultural drift, as in the song dialects of some birds (reviewed in Laiolo 2010). Cultural drift refers to a process analogous to genetic drift, whereby population reduction and isolation can erode signal diversity and increase differentiation among local song dialects. Cultural drift results from random errors in dialect transmission, but can be accelerated through increases in the rate of learning mistakes and reduced opportunities for learning from models (Laiolo 2010).

Disturbance can also cause changes in signal distributions that influence signal honesty and reliability and, consequently, selection on those signals. For example, the vastly increased availability of blue objects to bowerbird *Ptilonorhynchus violaceus* males should result in reduced variation among males in their ability to acquire blue ornaments. Due to its lower cost, blue bower ornamentation may become an unreliable indicator of male quality, leading to reduced sexual selection on this signal. Proximity to dense concentrations of carotenoids (e.g. orchards, berry farms) might similarly reduce selection for long-wavelength visual ornaments due to increased availability and reduced cost. Changes to signal honesty and reliability can also result from changes to transmission properties of the environment. For example, Wong et al. (2007) showed that in turbid water, visual signals become unreliable predictors of condition in sticklebacks.

At the population level, a widespread driver of signal evolution is reproductive character displacement, whereby signals diverge from those of closely related, sympatric heterospecifics. As pointed out by Servedio (2004), invasive species have the potential to exert selection on native signallers (see Chapter 14). Such signal divergence would serve to minimize gene flow between the introduced and native species. Further, as has been recognized for most of a century (Mayr 1942), native species are more likely to hybridize in disturbed environments. This may be due to increased opportunities for contact with introduced species or impacts on the way that receivers perceive signals.

Disturbance effects on communication systems can have evolutionary effects that extend far beyond their original impacts. We earlier discussed how humic acids (HA) impair chemical communication in swordtails. HA are found in any natural freshwater body and are considered benign substances; their discharge is entirely unregulated by the US Environmental Protection Agency. Yet, HA have caused irreversible evolutionary changes via their effects on communication. Specifically, by breaking down prezygotic isolating mechanisms between *X. malinche* and *X. birchmanni* (Fisher et al. 2006), HA are responsible for rampant hybridization and introgression between these species. Though very few first-generation hybrids are produced in contemporary populations, later-generation hybrids and backcrosses abound (Culumber et al. 2011), long after disturbance has passed. Via its effects on communication, a non-toxicological disturbance can therefore have irrecoverable evolutionary and ecological consequences.

2.6 Conclusions

Efficient communication is central to the health of animal populations. Given the importance of communication to biodiversity conservation, a more expansive approach to taxa and modalities is in order. Most studies to date have focused on acoustic effects in terrestrial habitats, with a heavy emphasis on birds (reviewed in Laiolo 2010; Patricelli and Blickley 2006). Among visual studies, most have centred on spectral changes to the light environment, rather than on how disturbance changes the spatiotemporal aspects of visual signals. Chemical communication in aquatic environments may be one of the most sensitive targets of environmental disturbance, since water chemistry can interact directly with signal transmission, production, and reception.

To our knowledge, there are no published studies that explicitly address anthropogenic effects on communication modalities other than vision, audition, and olfaction. Since vibrational communication depends on the propagation characteristic of specific host plants (Cocroft et al. 2010) and is sensitive to wind conditions (McNett et al. 2010), human-induced alterations of plant communities or habitat structure could have potentially deleterious effects on substrate-borne communication. Mining and other subterranean activities have the potential to disrupt communication in fossorial species (e.g. Nevo et al. 1991). Similarly, transmission of electrical signals is restricted to freshwaters of the rainy tropics; electrocommunication depends on water conductivity, which can change according to the concentration of salts and other substances. The relatively limited environments in which vibrational and electrical communication are possible may be an indication that these modalities are particularly susceptible to disturbance effects.

Finally, there is a dearth of knowledge in general about the mechanisms involved in response to signals beyond low-level properties of sensory systems. Recent studies (e.g. Heuschele and Candolin 2007) highlight the likely importance of changes in multimodal integration, both in terms of signaller behaviour and the way that receivers process and attend to information. Increased attention to the neural and psychological mechanisms underlying signaller decisions and receiver response is of primary importance both to our basic understanding of how animals communicate, and the applied challenge of understanding how disturbance impacts communication.

Acknowledgements

We thank B. B. M. Wong and U. Candolin for their helpful comments and tireless work in putting together this volume. H. Slabbekoorn provided feedback on an earlier draft, which greatly improved the manuscript. We are indebted to N. Ratterman, R. Cui, L. Dieckman, Z. Culumber, C. Carlson, and J. B. Johnson for valuable comments.

References

Ahtiainen, J. J., Alatalo, R. V., Mappes, J., and Vertainen, L. (2004). Decreased sexual signalling reveals reduced viability in small populations of the drumming wolf spider *Hygrolycosa rubrofasciata*. *Proceedings of the Royal Society of London B, Biological Sciences*, 271, 1839–45.

Allendorf, F. W. and Hard, J. J. (2009). Human-induced evolution caused by unnatural selection through harvest of wild animals. *Proceedings of the National Academy of Sciences of the USA*, 106, 9987–94.

Arellano-Aguilar, O. and Garcia, C. M. (2008). Exposure to pesticides impairs the expression of fish ornaments reducing the availability of attractive males. *Proceedings of the Royal Society of London B, Biological Sciences*, 275, 1343–50.

Baker, B. J. and Richardson, J. M. L. (2006). The effect of artificial light on male breeding-season behaviour in green frogs, *Rana clamitans melanota*. *Canadian Journal of Zoology-Revue Canadienne De Zoologie*, 84, 1528–32.

Barber, J. R., Crooks, K. R., and Fristrup, K. M. (2010). The costs of chronic noise exposure for terrestrial organisms. *Trends in Ecology & Evolution*, 25, 180–9.

Bayne, E. M., Habib, L., and Boutin, S. (2008). Impacts of chronic anthropogenic noise from energy-sector activity on abundance of songbirds in the boreal forest. *Conservation Biology*, 22, 1186–93.

Beeton, A. M. (2002). Large freshwater lakes: present state, trends, and future. *Environmental Conservation*, 29, 21–38.

Beyers, D. W. and Farmer, M. S. (2001). Effects of copper on olfaction of Colorado pikeminnow. *Environmental Toxicology and Chemistry*, 20, 907–12.

Bradbury, J. W. and Vehrencamp, S. L. (1998). *Principles of Animal Communication*, Sunderland, MA, Sinauer.

Brown, G. E., Adrian, J. C., Lewis, M. G., and Tower, J. M. (2002). The effects of reduced pH on chemical alarm signalling in ostariophysan fishes. *Canadian Journal of Fisheries and Aquatic Sciences*, 59, 1331–8.

Brumm, H. and Slater, P. J. B. (2006). Ambient noise, motor fatigue, and serial redundancy in chaffinch song. *Behavioral Ecology and Sociobiology*, 60, 475–81.

Buchanan, K. L. (2000). Stress and the evolution of condition-dependent signals. *Trends in Ecology & Evolution*, 15, 156–60.

Carreau, N. D. and Pyle, G. G. (2005). Effect of copper exposure during embryonic development on chemosensory function of juvenile fathead minnows (*Pimephales promelas*). *Ecotoxicology and Environmental Safety*, 61, 1–6.

Chang, E. F. and Merzenich, M. M. (2003). Environmental noise retards auditory cortical development. *Science*, 300, 498–502.

Clark, C. W., Ellison, W. T., Southall, B. L., Hatch, L., Van Parijs, S. M., Frankel, A., and Ponirakis, D. (2009). Acoustic masking in marine ecosystems: intuitions, analysis, and implication. *Marine Ecology-Progress Series*, 395, 201–22.

Clark, H. O. (2001). Use of a car alarm sequence in the northern mockingbird repertoire. *California Fish and Game*, 87, 115–16.

Cocroft, R. B., Rodriguez, R. L., and Hunt, R. E. (2010). Host shifts and signal divergence: mating signals covary with host use in a complex of specialized plant-feeding insects. *Biological Journal of the Linnean Society*, 99, 60–72.

Culumber, Z. W., Fisher, H. S., Tobler, M., Mateos, M., Barber, P. H., Sorenson, M. D., and Rosenthal, G. G. (2011). Replicated hybrid zones of *Xiphophorus* swordtails along an elevational gradient. *Molecular Ecology*, 20, 342–56.

Cummings, M. E. (2007). Sensory trade-offs predict signal divergence in surfperch. *Evolution*, 61, 530–45.

Cunnington, G. M. and Fahrig, L. (2010). Plasticity in the vocalizations of anurans in response to traffic noise. *Acta Oecologica-International Journal of Ecology*, 36, 463–70.

Dauwe, T., Janssens, E., and Eens, M. (2006). Effects of heavy metal exposure on the condition and health of adult great tits (*Parus major*). *Environmental Pollution*, 140, 71–8.

Drayton, J. M., Hunt, J., Brooks, R., and Jennions, M. D. (2007). Sounds different: inbreeding depression in sexually selected traits in the cricket *Teleogryllus commodus*. *Journal of Evolutionary Biology*, 20, 1138–47.

Dugatkin, L. A. and Godin, J.-G. J. (1992). Reversal of female mate choice by copying in the guppy (*Poecilia reticulata*). *Proceedings of the Royal Society of London B, Biological Sciences*, 249, 179–84.

Endler, J. A. (1978). A predator's view of animal color patterns. *Evolutionary Biology*, 11, 319–54.

Endler, J. A. (1992). Signals, signal conditions, and the direction of evolution. *American Naturalist*, 139, S125–53.

Endler, J. A., Westcott, D. A., Madden, J. R., and Robson, T. (2005). Animal visual systems and the evolution of color patterns: Sensory processing illuminates signal evolution. *Evolution*, 59, 1795–818.

Fabian, N. J., Albright, L. B., Gerlach, G., Fisher, H. S., and Rosenthal, G. G. (2007). Humic acid interferes with species recognition in zebrafish (Danio rerio). *Journal of Chemical Ecology*, 33, 2090–6.

Fisher, H. S., Wong, B. B. M., and Rosenthal, G. G. (2006). Alteration of the chemical environment disrupts communication in a freshwater fish. *Proceedings of the Royal Society of London B, Biological Sciences*, 273, 1187–93.

Fleishman, L. J. (1986). Motion detection in the presence and absence of background motion in an Anolis lizard. *Journal of Comparative Physiology*, 159, 711–20.

Fonken, L. K., Workman, J. L., Walton, J. C., Weil, Z. M., Morris, J. S., Haim, A., and Nelson, R. J. (2010). Light at night increases body mass by shifting the time of food intake. *Proceedings of the National Academy of Sciences of the USA*, 107, 18664–9.

Foote, A. D., Osborne, R. W., and Hoelzel, A. R. (2004). Whale-call response to masking boat noise. *Nature*, 428, 910–910.

Fuller, R. A., Warren, P. H., and Gaston, K. J. (2007). Daytime noise predicts nocturnal singing in urban robins. *Biology Letters*, 3, 368–70.

Fuller, R. C., Carleton, K. L., Fadool, J. M., Spady, T. C., and Travis, J. (2005). Genetic and environmental variation in the visual properties of bluefin killifish, *Lucania goodei*. *Journal of Evolutionary Biology*, 18, 516–23.

Geens, A., Dauwe, T., and Eens, M. (2009). Does anthropogenic metal pollution affect carotenoid colouration, antioxidative capacity and physiological condition of great tits (Parus major)? *Comparative Biochemistry and Physiology C-Toxicology, and Pharmacology*, 150, 155–63.

Gorissen, L., Snoeijs, T., Duyse, E., and Eens, M. (2005). Heavy metal pollution affects dawn singing behaviour in a small passerine bird. *Oecologia*, 145, 504–9.

Helfman, G. E. (2009). *The Diversity of Fishes: Biology, Evolution, and Ecology*, Hoboken, NJ, Blackwell.

Herbert-Read, J. E., Logendran, D., and Ward, A. J. W. (2010). Sensory ecology in a changing world: salinity alters conspecific recognition in an amphidromous fish, *Pseudomugil signifer*. *Behavioral Ecology and Sociobiology*, 64, 1107–15.

Heuschele, J. and Candolin, U. (2007). An increase in pH boosts olfactory communication in sticklebacks. *Biology Letters*, 3, 411–13.

Heuschele, J., Mannerla, M., Gienapp, P., and Candolin, U. (2009). Environment-dependent use of mate choice cues in sticklebacks. *Behavioral Ecology*, 20, 1223–7.

Holt, M. M., Noren, D. P., Veirs, V., Emmons, C. K., and Veirs, S. (2009). Speaking up: Killer whales (*Orcinus orca*) increase their call amplitude in response to vessel noise. *Journal of the Acoustical Society of America*, 125, EL27–EL32.

Hu, Y. and Cardoso, G. C. (2009). Are bird species that vocalize at higher frequencies preadapted to inhabit noisy urban areas? *Behavioral Ecology*, 20, 1268–73.

Hubbard, P. C., Barata, E. N., and Canario, A. V. M. (2002). Possible disruption of pheromonal communication by humic acid in the goldfish, *Carassius auratus*. *Aquatic Toxicology*, 60, 169-183.

Hunter, C. P. and Dwyer, P. D. (1997). The value of objects to Satin Bowerbirds *Ptilonorhynchus violaceus*. *Emu*, 97, 200–6.

Jepson, P. D., Arbelo, M., Deaville, R., et al. (2003). Gas-bubble lesions in stranded cetaceans. *Nature*, 425, 575–6.

Johnson, J. B. and Basolo, A. L. (2003). Predator exposure alters female mate choice in the green swordtail. *Behavioral Ecology*, 14, 619–25.

Kappler, A. and Haderlein, S. B. (2003). Natural organic matter as reductant for chlorinated aliphatic pollutants. *Environmental Science & Technology*, 37, 2714–19.

Kempenaers, B., Borgstrom, P., Loes, P., Schlicht, E., and Valcu, M. (2010). Artificial night lighting affects dawn song, extra-pair siring success, and lay date in songbirds. *Current Biology*, 20, 1735–9.

Laiolo, P. (2010). The emerging significance of bioacoustics in animal species conservation. *Biological Conservation*, 143, 1635–45.

Laiolo, P. and Tella, J. L. (2008). Social determinants of songbird vocal activity and implications for the persistence of small populations. *Animal Conservation*, 11, 433–41.

Leduc, A., Kelly, J. M., and Brown, G. E. (2004). Detection of conspecific alarm cues by juvenile salmonids under neutral and weakly acidic conditions: laboratory and field tests. *Oecologia*, 139, 318–24.

Leduc, A., Roh, E., and Brown, G. E. (2009). Effects of acid rainfall on juvenile Atlantic salmon (Salmo salar) antipredator behaviour: loss of chemical alarm function and potential survival consequences during predation. *Marine and Freshwater Research*, 60, 1223–30.

Li, W., Twohey, M., Jones, M., and Wagner, M. (2007). Research to guide use of pheromones to control sea lamprey. *Journal of Great Lakes Research*, 33, 70–86.

Longcore, T. and Rich, C. (2004). Ecological light pollution. *Frontiers in Ecology and the Environment*, 2, 191–8.

Lurling, M. and Scheffer, M. (2007). Info-disruption: pollution and the transfer of chemical information between organisms. *Trends in Ecology & Evolution*, 22, 374–9.

Madden, J. R. and Tanner, K. (2003). Preferences for coloured bower decorations can be explained in a nonsexual context. *Animal Behaviour*, 65, 1077–83.

Marchetti, K. (1993). Dark habitats and bright birds illustrate the role of the environment in species divergence. *Nature*, 362, 149–52.

Markman, S., Leitner, S., Catchpole, C., Barnsley, S., Muller, C. T., Pascoe, D., and Buchanan, K. L. 2008. Pollutants increase song complexity and the volume of the brain area HVC in a songbird. *PLOS One*, 3. e1674.

Marshall, A. J. (1954). *Bower-birds: Their Displays and Breeding Cycles*, Oxford, U.K., Clarendon Press.

Marshall, N. J. (2000). Communication and camouflage with the same 'bright' colours in reef fishes. *Philosophical Transactions of the Royal Society B: Biological Sciences*, 355, 1243–8.

Mayr, E. (1942). *Systematics and the Origin of Species*, New York, Columbia University Press.

Mcnett, G. D., Luan, L. H., and Cocroft, R. B. 2010. Wind-induced noise alters signaler and receiver behavior in vibrational communication. *Behavioral Ecology and Sociobiology*, 64, 2043–51.

Mesquita, R. M. R. S., Canario, A. V. M., and Melo, E. (2003). Partition of fish pheromones between water and aggregates of humic acids. Consequences for sexual signaling. *Environmental Science & Technology*, 37, 742–6.

Miller, P. J. O., Biassoni, N., Samuels, A., and Tyack, P. L. 2000. Whale songs lengthen in response to sonar. *Nature*, 405, 903–903.

Navara, K. J. and Nelson, R. J. (2007). The dark side of light at night: physiological, epidemiological, and ecological consequences. *Journal of Pineal Research*, 43, 215–24.

Nemeth, E. and Brumm, H. (2010). Birds and anthropogenic noise: are urban songs adaptive? *American Naturalist*, 176, 465–75.

Nevo, E., Heth, G., and Pratt, H. (1991). Seismic communication in a blind subterranean mammal—a major somatosensory mechanism in adaptive evolution underground. *Proceedings of the National Academy of Sciences of the USA*, 88, 1256–60.

Olivier, A., Leduc, H. C., Roh, E., Harvey, M. C., and Brown, G. E. (2006). Impaired detection of chemical alarm cues by juvenile wild Atlantic salmon (Salmo salar) in a weakly acidic environment. *Canadian Journal of Fisheries and Aquatic Sciences*, 63, 2356–63.

Ord, T. J., Peters, R. A., Clucas, B., and Stamps, J. A. (2007). Lizards speed up visual displays in noisy motion habitats. *Proceedings of the Royal Society of London B, Biological Sciences*, 274, 1057–62.

Ostlund-Nilsson, S. and Holmlund, M. (2003). The artistic three-spined stickleback (*Gasterosteus aculeatus*). *Behavioral Ecology and Sociobiology*, 53, 214–20.

Parris, K. M. and Schneider, A. (2009). Impacts of traffic noise and traffic volume on birds of roadside habitats. *Ecology and Society*, 14, 29.

Patricelli, G. L. and Blickley, J. L. (2006). Avian communication in urban noise: Causes and consequences of vocal adjustment. *Auk*, 123, 639–49.

Patricelli, G. L., Uy, J. A. C., and Borgia, G. (2003). Multiple male traits interact: attractive bower decorations facilitate attractive behavioural displays in satin bowerbirds. *Proceedings of the Royal Society of London B, Biological Sciences*, 270, 2389–95.

Pérez, C., Lores, M., and Velando, A. (2010a). Oil pollution increases plasma antioxidants but reduces coloration in a seabird. *Oecologia*, 163, 875–84.

Pérez, C., Munilla, I., Lopez-Alonso, M., and Velando, A. (2010b). Sublethal effects on seabirds after the Prestige oil-spill are mirrored in sexual signals. *Biology Letters*, 6, 33–5.

Reichert, M. S. (2010). Aggressive thresholds in *Dendropsophus ebraccatus*: habituation and sensitization to different call types. *Behavioral Ecology and Sociobiology*, 64, 529–39.

Rendall, D., Owren, M. J., and Ryan, M. J. (2009). What do animal signals mean? *Animal Behaviour*, 78, 233–40.

Ridgway, R. L., Silverstein, R. M., and Inscoe, M. N. (eds) (1990). *Behavior-Modifying Chemicals for Insect Management: Applications of Pheromones and Other Attractants*, New York, M. Dekker.

Rosenthal, G. G. (2007). Spatiotemporal dimensions of visual signals in animal communication. *Annual Review of Ecology Evolution and Systematics*, 38, 155–78.

Rushbrook, B. J., Head, M. L., Katsiadaki, I., and Barber, I. (2010). Flow regime affects building behaviour and nest structure in sticklebacks. *Behavioral Ecology and Sociobiology*, 64, 1927–35.

Ryan, M. J. (1998). Sexual selection, receiver biases, and the evolution of sex differences. *Science*, 281, 1999–2003.

Ryan, M. J., Akre, K. L., and Kirkpatrick, M. (2009). Cognitive mate choice. In: R. Dukas, J. R. (ed.) *Cognitive Ecology II*. Chicago: University of Chicago Press.

Ryan, M. J. and Brenowitz, E. A. (1985). The role of body size, phylogeny, and ambient noise in the evolution of bird song. *American Naturalist*, 126, 87–100.

Ryan, M. J. and Keddy-Hector, A. (1992). Directional patterns of female mate choice and the role of sensory biases. *American Naturalist*, 139, S4–S35.

Sandell, M. I., Smith, H. G., and Bruun, M. (1996). Paternal care in the European starling, *Sturnus vulgaris*: nestling provisioning. *Behavioral Ecology and Sociobiology*, 39, 301–9.

Scott, G. R. and Sloman, K. A. (2004). The effects of environmental pollutants on complex fish behaviour: integrating behavioural and physiological indicators of toxicity. *Aquatic Toxicology*, 68, 369–92.

Secondi, J., Aumjaud, A., Pays, O., Boyer, S., Montembault, D., and Violleau, D. (2007). Water turbidity affects the

development of sexual morphology in the palmate newt. *Ethology*, 113, 711–20.
Secondi, J., Hinot, E., Djalout, Z., Sourice, S., and Jadas-Hecart, A. (2009). Realistic nitrate concentration alters the expression of sexual traits and olfactory male attractiveness in newts. *Functional Ecology*, 23, 800–8.
Seehausen, O., Van Alphen, J. J. M., and Witte, F. (1997). Cichlid fish diversity threatened by eutrophication that curbs sexual selection. *Science*, 277, 1808–11.
Sergio, F., Blas, J., Blanco, G., Tanferna, A., López, L., Lemus, J. A., and Hiraldo, F. (2011). Raptor nest decorations are a reliable threat against conspecifics. *Science*, 331, 327–30.
Servedio, M. R. (2004). The what and why of research on reinforcement. *PLOS Biology*, 2, e420.
Shashar, N., Sabbah, S., and Cronin, T. W. (2004). Transmission of linearly polarized light in seawater: implications for polarization signaling. *Journal of Experimental Biology*, 207, 3619–28.
Slabbekoorn, H., Bouton, N., Van Opzeeland, I., Coers, A., Ten Cate, C., and Popper, A. N. (2010). A noisy spring: the impact of globally rising underwater sound levels on fish. *Trends in Ecology & Evolution*, 25, 419–27.
Slabbekoorn, H., Ellers, J., and Smith, T. B. (2002). Birdsong and sound transmission: The benefits of reverberations. *Condor*, 104, 564–73.
Smith, J. J., Leduc, A., and Brown, G. E. (2008). Chemically mediated learning in juvenile rainbow trout. Does predator odour pH influence intensity and retention of acquired predator recognition? *Journal of Fish Biology*, 72, 1750–60.
Sun, J. W. C. and Narins, P. A. (2005). Anthropogenic sounds differentially affect amphibian call rate. *Biological Conservation*, 121, 419–27.
Ten Cate, C., Verzijden, M. N. , and Etman, E. (2006). Sexual imprinting can induce sexual preferences for exaggerated parental traits. *Current Biology*, 16, 1128–32.
Thomas, J. D. (1997). The role of dissolved organic matter, particularly free amino acids and humic substances, in freshwater ecosystems. *Freshwater Biology*, 38, 1–36.
Turner, A. M. and Chislock, M. F. (2010). Blinded by the stink: nutrient enrichment impairs the perception of predation risk by freshwater snails. *Ecological Applications*, 20, 2089–95.
Vasconcelos, R. O., Amorim, M. C. P., and Ladich, F. (2007). Effects of ship noise on the detectability of communication signals in the Lusitanian toadfish. *Journal of Experimental Biology*, 210, 2104–12.
Velez, Z., Hubbard, P. C., Barata, E. N., and Canario, A. V. M. (2009). Adaptation to reduced salinity affects the olfactory sensitivity of Senegalese sole (Solea senegalensis Kaup 1858) to Ca 2 + and Na + but not amino acids. *Journal of Experimental Biology*, 212, 2532–40.
Verzijden, M. N., Ripmeester, E. A. P., Ohms, V. R., Snelderwaard, P., and Slabbekoorn, H. (2010). Immediate spectral flexibility in singing chiffchaffs during experimental exposure to highway noise. *Journal of Experimental Biology*, 213, 2575–81.
Ward, A. J. W., Duff, A. J., Horsfall, J. S., and Currie, S. (2008). Scents and scents-ability: pollution disrupts chemical social recognition and shoaling in fish. *Proceedings of the Royal Society of London B, Biological Sciences*, 275, 101–5.
Warren, P. S., Katti, M., Ermann, M., and Brazel, A. (2006). Urban bioacoustics: it's not just noise. *Animal Behaviour*, 71, 491–502.
Wojcieszek, J. M., Nicholls, J. A., and Goldizen, A. W. (2007). Stealing behavior and the maintenance of a visual display in the satin bowerbird. *Behavioral Ecology*, 18, 689–95.
Wong, B. B. M., Candolin, U., and Lindstrom, K. (2007). Environmental deterioration compromises socially enforced signals of male quality in three-spined sticklebacks. *American Naturalist*, 170, 184–9.
Zanette, L., Clinchy, M., and Sung, H. C. (2009). Food-supplementing parents reduces their sons' song repertoire size. *Proceedings of the Royal Society of London B, Biological Sciences*, 276, 2855–60.
Zang, X., Van Heemst, J. D. H., Dria, K. J., and Hatcher, P. G. (2000). Encapsulation of protein in humic acid from a histosol as an explanation for the occurrence of organic nitrogen in soil and sediment. *Organic Geochemistry*, 31, 679–95.
Zhang, J., Wang, T., Chameides, W. L., Cardelino, C., Kwok, J., Blake, D. R., Ding, A., and So, K. L. (2007). Ozone production and hydrocarbon reactivity in Hong Kong, Southern China. *Atmospheric Chemistry and Physics*, 7, 557–73.

CHAPTER 3

The endocrine system: can homeostasis be maintained in a changing world?

Katherine L. Buchanan and Jesko Partecke

⊃ Overview

In a changing world the endocrine system maintains internal physiological systems, ensuring survival and reproduction are maximized despite environmental perturbations. Flexibility is a key characteristic of the endocrine system, but when environmental parameters move outside the normal range, the system may be stretched beyond its capacity to cope. With particular reference to birds, we discuss the potential for anthropogenic change to affect endocrine mechanisms and the capacity of the endocrine system to maintain a homeostatic balance. In particular, we discuss the impact of climate change on phenology, the impacts of urbanization on endocrine systems, and the role of endocrine disrupting chemicals in effecting change. It is clear that the endocrine system is vulnerable to disruption and, as the system controls behaviour and reproduction, the effects could be potentially catastrophic. The response to anthropogenic change, as for other traits, will be determined by the magnitude of the anthropogenic effects and the speed of change.

3.1 Introduction

The world has always been in a state of flux. However, in recent times the speed of environmental change has increased substantially, with increased urbanization, the destruction and degradation of habitats, the reduction of food supplies through farming and fishing, the release of invasive species, pollution and, of course, climate change. Humans are dramatically altering environmental conditions throughout the world (Vitousek et al. 1997; Clark et al. 1993). Animals can potentially adapt, but the question here is whether they are flexible enough to be able to keep up with the speed of change. Physiological constraints are likely to play a role in determining the potential for adaptation. In this chapter, focusing mostly (but not exclusively) on avian examples, we seek to examine why a changing world affects the endocrine response, how the endocrine system is likely to respond, and what consequences this may have for species survival and the evolutionary process.

3.2 The endocrine system

The body has three intimately and intrinsically linked physiological systems: the endocrine system, the nervous system, and the immune system, each of which have their chemical messenger systems, some of which are produced by more than one system. Chemical messengers within the endocrine system are the classic examples of what many envisage as a hormone, but they are only a subset of the many chemical messengers produced by the body. By definition, endocrine hormones are pro-

Behavioural Responses to a Changing World. First Edition. Edited by Ulrika Candolin and Bob B.M. Wong.
© 2012 Oxford University Press. Published 2012 by Oxford University Press.

duced by endocrine glands, released into the bloodstream and, in only minute concentrations, effect changes in a tissue at some distance from the site of production (Brown 1994). However defining a 'true' hormone can be problematic due to their multiple sites of production, range of action, and modes of transfer through the body. Other chemical messengers include autocrine communication where the product directly affects the cells which produce it (e.g. neurotransmitters regulating their own release) and paracrine communication where the messenger affects adjacent cells (e.g. testosterone stimulating the Sertoli cells within the testes). Neuroendocrine secretion occurs when neurohormones are secreted from the nervous system into the peripheral circulatory system to act at some distance from the source of secretion (e.g. oxytocin produced by the posterior pituitary acting within the uterus). This variation can be found even within a single hormone. For example, 17β-estradiol has multiple sites of manufacture and modes of delivery. The brain produces many different neurotransmitters which modulate neural activity on site. Vertebrate hormones are highly conserved and generally show common pathways for biosynthesis. This is an important point, because it means synthetic substances which interact with the endocrine system (see below) in one taxon, are likely to have the same effect in another taxon (Van der Kraak et al. 1998). The main sites of endocrine production include the pineal gland (melatonin), the pituitary gland (follicle stimulating hormones, lutenizing hormones, adrenocorticotrophic hormone, oxytocin, vasopressin, thyroid stimulating hormone, growth hormone, prolactin), thyroid gland (thyroxine), pancreas (insulin), adrenal cortex (glucocorticoids), adrenal medulla (adrenaline, noradrenaline), testes (androgens), and ovaries (progesterones and estrogens) (Nelson 2011). However the heart, stomach and duodenum also act as endocrine glands and secrete hormones. Together this suite of chemical messengers act to maintain a physiological balance within the body, regulate growth and reproduction, as well as appropriate behavioural responses. Relevant to considering the networks of control is the hypothalamic–pituitary (HPA) axis, which determines the physiological response to external stressors, and the hypothalamic–pituitary–gonadal (HPG) axis, which regulates the reproductive system and also influences the immune response.

3.3 Environmental disruption of the endocrine response

One of the principle roles of the endocrine system is to maintain homeostasis, meaning that the internal state is maintained in constancy, despite fluxes in external conditions. It seems obvious, therefore, that the endocrine system should be flexible and able to adapt to changes in the external environment. The endocrine system also regulates adaptive physiological changes, such as the onset and maintenance of a reproductive state, aggression, fat stores, metabolic turnover, and secondary sexual traits amongst other factors (Nelson 2011). The magnitude and timing of these physiological changes are regulated by the endocrine response. The endocrine system involves a series of secretions which activate a cascade with a behavioural or physiological change as an endpoint (Fig. 3.1). Endocrine cascades are also characterized by positive and negative feedback loops which allow the system to be self-controlling. There is considerable flexibility within this system at all levels, which allows the animal to function despite external influences that can 'stress' the organism by challenging its ability to function within its normal physiological range (Fig. 3.1).

The classic work of Hans Selye first developed the concept of stress as we see it today: that similar physiological responses can be caused by a range of non-specific challenges (Ursin and Olff 1993), which function to promote survival. We now understand stress as a complex multidimensional concept that involves the stressor (the stimulus causing the reaction), the processing of the stimulus, and the physiological stress response which minimizes the damage to the organism when experiencing suboptimal conditions. The effects of environmental challenges can be additive, and habituation can occur over prolonged exposure. The physiological system is able to respond, therefore, to both predictable and unpredictable environmental changes, as well as to severe, but rare, scenarios (Wingfield 2008). This

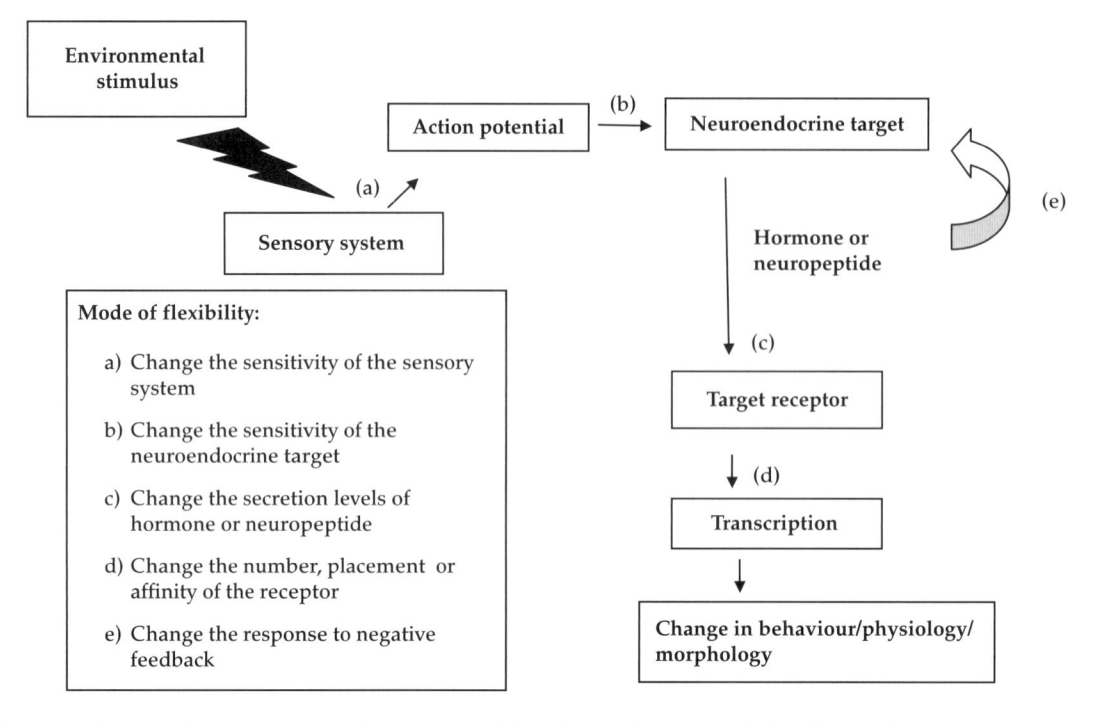

Figure 3.1 Overview of the hormonal signal transduction pathway and the main potential regulatory modes by which the endocrine system can be adjusted to novel environmental conditions. It is worth noting that many if not all parts of the system are subject to short-term adaptive change, as well as longer-term evolutionary adaptation.

Mode of flexibility:
a) Change the sensitivity of the sensory system
b) Change the sensitivity of the neuroendocrine target
c) Change the secretion levels of hormone or neuropeptide
d) Change the number, placement or affinity of the receptor
e) Change the response to negative feedback

means that the response is very flexible to even substantial changes in environmental conditions that fall within the 'normal' range, thus allowing homeostasis to be maintained. However, under extreme or novel conditions the 'normal' range may not apply. Anthropogenic change causes the normal boundaries to be reset by altering environmental stimuli such as food supply, temperature, availability of reproductive opportunities, and exposure to predators, pathogens, or parasites. Anthropogenic change produces stressors which lead to changes within the HPA axis for as long as the stimulus is present. For example, a prey species experiencing increased exposure to a new species of invasive predator will elevate its glucocorticoid levels to allow effective behavioural adjustments and mobilize fat reserves to allow energy for escape. The animal is likely to show some habituation, decreasing the endocrine response on each repeated exposure event, but the chronic stress of repeated exposure is likely to cause long term physiological damage as well as behavioural changes (e.g. reduced foraging) which may contribute to a reduction in condition. This is but one example of how anthropogenic change might push animals to their boundaries of 'coping'—a phenomenon which is crucially mediated by the functioning of the endocrine system. If the stimulus continues to operate outside of 'normal' levels, physiological changes will ensue in the longer term, which enable the system to adjust (e.g. alterations in receptor number or placement) without necessarily eliciting a genetic change. The ability of individuals to respond to novel challenges by means of phenotypic plasticity may reduce the strength of natural selection on physiological characters. In the longer term, selection on trait expression (e.g. rate of hormone production, receptor number, receptor affinity) may occur to allow the endocrine system to function within the new range of the stimulus intensity. The most crucial question, of course, is whether the speed of environmental change is too swift or of a magnitude which is too great to allow selection to occur. Therefore, although the endocrine system lends itself to coping with environmental change, the nature, speed, and magnitude of change are essential for determining the ultimate impact on the animal.

3.4 Photoperiodism and climate change

For most animals, the timing of breeding is essentially determined by long-term predictive cues such as photoperiod, which initiate a cascade of neuroendocrine and endocrine events that lead to the development of gonads and the activation of reproductive behaviours. In addition, supplementary cues such as food supply, temperature, weather conditions, and behavioural stimuli can modify the effects of photoperiod to advance or delay the timing of reproduction (Wingfield et al. 1993; Wingfield et al. 1992). This allows the exact timing of egg-laying to integrate multiple cues, such that offspring demands coincide with food availability. The advent of elevated temperatures and changes to rainfall patterns associated with global climate change may have important implications for the timing of life history stages of animals, resulting in alterations of the onset, duration, or termination of events, such as reproductive periods or migration (Wingfield 2008). It has long been recognized that male birds show seasonal development of the testes under photoperiodic influence (Farner and Wilson 1957), and female birds also show a strong photoperiodic influence to the initiation of breeding due to follicular stimulation (Farner et al. 1966). If the initiation of breeding was under purely photoperiod control, then lay dates would not alter with increased temperatures due to climate change. However, follicular development is also influenced by male behaviour, and other environmental cues (Hinde and Steele 1976), potentially including temperature per se. This would allow the exact timing of egg laying to integrate multiple cues, such that offspring demands coincide with food availability. Recent data suggest, however, that temperature does not directly affect initial follicular development or concentrations of either prolactin or luteinizing hormone (Visser et al. 2011), but that temperature plays an important role in the timing of the cessation of breeding. There are good reasons to suppose that the mechanisms controlling the timing of reproduction in the two sexes should differ, as selection on the timing of individual reproductive investment may differ between males and females (Ball and Ketterson 2008). As reproduction in females is much more energetically costly than males, it is crucial that females time egg production correctly. Female reproductive development is therefore tied to a number of supplementary cues, including food availability and temperature (Wingfield et al. 1996; Wingfield et al. 1997), whilst male gonadal development appears to rely on photoperiodic stimulation alone (Ball and Ketterson 2008). A series of studies have addressed whether elevated temperatures, which are predicted under climate change, have resulted in altered dates for the onset of breeding in a number of species (Visser et al. 2003), with the conclusion that in some avian populations at least there is strong support. Much of this effect is probably due to cues associated with food availability, or indeed direct effects of food intake on female condition. But it also seems that temperature may play a direct role in determining avian lay date separately from any effect of food availability. Under controlled photoperiod, female white-crowned sparrows *Zonotrichia leucophrys* show greater follicular development at higher temperatures (Wingfield et al. 1997). In addition, female great tits *Parus major* lay eggs earlier in elevated temperatures with a much smaller temperature difference between groups (Visser et al. 2009), whilst testis growth is probably not constrained by the associated energetic issues (Caro and Visser 2009). Here is also convincing evidence that temperature plays a crucial role in determining the length of the breeding season. Testicular regression has been shown to occur earlier at higher temperatures in both starlings *Sturnus vulgaris* (Dawson 2005) and great tits (Silverin et al. 2008), an effect which would predict a shortening of the breeding season. This effect has been inferred from recent reductions in the occurrence of second broods in great tits, suggesting that climate change may have fundamental effects on life history traits (Husby et al. 2009). This effect may well occur as a result of the mechanisms within the neuroendocrine system which induce photorefractoriness (Dawson and Sharp 2010). The endocrine mediators are not known, although this effect is likely to be associated with the timing of moult.

Photoperiodic responses are widespread and photoperiodism is crucial for initiating physiological and developmental processes across a range of taxa, including molluscs, anthropods, fish, frogs, birds, and mammals (Bradshaw and Holzapfel 2007). In mammals, photoperiod cues are essential for synchronizing seasonal changes in physiology and behaviour (Bradshaw and Holzapfel 2007) and may be particularly important in high-latitude species where the opportunity for breeding is short-lived. However, large bodied, longer-lived mammals tend to show a higher degree of photoperiodic cueing of reproduction, whilst smaller bodied mammals respond more to environmental cues, such as temperature or rainfall (Bronson 2009). In a situation where there is a mismatch between seasonal environmental cues and photoperiodic cues, as is predicted under climate change forecasts, it may be these longer-lived mammalian species which are more affected (Bronson 2009).

3.5 Urbanization and its ecological effects

The existence and rapid spread of urban areas represents one of the most striking human-induced changes to the environment (Johnston 2001). Besides the fact that urbanization results in a loss of species diversity, many native and non-native species have successfully colonized urban areas. Other species are on the march to colonize cities (McKinney 2006). A key issue is how species respond to, and cope with, a rapidly urbanizing world.

The urban environment as a new type of ecosystem significantly differs from nearby non-urban 'natural' habitats in a variety of abiotic and biotic factors. Urban populations, for instance, may suffer less from climatic stress especially during the winter months, due to the warmer microclimate ('heat island effect') or from lower predations risk. In addition, food in urban habitats is often abundant year round (Shochat et al. 2006). On the other hand, animals living in urban areas are confronted with many novel and potentially stressful conditions rarely experienced in their original, 'natural' environments, such as unfamiliar food sources, elevated anthropogenic disturbance, permanent presence and high density of humans, dogs, and cats, and increased levels of artificial lighting and noise. The combination of these factors creates a new environment favouring individuals that are able to cope behaviourally and physiologically with altered and 'novel' conditions.

Recent studies suggest that urbanization does not only affect the overall species composition and aspects of their phenology (Partecke et al. 2004; Partecke et al. 2005; McKinney 2006), but may change the general behavioural and physiological disposition of individuals thriving in urban areas (Bowers and Breland 1996; Slabbekoorn and Peet 2003; Brumm 2004; Shochat 2004; Miller 2006; Evans et al. 2010; Partecke and Gwinner 2007; see also Chapters 2 and 8). Because hormones often have pleiotropic effects on behaviour, physiology, and morphology, they may play an important role in adaptive phenotypic responses following colonization of novel environments such as urban areas.

3.6 What do we know about the effects of urbanization on hormonal responses?

To date, we still know very little about how urban life has altered the endocrine mechanisms of organisms thriving in urban areas. There is one specific hormone class, the glucocorticoid hormones, which has attracted a lot of attention in the last few years in urban studies. The release of glucocorticoid steroid hormones plays a major role in ensuring survival under adverse environmental conditions. The acute short-term secretion of these hormones is considered beneficial in that it helps to mediate adaptive responses, such as stimulating gluconeogenesis, inhibiting glucose utilization in nonessential tissues, mobilizing fat stores, and redirecting behaviour (Sapolsky et al. 2000; Romero 2004). In prolonged stress situations, however, chronically elevated levels of circulating glucocorticoids can impair reproductive, immune, and brain functions

(Sapolsky 1992). Hence, if urban environments create novel challenges, glucocorticoids might play an important role mediating behavioural and physiological responses of organisms to those challenges.

There is evidence that animals which thrive in urban areas are often noticeably bold or tame in the presence of humans (Evans et al. 2010; Møller 2008; see also Chapter 8). In line with these behavioural changes, reduced acute corticosterone stress response in urban European blackbirds *Turdus merula* compared to rural blackbirds has been interpreted as a result of local adaptation to the urban-specific environmental condition (Partecke et al. 2006). Although the results of this common garden study argue in favour of microevolutionary changes, early environmental or maternal effects could not be excluded. In contrast, existing field studies on stress physiology—including six bird species and one lizard species—are quite inconsistent (Bonier et al. 2007; French et al. 2008; Fokidis et al. 2009; Schoech et al. 2004). Thus, there seems to be no uniform pattern with regard to the effect of urban life on the secretion of glucocorticoids, either for baseline levels or for stress-induced corticosterone levels when comparing urban and rural populations. It is conceivable that these inconsistencies between different species may be explained by differences in life histories between species or between urban habitats on large geographical scales. There is a clear need for more data from paired urban and rural populations of different species and data from different parts of the world, before firm conclusions can be drawn as to what extent urbanization affects stress physiology.

Lack of knowledge is even larger when it comes to other hormones associated, for example, with the hypothalamic–pituitary–gonadal axis, such as androgens or estrogens. This is quite remarkable because we know that these steroid hormones are involved in the expression of behaviours, such as aggression, courtship, or sociality. Indications for potential differences in the secretion of these hormones between urban and rural animals comes from a study on free-living European blackbirds (Partecke et al. 2005). Urban males had overall lower luteinizing hormone and testosterone plasma levels during gonadal growth than forest males, while females of both populations did not differ in luteinizing hormone and estradiol plasma levels. These differences in hormone plasma levels between urban and forest males are intriguing and may reflect specific hormonal responses of urban and forest blackbirds exposed to different social and environmental conditions. However, more studies are definitely needed to confirm these conclusions.

Another potentially important anthropogenic factor affecting the ecology of animals thriving in urban areas is artificial night lighting, or 'light pollution'. The effects of artificial night lighting on the ecology of animals may be quite widespread and biologically significant. Altered light irradiance in urban areas has been suggested to influence daily and seasonal organization, such as foraging behaviour, reproduction, migration, and communication (Longcore and Rich 2004; Miller 2006; Kempenaers et al. 2011; Rowan 1937; see also, for example, Chapters 2 and 6). These major effects of artificial night lighting are due to the pivotal role natural light plays in the synchronization of daily and seasonal biological processes. In an urban environment, however, organisms are exposed to a combination of natural photoperiod and artificial night lighting. The underlying physiological mechanisms which may trigger these behavioural responses when exposed to artificial night lighting are still unclear. One potential candidate could be the hormone melatonin, which is considered to be part of the central pacemaking system and thus may mediate these urban-specific behavioural responses (Gwinner and Brandstatter 2001).

3.7 Chemical pollution and endocrine disruption

Endocrine disrupters represent a broad class of chemicals of diverse form, which have in common the capacity to modulate or disrupt endocrine function. The fact that certain synthetic compounds could functionally mimic natural hormone structures was recognized in the early part of the twentieth century (Dodds et al. 1938), but the significance of this only became apparent later in the century when addressing the effects of the pesticide DDT on the sexual development and welfare of both animals and humans (Carlson 1962; Bitman et al. 1968). Since that time, considerable effort has been made to identify the compounds of most environmental concern. Much attention has been focused on compounds deemed to have 'estrogenic activity' (i.e. estrogen agonists or antagonists). These include natural chemicals (e.g. genistein), pollutants (e.g. PCBs, dioxins, DDT), pharmaceuticals (e.g. 17a ethinylestradiol), and industrial compounds (e.g. Bisphenol A, phthalates, alkylphenols). However, many compounds modulate multiple endocrine pathways (e.g. o,p'-DDT is both weakly estrogenic and strongly antiandrogenic), and often the mechanism of action is unclear (Guillette Jnr 2006). Whilst progress has been made in identifying a range of chemicals which, at least in developed countries, are now banned or restricted (e.g. PCBs), many chemicals which are known to modulate endocrine function continue to be routinely used (phthalates), as the exposure levels for wildlife or humans is thought to be low enough to be biologically irrelevant. The most common route of exposure of wildlife has been through the leaching of chemicals into water sources to which aquatic organisms are then exposed. As a consequence, there is an immense body of literature on the potential effects of endocrine disrupting chemicals on aquatic invertebrates and vertebrates (reviewed in, for example, Tyler et al. 1998; Ankley and Giesy 1998). By contrast, there has been substantially less research examining the potential exposure of terrestrial organisms.

The range of chemicals hypothesized to modulate endocrine function in vertebrates is considerable (Giesy and Snyder 1998), but one group of primary interest is the chemicals used to produce plastics. The most widely used plasticizers include a range of phthalate compounds and bisphenol A (BPA), all of which are known to have estrogenic activity (Oehlmann et al. 2009; Vandenberg et al. 2009). Globally, 2.7 million metric tonnes of phthalates are produced each year, with phthalates constituting up to 50% of the current PVC production by weight (Bauer and Herrmann 1997). BPA worldwide production is 2.5 million metric tonnes annually (Staples et al. 2002). Current evidence suggests that at very low exposure levels (ng/l–ug/l range) these compounds can significantly affect molluscs, crustaceans, and amphibians by altering endocrine function (Oehlmann et al. 2009). There is considerable evidence that these compounds also affect fish, but less is known about how other vertebrate taxa might be affected (Tyler et al. 1998b; Oehlmann et al. 2009).

The hypothesized modes of endocrine modulation include changes in hormone biosynthesis, hormone transport, hormone metabolism, and hormone action, with much more known about the effects on steroid and thyroid hormones than the physiology of peptide and catecholamine hormones (Van der Kraak et al. 1998). Hormone biosynthesis can be affected by xenobiotic chemicals influencing the availability of hormone precursors, or by inhibiting the enzymatic pathways involved in hormone synthesis, although the exact mechanisms are unclear. Hormone transport is controlled by specific binding globulins in the bloodstream, which have high affinity for certain hormones. Unbound hormones are more quickly cleared from the bloodstream, so the presence of globulins has an important effect on hormone efficacy. One action of xenobiotics is to displace hormones from their natural binding globulins, causing an increased clearance rate of the hormone from the bloodstream. Hormone metabolism is affected by xenobiotic compounds through alterations to the enzyme pathways that effect hormone metabolism and clearance. The majority of hormones produced are removed through metabolism in the liver, kidneys, and intestine, rather than metabolism at the target site, so chemicals which affect enzyme activity in these tissues are likely to be relevant. Hormone action is affected when the binding of the hormone with its target receptor is affected. Although some chemicals are strongly suspected to act as endocrine disrupters, their mode of action is unknown. For example, atrazine is one of the most commonly used garden herbicides, with global production exceeding 70,000 tonnes. It has been shown to be a teratogen and also to cause reproductive abnormalities in amphibians (Hayes et al. 2003). Although the mechanism of action is unknown and controversial, modulation of the endocrine system is implicated.

The effects of endocrine disrupting chemical are extensively reviewed elsewhere (Giesy and Snyder

1998; Oehlmann et al. 2009). In aquatic organisms, most effects are hypothesized to occur through the constant exposure of individuals to low levels of these chemicals in their surrounding medium. There is ample evidence that fish populations have been adversely affected by endocrine disrupters through effects on individual physiology, anatomy and, less often, behaviour (Tyler et al. 1998b; Giesy and Snyder 1998; Tyler et al. 1998a; Jobling et al. 1998). In birds, dramatic effects were witnessed in the 1950–60s, in wild populations of birds of prey as a result of egg shell thinning. This occurred as a result of the direct effects of contaminants, principally p,p' –DDE (a metabolite of DDT) on the egg shell gland, rather than through endocrine disruption *per se*. DDT is a recognized endocrine disrupter which has complex effects on the endocrine system, but p,p' –DDE is a recognized estrogen antagonist which inhibits the binding of estrogen with its receptor. Although DDT is banned in many developed countries, persistent levels of p,p' –DDE have been found in avian eggs in areas that previously received DDT treatment, with associations between organochlorine levels and hatching rates (Bishop et al. 2000). In addition, DDT is still used in many developing parts of the world. This can have potentially dire consequences for migratory birds and their predators due to the transport of the chemicals to unpolluted areas during migration.

As mentioned above, the risk of endocrine disruption to aquatic animals has long been recognized (Giesy and Snyder 1998). In particular, fish, amphibians, and reptiles, which have a high degree of reproductive plasticity, are at risk of abnormal development, reduced fertility, and morphological disorders (Edwards et al. 2006). In contrast, the evidence for endocrine disruption in terrestrial wild birds has been rather scant, in part due to credible routes of terrestrial exposure, or due to a lack of evidence that the endocrine system is involved in the adverse effects (Dawson 2000; Giesy et al. 2003). In attempting to understand the potential for these chemicals to adversely affect wildlife, it is important to quantify the *in vitro* effects, the dose levels, the period of dosage, and then the *in vivo* exposure and end points. Considerable effort has been invested in quantifying *in vitro* effects. A range of studies has been involved in quantifying dose levels by sampling animals to obtain estimates of the level of contamination over time. Much less work has been focused on ecologically relevant estimations of exposure due to the difficulties associated with this work. However, for good estimates of wildlife exposure levels, it is essential that an interdisciplinary approach is taken integrating behavioural, ecological, and toxicological studies (Clotfelter et al. 2004). Although there is now an increasing number of studies documenting the effects of EDCs on behaviour, behaviour is rarely assessed in ecotoxicological studies and behavioural information is rarely integrated into estimates of exposure levels (Clotfelter et al. 2004). By using behavioural observations to estimate exposure levels, it has recently been demonstrated that birds foraging on invertebrates within sewage treatment works can be vulnerable to significant exposure to a range of EDCs (Markman et al. 2007). Dods et al. (2005) found that tree swallows *Tachycineta bicolor* breeding at a wastewater treatment plant had lower fledging success, with fledging liver mass being higher, indicating a possible response in relation to contamination. This was supported by the finding that insects at the treatment plant contained substantially higher levels of 4-nonylphenol compared to those trapped at a wildlife refuge control site, although the causal link is far from demonstrated. 4-nonylphenol is a surfactant which has a number of industrial uses (pulp and textile processing), and also enters the wastewater system from municipal sources due to its use in detergents. It has a range of endocrine (Dussault et al. 2005) and non-endocrine-related effects (Staples et al. 2004) that are associated with reduced reproductive performance in animals above certain levels of exposure. This raises the possibility that the sewage treatment process may offer a route by which these chemicals are able to exert their effects through the food chain to taxa such as birds, which have previously been assumed to be unaffected.

From an evolutionary perspective, there has been recent interest in the potential of endocrine disrupting chemicals to alter signalling strategies in animals (Shenoy and Crowley 2010). There is a plethora of studies demonstrating that chemicals which can act

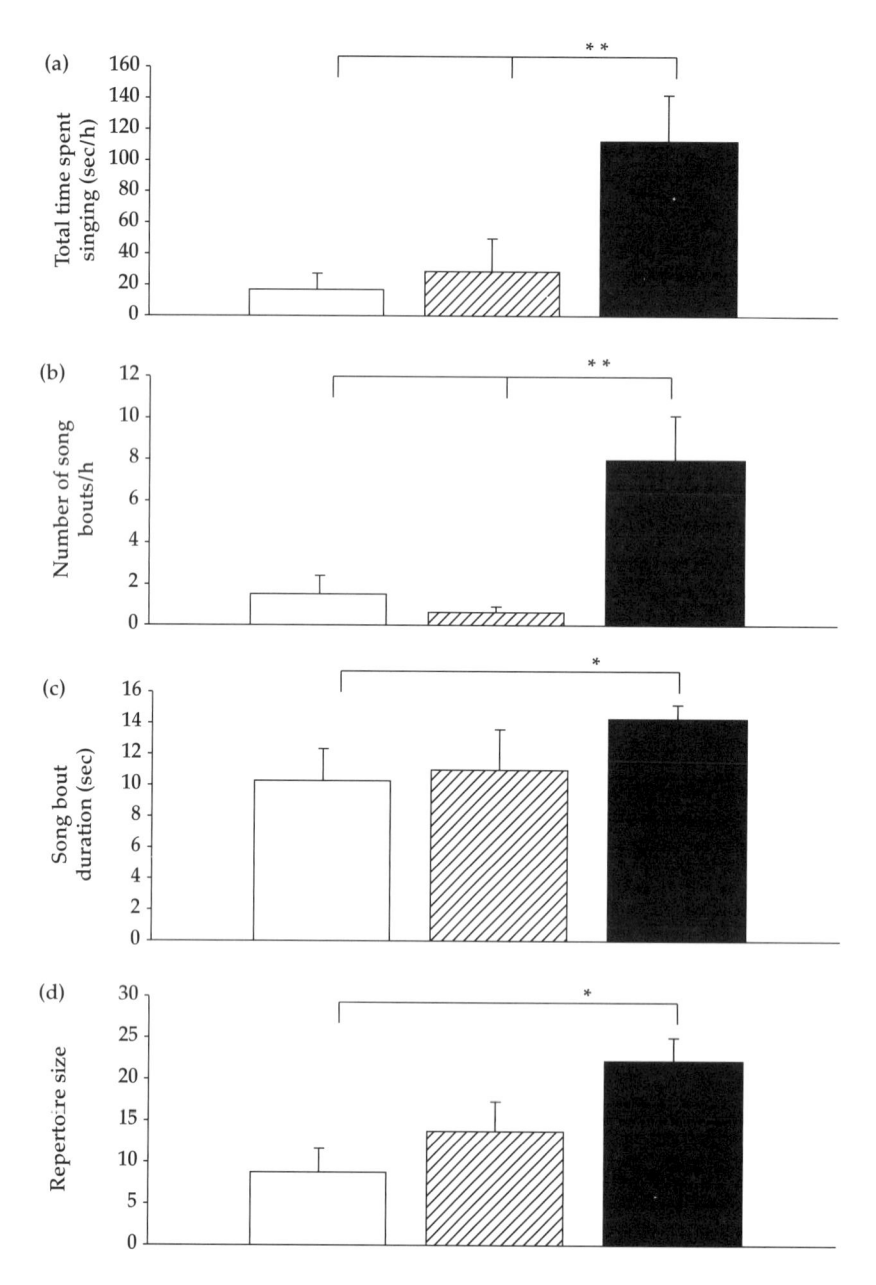

Figure 3.2 Song production in male starlings exposed to estrogenic chemicals. The song production of male starlings in three treatment groups: control (open bars); 17 β estradiol (E2), alone dosed (hatched bars); and chemical mixture: E2, dibutylphthalate, dioctylphthalate and bisphenol A dosed (black bars). Dose levels given at ecologically relevant levels after field sampling. (a) Total time spent singing (sec/h). (b) Number of song bouts per hour. (c) Song bout duration. (s) (d) Repertoire size. Graphs show means + SEM $* = P < 0.05$; $** = P < 0.01$. Reproduced from Markman et al. (2008).

as endocrine disrupters affect the reproductive traits of a range of animals (Shenoy and Crowley 2010), although little evidence that these effects are mediated by endocrine disruption *per se*. This raises the question whether endocrine disrupters can have relevant effects on the evolution of sexual signals. If signalling strategies can change fast enough it would be predicted that, within exposed populations, resistance to chemical pollutants would be selected, or a change in signalling strategy would occur, with a switch to using signals that are resistant to exposure (Shenoy and Crowley 2010). Female preferences can evolve quickly, but the effects are difficult to predict because preferences—as well as male traits—may show changes under exposure to EDCs. Markman et al. (2008) have recently shown that male European starlings, experimentally exposed to a suite of endocrine disrupting chemicals delivered at ecologically relevant levels, increased their song production. Male starlings were dosed with the full spectrum of chemicals found in the invertebrate tissue taken from the sewage treatment works, at the dose level estimated from foraging rates in the field. Birds in the group which received the dose with all the estrogenic constituents produced longer, more complex song bouts (Fig. 3.2) (Markman et al. 2008). This was found to be due to the hypermasculinization of the brain and enlargement of the HVC (used as a proper name), the nucleus in the brain associated with the production of complex song (Markman et al. 2008). The song of these males was preferred by females not exposed to any chemical pollutants, suggesting that sexual selection favours males feeding in polluted areas (Markman et al. 2008). There is evidence that natural selection may play an opposing role, as pollutant-exposed males had suppressed immune responses to experimental challenges (Markman et al. 2008). In addition, nestling starlings fed similar ecologically relevant dose levels of the same chemicals were found to have reduced growth rates and suppressed immune function (Markman et al. 2010).

3.8 Conclusions

Understanding how the endocrine system responds to long term environmental change is crucial for understanding how, when, and to what degree animals are able to adapt to a changing world (Wingfield 2008). The endocrine system is intrinsically flexible through its design, with the cascades and feedback loops providing the capacity for short and longer-term maintenance of essential biological functions, despite a changing external environment. In some cases, it is conceptually possible to visualize how environmental change might lead to changes in either the initial sensitivity of the neuroendocrine target or in receptor number, affinity, or placement (Fig. 3.1). In terms of the effects of climate change on the timing of breeding, it seems likely that selection may act rapidly, particularly in species which integrate different environmental cues to determine either the onset or termination of breeding (Wingfield 2008). Urbanization is likely to initially bring additional stressors, however some species have been highly successful in adapting to these, suggesting that effects of continual stressors may have fundamental consequences for the endocrine system on the move to an urbanized environment. However, we currently have little hard evidence for the physiological changes that occur in relation to long term environmental change. In particular, in wild animals, there is no information on what parts of the endocrine response (Fig. 3.1) have the strongest environmental or genetic control, and so the effects of environmental change on basic physiological parameters are difficult to predict. When examining the potential consequences of endocrine disruption, the evolutionary routes available to animals exposed to EDCs appear to be much more limited, and are ultimately determined by exposure levels and the detrimental effects on fitness. Evading endocrine disruption relies on redesigning the hormone receptor structure to allow binding with the natural hormone, but not its mimic. This may well be possible in some cases, but does not seem likely to be biologically easy, if possible, in many cases. The solution here probably lies in legislation to reduce exposure levels where effects are demonstrable. The key driver may be recognition that humans, similar to wildlife, are susceptible to the effects of endocrine disrupters such as BPA, motivating us to monitor and reduce our impact (Nelson 2011).

Within the field of environmental endocrinology, there is a long history of iconic work on a few model species identifying many of the endocrine mechanisms underlying behavioural changes (Bradshaw 2007). More recently, attempts have been made to integrate these results in conceptual frameworks to explain the evolution of the endocrine system within the constraints imposed (Hau 2007; McNamara and Buchanan 2005). Integrating empirical information on the mechanisms involved with theoretical approaches to understanding the evolutionary pathways will allow us to understand the limitations of the endocrine system and make predictions about degrees of vulnerability of wildlife to environmental change. Some crucial information is still lacking. There is still relatively little information on the heritability of endocrine responses in terms of understanding the potential for rapid adaptive change. Through habitat destruction and removal, humans have had fundamental effects on many vertebrate species, presumably impacting on the endocrine systems of the successful survivors. The historic effects that these changes may have had on the endocrine systems of the remaining fauna are surely difficult to interpret, but could be somewhat quantified by examining the endocrine system and its response to standardized conditions of animals from either more or less degraded habitats. A vast amount of time, energy, and funding has been invested in attempting to understand the phenomenon of endocrine disruption, and yet we understand rather little about ecologically relevant dose levels of many common chemicals and the risks these doses pose for wildlife. Priorities in this area lie in combining behavioural studies with ecotoxicology and studies of gene expression to obtain this type of information. Developing a realistic interpretation of how anthropogenic change is likely to impact on animals involves integrating interdisciplinary interpretations of the 'response to change' and control of seasonal effects (Visser et al. 2010). The answers are unlikely to be simple, but answers are necessary in order to understand both the mechanisms and the risks to wildlife.

Acknowledgements

This chapter benefitted from earlier discussions with Alistair Dawson, Marcel Visser, and Samuel Caro. We would like to thank Bob Wong and Ulrika Candolin, as well as two anonymous reviewers for their comments on an earlier draft of this manuscript. Jesko Partecke was supported by the Volkswagen-Foundation.

References

Ankley, G. T. and Giesy, J. (1998). Endocrine disruptors in wildlife: a weight-of-evidence perspective. *In:* Kendall, R., Dickerson, R., Giesy, J., and Suk, W. (eds) *Principles and Processes for Evaluating Endocrine Disruption in Wildlife.* Pensacola, SETAC.

Ball, G. F. and Ketterson, E. D. (2008). Sex differences in the response to environmental cues regulating seasonal reproduction in birds. *Philosophical Transactions of the Royal Society B: Biological Sciences*, 363, 231–46.

Bauer, M. and Herrmann, R. (1997). Estimation of the environmental contamination by phthalic acid esters leaching from household wastes. *Science of the Total Environment*, 208, 49–57.

Bishop, C., Collins, B., Mineau, P., Burgess, N., Read, W., and Risley, C. (2000). Reproduction of cavity-nesting birds in pesticide sprayed apple orchards in southern Ontario, Canada (1998–1994). *Environmental Toxicology and Chemistry*, 19, 588–99.

Bitman, J., Cecil, H., et al. (1968). Estrogenic activity of o,p'-DDT in the mammalian uterus and avian oviduct. *Science*, 162, 371.

Bonier, F., Martin, P.R., Sheldon, K. S., Jensen, J. P., Foltz, S. L., and Wingfield, J. C. (2007). Sex-specific consequences of life in the city. *Behavioral Ecology*, 18, 121–9.

Bowers, M. A. and Breland, B. (1996). Foraging of Gray squirrels on an urban-rural gradient: use of the GUD to assess anthropogenic impact. *Ecological Applications*, 6, 1135–42.

Bradshaw, D. (2007). Environmental endocrinology. *General and Comparative Endocrinology*, 152, 125–41.

Bradshaw, W. E. and Holzapfel, C. M. (2007). Evolution of animal photoperiodism. *Annual Review of Ecology Evolution and Systematics*, 38, 1–25.

Bronson, F. H. (2009). Climate change and seasonal reproduction in mammals. *Philosophical Transactions of the Royal Society B: Biological Sciences*, 364, 3331–40.

Brown, R. (1994). *An Introduction to Neuroendocrinology*, Cambridge, Cambridge University Press.

Brumm, H. (2004). The impact of environmental noise on song amplitude in a territorial bird. *Journal of Animal Ecology*, 73, 434–40.

Carlson, R. (1962). *Silent Spring*, London.

Caro, S. P. and Visser, M. E. (2009). Temperature-induced elevation of basal metabolic rate does not affect testis

growth in great tits. *Journal of Experimental Biology*, 212, 1994–8.
Clark, W. C., Kates, R. W., Turner, B. L., Richards, J. F., and Meyer, W. B. (eds) 1993. *The Earth as Transformed by Human Action*, Cambridge, UK, Cambridge Universtiy Press.
Clotfelter, E., Bell, A., and Levering, K. (2004). The role of animal behaviour in the study of endocrine-disrupting chemicals. *Animal Behaviour*, 68, 665–76.
Dawson, A. (2000). Mechanisms of endocrine disruption with particular reference to occurrence in avian wildlife: A review. *Ecotoxicology*, 9, 59–69.
Dawson, A. (2005). The effect of temperature on photoperiodically regulated gonadal maturation, regression and moult in starlings—potential consequences of climate change. *Functional Ecology*, 19, 995–1000.
Dawson, A. and Sharp, P. J. (2010). Seasonal changes in concentrations of plasma LH and prolactin associated with the advance in the development of photorefractoriness and moult by high temperature in the starling. *General and Comparative Endocrinology*, 167, 122–7.
Dodds, E., Goldberg, L., Lawson, W., and Robinson, R. (1938). Oestrogenic activity of certain synthetic compounds. *Nature*, 141, 247.
Dods, P., Birmingham, E., Williams, T., Ikonomou, M., Bennie, D., and Elliott, J. (2005). Reproductive success and contaminants in tree swallows (Tachycineta bicolor) breeding at a wastewater treatment plant. *Environmental Toxicology and Chemistry*, 24, 3106–12.
Dussault, E., Sherry, J., Lee, H., Burnison, B., Bennie, D., and Servos, M. (2005). In vivo estrogenicity of nonylphenol and its ethoxylates in the Canadian environment. *Human and Ecological Risk Assessment*, 11, 353–64.
Edwards, T., Moore, B., and Guillette Jnr, L. (2006). Reproductive dysgenesis in wildlife: a comparative review. *International Journal of Andrology*, 29, 109–20.
Evans, J., Boudreau, K., and Hyman, J. (2010). Behavioural syndromes in urban and rural populations of song sparrows. *Ethology*, 116, 588–95.
Farner and Wilson (1957). A quantitative examination of testicular growth in the white-crowned sparrow *Biological Bulletin*, 113, 254–67.
Farner, D., Follett, B., King, J., and Morton, M. (1966). A quantitative examination of ovarian growth in the white-crowned sparrow. *Biological Bulletin*, 130, 67–75.
Fokidis, H. B., Orchinik, M., and Deviche, P. (2009). Corticosterone and corticosteroid binding globulin in birds: Relation to urbanization in a desert city. *General and Comparative Endocrinology*, 160, 259–70.
French, S.S., Fokidis, H. B., and Moore, M. C. (2008). Variation in stress and innate immunity in the tree lizard (*Urosaurus ornatus*) across an urban-rural gradient. *Journal of Comparative Physiology—B, Biochemical, Systemic, and Environmental Physiology*, 178, 997–1005.
Giesy, J., Feyk, L., Jones, P., Kannan, K., and Sanderson, T. (2003). Review of the effects of endocrine disrupting effects in birds. *Pure and Applied Chemistry*, 75, 2287–303.
Giesy, J. and Snyder, E. (1998). Xenobiotic modulation of endocrine function in fishes. *In:* Kendall, R., Dickerson, R., Giesy, J., and Suk, W. (eds) *Principles and Processes for Evaluating Endocrine Disruption in Wildlife*. Kiwah Island SC. Pensacola FL, SETAC.
Guillette Jnr, L. (2006). Endocrine disrupting contaminants—Beyond the dogma. *Environmental Health Perspectives*, 114, 9–12.
Gwinner, E. and Brandstatter, R. (2001). Complex bird clocks. *Philosophical Transactions of the Royal Society B: Biological Sciences*, 356, 1801–10.
Hau, M. (2007). Regulation of male traits by testosterone: implications for the evolution of vertebrate life histories. *Bioessays*, 29, 133–44.
Hayes, T., Haston, K., Tsui, M., Hoang, A., Haeffele, C., and Vonk, A. (2003). Atrazine-induced hermaphroditism at 0.1 ppb in American leopard frogs (Rana pipiens): Laboratory and field evidence. *Environmental Health Perspectives*, 111, 568–75.
Hinde, R. and Steele, E. (1976). The effect of male song on an oestrogen dependent behaviour in the female canary Serinus canaria. *Hormones and Behavior*, 7, 293–304.
Husby, A., Kruuk, L. E. B., and Visser, M. (2009). Decline in the frequency and benefits of multiple brooding in great tits as a consequence of a changing environment. *Proceedings of the Royal Society of London B, Biological Sciences*, 276, 1845–54.
Jobling, S., Nolan, M., Tyler, C., Brighty, G., and Sumpter, J. (1998). Widespread sexual disruption in wild fish. *Environmental Science and Technology*, 32, 2498–506.
Johnston, R. F. (2001). Synanthropic birds of North America. In: Marzluff, J. M., Bowman, R., and Donnelly, R. E. (eds) *Avian Ecology and Conservation in an Urbanizing World*. Norwell, MA, Kluwer Academic.
Kempenaers, B., Borgström, P., Loës, P., Schlicht, E., and Valcu, M. (2011). Artificial night lighting affects dawn song, extra-pair siring success, and lay date in songbirds. *Current Biology*, 20, 1735–9.
Longcore, T. and Rich, C. (2004). Ecological light pollution. *Frontiers in Ecology and the Environment*, 2, 191–8.
Markman, S., Guschina, I., Barnsley, S., Buchanan, K., Pascoe, D., and Muller, C. (2007). Endocrine disrupting chemicals accumulate in earthworms exposed to sewage effluent. *Chemosphere*, 70, 119–25.

Markman, S., Leitner, S., Catchpole, C., Barnsley, S., Muller, C., Pascoe, D., and Buchanan, K. (2008). Pollutants increase song complexity and the volume of the brain area HVC in a songbird. *PLOS ONE*, 3, e1674.

Markman, S., Muller, C., Pascoe, D., Dawson, A., and Buchanan, K. (2010). Pollutants affect development in nestling starlings Sturnus vulgaris. *Journal of Applied Ecology*.

McKinney, M. L. (2006). Urbanization as a major cause of biotic homogenization. *Biological Conservation*, 127, 247–60.

McNamara, J. and Buchanan, K. (2005). Stress, resource allocation and mortality. *Behavioral Ecology*, 16, 1008–17.

Miller, M. W. (2006). Apparent effects of light pollution on singing behavior of American robins. *The Condor*, 108, 130–9.

Møller, A. P. (2008). Flight distance of urban birds, predation and selection. *Behavioral Ecology and Sociobiology*, 63, 63–75.

Nelson, R. J. (2011). *An Introduction to Behavioural Endocrinology*, Sunderland, Sinauer.

Oehlmann, J., Schulte-Oehlmann, U., Kloas, W., Jagnytsch, O., Lutz, I., Kusk, K. O., Wollenberger, L., Santos, E. M., Paull, G. C., Van Look, K. J. W., and Tyler, C. R. (2009). A critical analysis of the biological impacts of plasticizers on wildlife. *Philosophical Transactions of the Royal Society B: Biological Sciences*, 364, 2047–62.

Partecke, J. and Gwinner, E. (2007). Increased sedentariness in European blackbirds following urbanization: A consequence of local adaptation? *Ecology*, 88, 882–90.

Partecke, J., Schwabl, I., and Gwinner, E. (2006). Stress and the city: Urbanization and its effects on the stress physiology in European blackbirds. *Ecology*, 87, 1945–52.

Partecke, J., Van't Hof, T., and Gwinner, E. (2004). Differences in the timing of reproduction between urban and forest European blackbirds (*Turdus merula*): result of phenotypic flexibility or genetic differences? *Proceedings of the Royal Society of London B, Biological Sciences*, 271, 1995–2001.

Partecke, J., Van't Hof, T., and Gwinner, E. (2005). Underlying physiological control of reproduction in urban and forest-dwelling European blackbirds *Turdus merula*. *Journal of Avian Biology*, 36, 295–305.

Romero, L. M. (2004). Physiological stress in ecology: Lessons from biomedical research. *Trends in Ecology and Evolution*, 19, 249–55.

Rowan, W. (1937). Effects of traffic disturbance and night illumination on London starlings. *Nature*, 139, 668–9.

Sapolsky, R. M. (1992). *Stress, the Aging, and the Mechanisms of Neuron Death*, Cambridge, Massachusetts, MIT Press.

Sapolsky, R. M., Romero, L. M., and Munck, A. U. (2000). How do glucocorticoids influence stress responses? Integrating permissive, suppressive, stimulatory, and preparative actions. *Endocrine Reviews*, 21, 55–89.

Schoech, S. J., Bowman, R., and Reynolds, S. J. (2004). Food supplementation and possible mechanisms underlying early breeding in the Florida Scrub-Jay (*Aphelocoma coerulescens*). *Hormones and Behavior*, 46, 565–73.

Shenoy, K. and Crowley, P. (2010). Endocrine disruption of male mating signals: ecological and evolutionary implications. *Functional Ecology*, 25, 433–48.

Shochat, E. (2004). Credit or debit? Resource input changes population dynamics of city-slicker birds. *Oikos*, 106, 622–6.

Shochat, E., Warren, P. S., Faeth, S. H., McIntyre, N. E., and Hope, D. (2006). From patterns to emerging processes in mechanistic urban ecology. *Trends in Ecology and Evolution*, 21, 186–91.

Silverin, B., Wingfield, J., Stokkan, K. A., Massa, R., Jarvinen, A., Andersson, N. A., Lambrechts, M., Sorace, A., and Blomqvist, D. (2008). Ambient temperature effects on photo induced gonadal cycles and lehormonal secretion patterns in Great Tits from three different breeding latitudes. *Hormones and Behavior*, 54, 60–8.

Slabbekoorn, H. and Peet, M. (2003). Birds sing at a higher pitch in urban noise. *Nature*, 424, 267.

Staples, C., Mihaich, E., Carbone, J., Woodburn, K., and Klecka, G. (2004). A weight of evidence analysis of the chronic ecotoxicity of nonylphenol ethoxylates, nonylphenol ether carboxylates and nonylphenol. *Human and Ecological Risk Assessment*, 10, 999–1017.

Staples, C., Woodburn, K., Caspers, N., Hall, A., and Klecka, G. (2002). A weight of evidence approach to the aquatic hazard assessment of bisphenol A. *Hum Ecol Risk Assess.*, 8, 1083–105.

Tyler, C., Jobling, S., and Sumpter, J. (1998). Endocrine disruption in wildlife: a critical review of the evidence. *Critical Reviews in Toxicology*, 28, 319–61.

Ursin, H. and Olff, M. (eds) (1993). *The Stress Response*, London, Academic Press.

Van der Kraak, G., AZacharewski, T., Janz, D., Sanders, B., and Gooch, J. (1998). Comparative endocrinology and mechanisms of endocrine modulation in fish and wildlife. In: Kendall, R., Dickerson, R., Giesy, J., and Suk, W. (eds.) *Principles and Processes for Evaluating Endocrine Disruption in Wildlife*. Kiwah Island SC. Pensacola FL, SETAC.

Vandenberg, L. N., Maffini, M. V., Sonnenschein, C., Rubin, B. S. and Soto, A. M. (2009). Bisphenol-A and the great divide: a review of controversies in the field of endocrine disruption. *Endocrine Reviews*, 30, 75–95.

Visser, M., Adriansen, F., van Balen, J. H., Blondel, J., Dhondt, A. A., van Dongen, S., du Feu, C., Ivankina, E., Kerimov, A., de Laet, j., Mathysen, E., McCleery, R., Orell, M. and Thomson, D. (2003). Variable responses to large-scale climate change in European Parus populations. *Proceedings of the Royal Society of London B, Biological Sciences*, 270, 367–72.

Visser, M., Schaper, S., Holleman, L. J. M., Dawson, A., Sharp, P., Gienapp, P., and Caro, S. P. (2011). Genetic variation in cue sensitivity involved in avian timing of reproduction. *Functional Ecology*, 25, 868–77.

Visser, M. E., Caro, S. P., van Oers, K., Schaper, S. V., and Helm, B. (2010). Phenology, seasonal timing and circannual rhythms: towards a unified framework. *Philosophical Transactions of the Royal Society B: Biological Sciences*, 365, 3113–27.

Visser, M. E., Holleman, L. J. M., and Caro, S. P. (2009). Temperature has a causal effect on avian timing of reproduction. *Proceedings of the Royal Society of London B, Biological Sciences*, 276, 2323–31.

Vitousek, P. M., Mooney, H. A., Lubchenco, J., and Melillo, J. M. (1997). Human domination of earth's ecosystems. *Science*, 277, 494–9.

Wingfield, J. C. (2008). Comparative endocrinology, environment and global change. *General and Comparative Endocrinology*, 157, 207–16.

Wingfield, J. C., Hahn, T. P., and Doak, D. (1993). Integration of environmental factors regulating transitions of physiological state, morphology and behaviour. In: Sharp, P. J. (ed.) *Avian Endocrinology.* Bristol, Society for Endocrinology.

Wingfield, J. C., Hahn, T. P., Levin, R., and Honey, P. (1992). Environmental predictability and control of gonadal cycles in birds. *Journal of Experimental Zoology*, 261, 214–31.

Wingfield, J. C., Hahn, T. P., Wada, M., Astheimer, L. B., and Schoech, S. (1996). Interrelationship of day length and temperature on the control of gonadal development, body mass and fat score in white-crowned sparrows, *Zonotrichia leucophrys gambelii. General and Comparative Endocrinology*, 101, 242–55.

Wingfield, J. C., Hahn, T. P., Wada, M., and Schoech, S. J. (1997). Effects of day length and temperature on gonadal development, body mass and fat depots in white-crowned sparrows, *Zonotrichia leucophrys pugetensis. General and Comparative Endocrinology,* 107, 44–62.

CHAPTER 4

Experience and learning in changing environments

Culum Brown

> ### ⮞ Overview
> This chapter examines how animals use learning to adapt to environmental change. It focuses on the consequences of anthropogenic activities for learning and the adjustment of animal behaviour. First, I introduce learning and its role in the development and evolution of behaviour. In particular, I identify the circumstances under which learning is likely to play a key role in the adaptation to novel environments and rapid environmental change. I then provide a number of examples of the way in which animals adapt to human induced habitat change, using urbanization and climate change as case studies. From this, we should be able to draw some general conclusions about the characteristics of species that enable them to cope with rapid environmental shifts.

4.1 Introduction

Over the years the ethology and comparative psychology paradigms have merged to form a modern synthesis that recognizes the role of both genes and the environment in shaping animal behaviour. The approach recognizes that animals display a large degree of phenotypic plasticity in response to variable environments and that much of this is mediated via learning. Learning is described as a change of behaviour induced by experience, which provides animals with an opportunity to adapt their behaviour to suit contemporary environments. It is the primary mechanisms animals use to cope with shifting environmental variables. Changes in behaviour as a result of experience can also be induced through mechanisms other than learning, for example by physiological shifts, but here we are primarily concerned with cognitive changes induced by neural plasticity.

In the modern world, the environment that animals are adapting to is more often than not the product of human disturbance. Human affected landscapes are characterized by large-scale land clearance for farming or mining, and by urbanization as human populations increasingly consolidate in cities. More recently, it has become clear that human impacts can operate at a global scale as evidenced by human induced climate change and acid rain. Rapid changes in the environment are typical of modern human activities and such changes often occur too rapidly for evolutionary processes to respond. Thus, in many circumstances, learning may provide the only avenue for animals to track environmental variation.

4.2 Learning and its role in the development of behaviour

The development of animal behaviour is controlled by two underlying mechanisms. The first is an automated response to specific stimuli within the environment that is entirely predetermined by specific genes. These innate behavioural responses are the result of natural selection acting over many

---

Behavioural Responses to a Changing World. First Edition. Edited by Ulrika Candolin and Bob B.M. Wong.
© 2012 Oxford University Press. Published 2012 by Oxford University Press.

generations. The second mechanism involves modification of behaviour through experience and interaction with stimuli in the environment during an individual's lifetime. Learning has the added benefit in that animals can change their response to specific cues or respond to new cues during ontogeny. Like innate behavioural responses, learning ability may also evolve over evolutionary time and there are multiple examples of differences in learning ability between closely related species and populations within species (e.g. Brown and Braithwaite 2005).

Whilst these two mechanisms (genetic inheritance and learning via experience) are often portrayed as polar opposites (the Nature–Nurture debate), in reality this is a false dichotomy as most behaviour is determined by some mixture of the two. Animals may have evolved an innate predisposition to perform a certain behaviour in a given context or pay attention to certain stimuli, but these behaviours or the response to the stimuli are then modified as a result of experience to produce an optimal response specific to local conditions. For example, innate predator recognition may result in a response to a generalized predatory threat (e.g. any large moving object), but the response is fine-tuned by learning as the animal comes to recognize which large objects are especially associated with danger in the current environment and which are not. In this manner the innate and learned responses interact to produce an adaptive response such that the animal will only display an antipredator response when it is appropriate to do so (Lima and Dill 1990).

It is important to recognize that both learning and innate responses contribute to animal behaviour. Even if predator recognition were entirely innate, the prey animal would still alter its behaviour when presented with a generalized predator cue, for example by seeking cover. Learning, however, provides an opportunity to fine-tune the response and enables the animal to choose from an infinitely larger set of behavioural responses to suit any given context (e.g. the animal may recognize that the predator is satiated and thus does not presently represent a risk; Licht 1989).

There are a number of criteria that must be met in order for learning to occur. First, animals must be able to determine which environmental cues are associated with important biological events. This is in fact a daunting task given the seemingly infinite number of cues and events that occur in the natural world. Much of the information emanating from the environment is effectively noise and has no relevance to the animal. Thus, determining the relevant cues in the environment must be one of the most important constraints on the learning process. Second, they must be able to determine the consequences of their behavioural response to environmental cues. In this regard, the immediate reward of performing the response is measured by reference to a change in internal state (e.g. hunger or fear level). A large number of studies have shown that if the gap between the animal performing an action and the consequences (reward or punishment) is separated by too long a time period then no association will be drawn (e.g. trace conditioning; Shettleworth 1998).

The combination of innate and learned behaviour that best tracks environmental change will result in the optimal behavioural phenotype. This will depend on the degree to which the environment is stable and, hence, predictable through space and time. Let us first consider spatial heterogeneity. When we consider spatial heterogeneity, we must examine it at several scales. At the individual level, the appropriate spatial scale is that of the home-range. That is the environment in which the individual operates and where the fitness benefits associated with performing certain behaviours are relevant. At the population level, we must also consider the potential for immigration and emigration from local microenvironments, where individuals may experience a range of alternative environmental variables. At the species level, the appropriate spatial scale is the entire species distribution. As we proceed through larger spatial scales it becomes apparent that the environment is less likely to be homogeneous and behaviour will need to be increasingly plastic to cope with this variability. Thus, learning plays a larger role in shaping the behaviour of species that occupy large geographical areas that encompass a variety of environmental conditions.

When we are dealing with temporal heterogeneity there are essentially two important levels to consider. Firstly, the degree to which the environment is temporally stable during the lifetime of an individual. If the environment is highly stable there will be little need for the individual to display behavioural plasticity and innate behaviours should dominate. At intermediate levels of temporal heterogeneity, learning should prevail since innate responses will not be sufficient to track environmental variability. Secondly, the intergenerational scale concerns the degree to which the environment is stable from one generation to the next. In circumstances where the environment is temporally stable over many generations, behavioural traits can become genetically fixed in the population. But, if there is temporal environmental variance between generations then plasticity will be favoured. Because environments are variable in space and time, no given phenotype will ever be consistently optimal and some degree of phenotypic plasticity is required. Thus learning tends to be a ubiquitous trait in virtually all animals. It is important to note that the relevant temporal and spatial scales will vary depending on the species we are concerned with.

The temporal and spatial dimensions may also interact when we are concerned with dispersal over multiple generations (see also Chapter 5). High dispersal rates will result in behavioural (and genetic) panmyxia at the species level, whereas low levels of dispersal will generate population diversity. In the former case, innate behavioural responses will be highly generalized because they will have to encapsulate the total environmental variance experienced by the species. In this context, behavioural plasticity would be best generated via learning to fine-tune behaviour to suit the specific location an individual finds itself in. In the latter case, innate behavioural responses may be more specific and closely match local ecological conditions and learning would be less important. Indeed, isolated populations on the fringe of the species range are more likely to develop novel behavioural phenotypes (Cassel-Lundhagen et al. 2011). However, behaviour is commonly controlled by multiple loci each with a small effect. Therefore, even small amounts of gene flow will likely erode local adaptations and, once again, learning would be favoured.

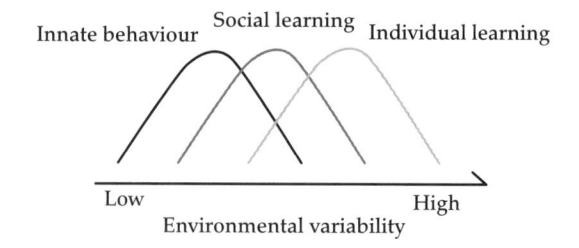

Figure 4.1 The relative contributions of three forms of plasticity generating mechanisms at varying levels of environmental variability.

To summarize, if the environment were completely homogenous then there would be no need for learning at all and innate behavioural responses would dominate. At the other extreme, the environment could be completely heterogeneous and there would be no opportunity to learn, because environmental cues would become too unstable to be of any predictive value. Needless to say, in the real world, the environment is never completely stable nor is it completely unstable so some degree of plasticity is required. We might generally predict that at low levels of heterogeneity, innate behaviours would dominate but as the environment becomes increasingly heterogenous learning would prevail (Fig. 4.1).

4.3 Social learning

Learning can occur both through individual trial and error learning (sometimes referred to as asocial learning), through social learning, or some mix of the two (Brown and Laland 2003). Social learning can be defined as incidences where individuals acquire new information about their environment or adopt a new behaviour as a result of observing or interacting with another individual or its products (Brown and Laland 2003). Individual learning requires the animal to interact directly with the environment and make associations between cues and events. It may take a considerable amount of time before individuals make appropriate associations and, therefore, individual learning can incur multiple costs as outlined below. In contrast, social learning enables animals to take short-cuts by using cues emanating from conspecifics, so called public information, to focus on particular cues or locations (stimulus or local enhancement) or to reproduce the outcome of the conspecifics' behaviour (goal emulation) (Boyd and Richerson 1985). In some species, individuals may even adopt new behavioural patterns by direct imitation (e.g. birds; Zentall 2004). Moreover, social learning may expose individuals to novel conditions or focus their attention on novel stimuli, after which individual learning can then act to generate novel behavioural repertoires. Thus, social learning can lead to the rapid adoption of novel behavioural phenotypes that are likely to be locally adaptive and potentially produce behavioural innovations.

Unlike individual learning, social learning enables information to be passed on through generations via vertical transmission. In this respect, it has similarities with innate behavioural responses. Thus, local behavioural traditions can form which are themselves subject to natural selection. Unlike heritable traits, however, behavioural traditions can rapidly transform via copying error or the introduction of novel behavioural patterns into the population within an individual's lifetime. Social learning can also incur costs if the environment shifts and the cultural norm becomes disconnected from current environmental conditions. For example, Day et al. (2001) trained guppies *Poecilia reticulata* to utilize one of two coloured doors to access a feeding site. When new individuals were added to the school, they continued to follow the trained demonstrators even when the food patch was moved closer to the alternative door. Thus social learning can favour the maintenance of maladaptive behavioural patterns if individuals do not sample the environment for themselves. Recent studies suggest, however, that most individuals rely on a mixture of private and public information to make decisions (van Bergen et al. 2004), which means that maladaptive behavioural traits are soon extinguished (Brown and Laland 2003).

In a broader context, individual learning is most beneficial in conditions subject to higher levels of environmental heterogeneity. Even though social learning can speed up the learning process, it is better suited to lower levels of environmental heterogeneity because local behavioural traditions are more likely to be out of date at high levels of environmental variability (Fig. 4.1).

4.4 Interaction between innate and learnt responses

The interaction between innate and learned behavioural responses is an interesting topic worthy of closer scrutiny. It is likely that the manner in which the two mechanisms interact is more than just additive. Innate mechanisms often act as filters to focus the attention of the animal on environmental cues that are likely to be biologically meaningful. Learning then enables the animal to make associations between cues or outcomes through classical or operant conditioning. It is important to note that in cases where a single cue may have different predictive outcomes in different environments or contexts, it is vital that the animal is flexible in the way it can respond. Similarly, displaying a given behaviour can have various outcomes depending on the context in which it is performed. For example, an animal may respond to a predatory cue (e.g. olfactory cues) by seeking cover in dense vegetation, but this would not be an appropriate response if the particular predator detected is commonly encountered in vegetation.

There are many examples of the interaction between innate and leant responses in the animal world. During operant conditioning experiments, for example, rats *Rattus norvegicus* are more likely to associate gastric illness with olfactory cues rather than mechanical ones (e.g. electric shock or flashing lights) (Garcia et al. 1974). Many fish have a strong preference for red coloured objects because red is associated with high quality foods. Guppies can be taught to find a food reward hidden behind a red partition far more quickly than if the partition is any other colour (C. Brown unpublished data). This preference for red has been secondarily adopted as a sexually selected trait to attract mates, thus, the colour red may be of predictive value in multiple contexts (Rodd et al. 2002).

One might view the interaction between innate and learned behaviours in different ways. To some extent, a predisposition to pay attention to some cues in favour of others may be viewed as a constraint because it limits the possibility that an animal will pay attention to novel cues when they become available, thereby limiting the development of innovative behavioural patterns. On the other hand, there is little doubt that paying attention to specific cues in the environment dramatically reduces the time it takes to learn associations between environmental cues and important biological events, thereby dramatically reducing the cost of learning. The problem with these combinations of innate dispositions acting in concert with individual learning is that the system breaks down if the environmental cues upon which the innate response depends shift, perhaps as a result of habitat disturbance or climate change. A good example of this is the historically strong connection between the leaf sprouting time of trees, the emergence of caterpillars, and the breeding time of insectivorous birds such as the great tit *Parus major* (Both et al. 2009). This mismatch is a reflection of the fact that the various different trophic levels are responding to different environmental cues to adjust their phenologies (see Chapter 6 for more detail). Thus, as the environment gets progressively warmer, there is a growing gap between the timing of the emergence of caterpillars and the hatching time of the birds, which leads to rapid population declines in the bird population. Similar observations have been made between migrating cuckoos and their resident hosts (Møller et al. 2011). In many ways an over reliance on genetic predispositions greatly enhances that probability that a species will fall victim to an evolutionary trap from which it is unlikely to emerge (see Chapter 16 for more information).

4.5 Costs associated with learning

It is quite apparent that animal behaviour is not infinitely plastic and learning is constrained by certain costs. The fact that we see differences in learning ability between different populations of the same species, or even between sexes, suggests that the evolution of learning depends on the relative costs and benefits of enhanced learning ability. While the costs associated with learning have been predicted by a range of models, there is still relatively little experimental evidence.

One obvious source of costs is that associated with the development and maintenance of neural tissues. There is no doubt that brain matter is by far and away the most expensive tissue in the body to maintain. Of course this argument relies on the fact that there is a link between brain size and learning ability. This is still a highly controversial topic, but there are a number of direct and indirect lines of evidence that suggest it may well be true. In terms of direct evidence, butterflies that have a high capacity to learn the location of host plants have larger mushroom bodies in their brains (Snell-Rood et al. 2009) and comparative studies have revealed that a larger hippocampus is associated with improved spatial learning in mammals and birds (reviewed in Sherry et al. 1992). Both of these brain structures are intimately associated with spatial learning. Indirect evidence also comes in many forms. Firstly animals reared in enriched environments tend to have larger brains and a higher capacity to learn. Salmon *Salmo salar* par reared in enriched environments, for example, are better able to generalize between food types and thus learn to forage on novel prey items more quickly than those reared in impoverished conditions (Brown et al. 2003). Secondly, changes in brain sizes have been observed in animals when they no longer require their high capacity for learning. One very nice example is that of the male meadow vole *Microtus pennsylvanicus*, whose brain size increases during the breeding season in line with the increasing size of his territory. Similarly, black-capped chickadees *Poecile atricapillus* increase their brain size during autumn when they are busy caching food items for the coming winter. Moreover, there is some evidence that larger brained species are more likely to develop behavioural innovations through their higher capacity for learning, both socially and asocially (Reader et al. 2003; Sol et al. 2002).

As alluded to above, there are two ways in which brain size can be enhanced. The first is a congenital (inherited) investment in brain size that arms the animal with a high capacity for learning. But this approach has the potential to incur a high cost because the energetic investment is made before learning occurs, thus the cost may be borne even if the animal doesn't fully utilize its learning capacity

(Dukas 1998). Alternatively, investment may be made in brain size during ontogeny as the animal faces particular environmental challenges (e.g. learning tasks; Clayton and Krebs 1994). In this instance, brain growth can be specifically targeted to those areas that are needed when required, and would greatly reduce the chances of learning capacity redundancy and associated costs. Reducing the costs of learning would encourage the persistent use of learning, both by individuals during ontogeny, and by species over evolutionary time.

Another cost of learning is often referred to as the cost of being naïve. Naïve individuals must first sample the environment to determine what the contemporary conditions are and the information they gather may not be entirely accurate. A small amount of environmental variability can generate a degree of uncertainty. Thus, during the following period of trial and error learning it takes a considerable amount of time for associations to be formed, and behaviour is therefore sub-optimal during this time. During both of these phases, there is a cost in terms of time that would otherwise be available to perform other behaviours, the energetic cost of performing the sub-optimal behaviour, as well as the costs associated with making errors. Some of the costs of being naïve are actually associated with picking an appropriate behaviour from a potentially large behavioural repertoire rather than the process of learning itself. Thus, the costs of being naïve can be dramatically reduced by enhancing learning speed. Even innate behavioural responses may incur a cost of being naïve because it is unlikely that the present environmental conditions are exactly the same as those over the proceeding generations, during which the innate behaviour was shaped. Naturally, the costs associated with a genetically fixed trait are huge should the animal find itself in an inappropriate environment.

There are a number of good examples in the literature that clearly illustrate the cost of being naïve. In fishes, it has been demonstrated that the handling time of novel prey items is relatively high and, in combination with the development of a new search image, represents a considerable cost to switching to novel prey items (Warburton 2003). Moreover, it is apparent that fish are aware of this cost. Thus, fish may delay the switch from their preferred prey item to a novel prey item even when the latter becomes increasingly abundant. Similar observations have been made in bees *Apis mellifera* (Laverty and Plowright 1988). Thus, the cost of being naïve can impede optimal foraging.

4.6 Learning and evolution

One intriguing question that is often fiercely debated is the extent to which learning facilitates or inhibits the evolution of novel behavioural phenotypes. This is particularly important in the context of rapid environmental change associated with human activities. To address this question, we must consider the interaction between innate predispositions and learning. After all, natural selection can only act on the heritable components of behaviour, which may include the heritability of plasticity or general learning ability itself.

A number of arguments suggest that learning would inhibit natural selection in response to environmental change. First, the interaction between innate and learnt behaviours may inhibit behavioural innovation by limiting the cues to which animals pay attention (see Section 4.4). This effectively restricts the animal's ability to respond to novel environmental stimuli and thereby prevent the production of behavioural innovations. Second, learning allows animals to fine-tune their innate behavioural responses and thus learning may act as a buffer against selective forces. This would have the effect of reducing selection on the underlying innate behavioural traits, thereby slowing down their evolution.

On the other hand, a number of models have suggested that learning increases the probability of invading novel habitats (Thibert-Plante and Hendry 2011; Chapter 14). Exposure to new environments may generate novel selective forces that then act by shifting the underlying genetic architecture by natural selection. Thus, learning, by facilitating invasion, may ultimately result in the rapid movement of innate behaviours towards a new optimum in novel environments (Price et al. 2003; Fig. 4.2). If this is the case, learning is likely to result in rapid shifts in the evolution of behaviour in response to large shifts in

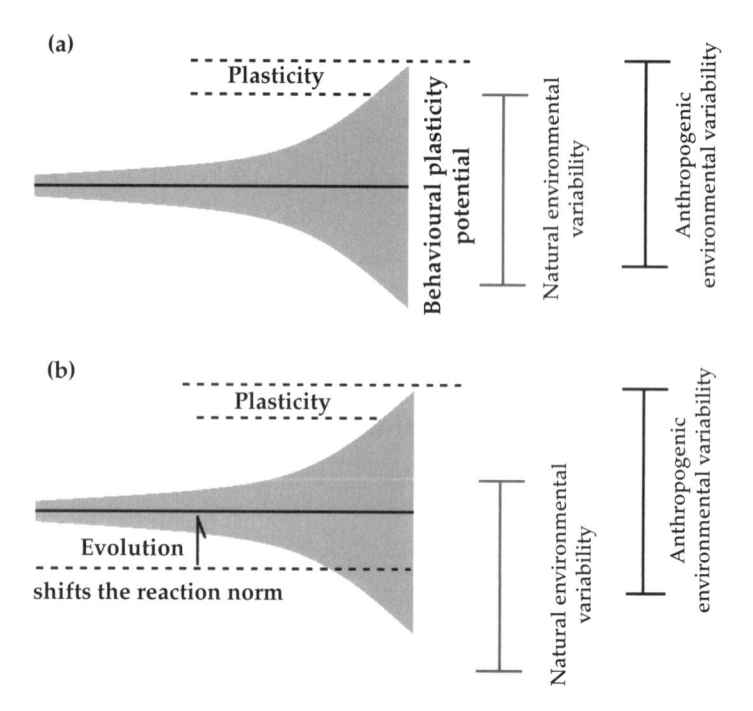

Figure 4.2 Degree of plasticity (y axis) displayed by a range of individuals in a population in response to a given environmental variable (x axis). Individuals with low levels of plasticity (left end of the x axis) cannot respond to environmental changes. An individual with a high degree of behavioural plasticity primarily mediated via learning (right end of the x axis) is confronted with a sudden environmental change. If the environmental change is small (a), these individuals will be able to cope by changing their phenotype. Alternatively, if the shift is large (b), then some of the required change in behaviour will be generated through plasticity, but this will need to be accompanied by an underlying, elevated shift in the reaction norm via selection on innate traits, perhaps via the Baldwin effect (arrow). Natural selection would ensure this new reaction norm spreads through the population.

environmental variables brought on by human activities. One such method by which this might occur is via the Baldwin effect (see also Chapter 1). However, quantitative genetic models suggest that the conditions under which learning facilitates evolution are highly restricted (Lande 2009).

Many of these models, however, do not consider the source of individual behavioural plasticity, but take as their starting point that some degree of plasticity exists. Plasticity itself may be heritable and can also be generated by physiological changes in addition to cognitive changes (i.e. learning). As mentioned above, innate behavioural responses can only generate limited plasticity but this is significantly enhanced through learning. The interaction between social learning and individual learning, however, may expose individuals to novel situations or direct attention to novel cues. Wilson (1985) suggested that through high rates of social and individual learning, individuals expose themselves to novel selective pressures, which can then act as a driver for evolution. In both birds and primates, there are correlations between the propensity to learn and rates of behavioural innovation (Sol et al. 2002; Reader 2003). But, innovative species and individuals are also characterized by high rates of neophilia. Thus, a combination of high learning capabilities generating plasticity and exposure to novelty likely plays a key role in the evolution of novel behavioural types.

4.7 Learned responses to human induced environmental variation

There is little doubt that humans are creating a degree of environmental change at a rate that has rarely been witnessed in Earth's geological history. Much of this damage has been inflicted by a few key processes including land clearing, urbanization, overharvesting of prey species, and pollution. Some of these processes have resulted in rapid and irreversible damage to the environment, providing species with little opportunity to adapt, whilst others, such as climate change, are more insidious.

Some of the less obvious examples of human induced environmental change are brought about indirectly. Whilst pollution has a very direct influence on species behaviour and survival, indirect effects include climate change (see subsection 4.6.2) and acid rain. In apparently undisturbed streams in North America, acid rain can have surprising effects on animal behaviour and the learning process in particular. Many young salmonids rely on chemical cues to recognize predators through associative learning, but acid rain has changed the configuration of key components of the alarm cues to the extent that fish no longer respond to them (Leduc et al. 2006). Changes in ocean acidification associated with climate change have also interrupted the chemosensory ability of clown fish *Amphiprion percula* both during settlement and during predator encounters (Munday et al. 2010). We do not know if these fish shift their behaviour to rely more on alternative cues (e.g. visual or auditory) to compensate, nor do we know if they will be able to adapt their chemical recognition capabilities to suit the new environment.

The commercial overharvesting of animals has also had a large impact on animal behaviour. Although much of this effect in terrestrial habitats occurred hundreds of years ago, this process is ongoing in the marine environment. Systematic hunting of cod has led to changes in a wide range of traits including size at maturity, fecundity and migration routes, the latter of which is likely maintained via cultural traditions (Ferno et al. 2006). The overharvesting of large marine fish species by commercial fishing operations has also significantly altered the behaviour of prey species both directly and indirectly. In both terrestrial and aquatic ecosystems, the systematic removal of predatory species (so called apex predators) or other key stone species can have flow-on effects throughout the food web (Crooks and Soulé 1999). Fish communities that are varyingly exposed to commercial fishing pressure differ in their predation risk, as large piscivores are removed, which, in turn, affects the foraging behaviour of multiple small, prey fish species (Madin et al. 2010). These changes are brought about by a mixture of immediate learned responses by individuals and long-term shifts in selective pressures.

I have argued that the key to responding to environmental variation is behavioural plasticity, primarily mediated via learning. In disturbed environments, generalists maintain their numbers (some may even flourish) whereas specialists rapidly decline. The success of generalists has been correlated with their high degree of behavioural flexibility and a lack of neophobia (Reader 2003). Habitat generalists can adjust their movement patterns and resource use to make the most of the modified habitat. Most strikingly, they are able to change their foraging behaviour, use of shelter, and so on to incorporate novel resources that are available in human environments, which helps mitigate against the loss of natural resources. It should come as no surprise that the characteristics of species likely to respond favourably to human induced environmental change match very closely those of invasive species (i.e. short generation time, rapid growth, large fecundity potential, etc: Sol et al. 2002).

Innovation frequency is a decent surrogate for behavioural flexibility, including learning and problem solving (Webster and Lefebvre 2001), and has been implicated in the invasion potential of birds (Sol et al. 2002). The blackbird *Turdus merula*, for example, is often portrayed as a successful invader and colonizer of a range of habitats around the world. Over 23 foraging innovations have been reported for this species, including hawking for insects and attacking skinks (Sol et al. 2002). This species was the fastest learner in comparative tests (Sasvari 1985) and has low levels of neophobia (Marples et al. 1998). A lack of fear when operating in new environments or in the presence of novel objects, combined with high rates of learning, is likely to lead to the discovery of new resources, such as novel prey items. Indeed the response of birds to urban environments is one of the most well studied examples of how animals adapt to human induced environmental changes.

4.7.1 Learned responses to urbanization

As one moves along the urbanization gradient from native bushland to cities, species richness tends to decline (Clergeau et al. 1998). The homogenization process of urbanization is similar all over the world, thus, the same species tend to be present in cities around the globe. The fauna is often dominated by a small number of non-native, urban specialists (Blair 1996). For most native species, living in urban areas is a stressful business, as these species are either not pre-adapted to live in urban environments or are not able to adapt. Indeed, recent evidence suggests that some species, which have been traditionally thought of as urban specialists, are not coping as well as it first appeared. For example, despite incorporating human food sources into their diet, urban starlings show reduced food provisioning at the nest, resulting in lower nestling weight compared to starlings in nonurban areas (Mennechez and Clergeau 2006).

The key question is what characterizes successful urban adapters? The potential list is long and includes: high sociality, the utilization of human resources, a lack of fear, tendency to be sedentary, and broad diet (Clergeau et al., 1998). Many of these traits, however, are likely to be pre-adaptations and have little to do with learning and behavioural plasticity *per se*. For example, birds that have a tendency to nest in trees are worse off in the cities than those that nest in crevices, such as pigeons *Columba livia*. Kark (2007) added another trait to this list: behavioural flexibility.

Foraging innovation, including the introduction of novel behaviours and the utilization of new food sources, appears to be one of the key characteristics of urban specialists. Urban adapters must be able to feed on a wide variety of prey items and readily adopt new resources. Thus, scavengers or generalist omnivores tend to do quite well in cities (Courtney and Fenton 1976). There are numerous accounts of feeding innovations in urban bird species (Lefebvre et al. 1997). Starlings picking off insects plastered to the front of arriving trains, sparrows operating bus station doors by setting off the motion censor, and birds entering shopping centres by waiting for the automated doors to open as shoppers pass through, providing them with access to food. It is unclear how these behaviours develop, but they are likely to be opportunistic in the first instance and reinforced by operant conditioning, and then spread through the population via social learning (e.g. Hinde and Fisher 1951).

A number of studies have reported significant shifts in the temporal and spatial activity patterns of urban animals, many of which occur via learning. In some instances, activity is extended because of a reduction in natural predators. In other instances, activity patterns are moved to avoid urban predators, such as cats or humans, or to benefit from new resources. For instance, black bears *Ursus americanus* in urban areas that have access to garbage have significantly different time budgets. They have also shifted to becoming nocturnal, are 30% bigger, have sex ratios skewed towards males, and increased female reproductive success (Beckmann and Berger 2004). Thus, it is apparent that shifts in activity times and diets associated with urban environments may have flow on effects on life history traits.

The increased availability of food in landfills and trash cans can increase local population densities. Animals soon learn the location of these stable food sources and are attracted to them. For example, the distributions of gray squirrels *Sciurus carolinensis* and chipmunks *Tamias striatus* in urban areas are determined by the proximity to artificial food sources rather than by other habitat characteristics (Flyger 1970; Ryan and Larson 1976). In urban areas, insectivorous bats spend a significant amount of time hunting around street lights, and subsequently densities are higher than elsewhere (Jung and Kalko 2009). Clumping around artificial food sources can have several secondary influences on urban animals. Increased social interactions can result in risk of disease/parasite transmission and changes in social structure. Raccoons *Procyon lotor* in urban areas have smaller and more stable home ranges than rural individuals (Prange et al. 2004). House geckos *Lepidodactylus lugubris* outcompete other native gecko species, but only when prey patches are clumped around lights (Petren and Case 1996).

Animals typically tradeoff foraging opportunity with predation risk. This balance can be affected by a large number of factors, including competition for food, perhaps mediated by conspecific densities, and perceived risk of predation. Thus, risk assessment in urban animals is influenced by a range of factors that are likely to vary from those experienced in the wild (Blumstein 2002). As populations habituate to humans, as part of their adaptation to urban environments, it is plausible that they equally reduce fear towards predators in general. Squirrels in urban environments show reduced responses to humans, compared to suburban and rural squirrels, as well as reduced responses to hawk and coyote vocalizations (McCleery 2009). McCleery (2009) suggested that this is evidence of a generalized decrease in antipredator behaviour, but it may also be indicative of behavioural plasticity, since both hawks and coyotes are rare in urban settings. Comparative analyses have shown similar processes in a wide range of birds, where shorter flight distances to approaching predators characterize urban species (Møller 2008). Like other behavioural adaptations to urban environments, it is not clear if these changes stem from phenotypic plasticity operating through individual habituation to humans, or from evolutionary responses over multiple generations (Diamond 1986).

Urban animals need to successfully communicate with one another, which is a relatively difficult task in a noisy environment. For passerines, this is highly problematic because the urban environment interferes with the transmission of their calls. There is plenty of evidence that urban birds respond to background noise in a number of ways, including shifting the frequency of their songs (see Patricelli and Brickley 2006 and Chapter 2 for reviews) or changing the time at which they sing (Fuller et al. 2007). Similar observations have been made in a variety of animals, ranging from whales to frogs (Parks et al. 2011; Sun and Nairns 2005). To date, however, it is unclear how animals achieve these shifts. Bermudez-Cuamatzin et al. (2011) found house finches *Carpodacus mexicanus* to instantaneously alter their song to avoid the dominant frequencies in the background noise, indicating high plasticity in song. Shifts in song characteristics of passerine birds might also be culturally inherited, because song is vertically transmitted from parent to offspring during a sensitive period early in ontogeny. If male urban birds are singing higher frequency songs, then young male birds in the vicinity are likely to produce high frequency songs (Luther and Baptista 2010). Noisy environments also exist in the wild (e.g. near streams, or in windy areas), so it should not surprise us that animals can adapt to a noisy urban environment (Dubois and Martens 1984). Nevertheless, behavioural plasticity in communication seems to be part of the urban dweller's survival kit (see Chapter 2 for full discussion).

4.7.2 Learned responses to climate change

It is widely accepted that recent changes in global climate have had numerous effects on the behaviour of animals. These changes include direct changes in phenology with respect to migration, development and reproduction (see Chapters 6 and 8). To date, observations of shifts in phenology dominate the climate change literature when it comes to animal behaviour. But, there are also far more subtle behavioural changes, which are induced by climate change indirectly, such as those resulting from shifting food availability or predator density. For example, during years of high snowfall, snowdrifts accumulate in Antarctica, allowing giant petrels *Macronectes giganteus* to safely crash land and feed on the Antarctic petrel's *Thalassoica antartica* chicks and eggs (Van Franeker et al. 2001). However, observations of behavioural innovations brought about by learning in response to climate change are very rare. Fortunately, we can rely on observations of animals that have undergone large shifts in climate in the past, such as prolonged droughts, to provide an insight into how animals are likely to react to future climate change scenarios.

Shifts in behaviour as a response to global warming are expected, given that behavioural adjustment is often a first means of coping with environmental variability. One of the basic underpinnings of behavioural ecology is that even short-term changes in climate can dramatically alter animal behaviour (Krebs and Davies 1991). Flying foxes *Pteropus poliocephalus* spend significantly more time fanning themselves in their roosts during hot days, rather than sleeping, which has significant implications for energy expenditure (C. Brown unpublished data). Similarly, during drought, impala *Aepyceros melampus* need to drink and their movement patterns become restricted to water courses where predation risk is high (Jarman and Jarman 1973). Both examples illustrate the typical behavioural flexibility illustrated by animals in response to climate variation.

Much of the evidence of behavioural shifts influenced by climate change involves movement in phenology. Mammals, for example, emerge from hibernation earlier in response to global warming, which has several impacts on their behaviour (Stirling et al. 1999; Inouye et al. 2000; Reale et al. 2003). While there is no doubt that changes in behaviour are occurring, far less is known about the mechanisms by which these changes arise. As discussed above, there are essentially two ways behaviour can track climate change: (1) microevolutionary responses via natural selection or (2) phenotypic plasticity, which may be mediated by learning. Recent studies have found evidence of both microevolutionary and plastic responses to climate change (Reale et al. 2003; Bradshaw and Holapzfel 2006). In an analysis of a long-term data set on great tits in the UK, Charmantier et al. (2008) used an individual mixed modelling approach to examine the reaction norms of individual females to changes in spring temperatures. She found the population to have closely tracked environmental change associated with global warming over the last 50 years, and much of this was due to individual plastic responses.

In these examples, the plastic response observed in animals is largely a direct physiological response to environmental cues. There is little suggestion that shifts in the behaviours are generated by learning, or that they cascade to other behavioural patterns that are influenced by learning or other cognitive decision making processes. Nevertheless, since the different members of an ecological community vary in their responses to climate change, multiple trophic interactions will become disconnected in space and time, and new relationships will be born. Thus, behaviour will have to alter accordingly. The development of novel behaviours will be particularly obvious in long lived species and in those that show range shifts to keep up with climate change, such as migratory birds. Highly mobile species will head polewards (or to higher elevations) at a far greater rate than less mobile species. As these animals encounter new communities, they will inevitably have to adjust their foraging and antipredator responses via learning (Walther et al. 2002). Bluefin tuna *Thunnus thynnus*, for example, migrate across the ocean searching for suitable prey species, which can be directly influenced by climate induced changes in upwelling and current circulation (Polovina 1996). If there are cultural elements to migratory patterns, shifts in migration patterns may be further exacerbated by human exploitation, if the largest and most knowledgeable individuals are selectively removed from the population (Ferno et al. 2006).

Long lived animals may be able to rely on previous experience and retreat to refugia during tough times, the location of which may be culturally inherited. Elephants *Loxodonta africana*, for example, are long lived and maintain a variety of traditions within family groups. During extreme drought events, large numbers of juveniles perish, but calves with mothers that had previous experience of extreme drought were more likely to survive (Foley et al. 2008). Similarly, elephant seals *Mirounga leonina* show a high degree of foraging site fidelity, even though the productivity of their preferred foraging site varies from year to year. It's thought that this long-term strategy ameliorates fluctuation in food availability in response to climate variables (Bradshaw et al. 2004).

The social organization of animals can also vary depending on the availability of prey items, the abundance of which is influenced by climate variation. As prey availability increases, the competition of food resources within the group is reduced and, thus, animals can afford to form larger social groups (Lusseau et al. 2004; Western and Lindsay 1984). Observations of Alaskan red foxes *Vulpes vulpes* show large changes in breeding behaviour during poor seasons brought about by El Nino. In good years (La Nina), the foxes breed polygynously. Males have many mates and female helpers at the den. But during El Nino, food supplies drop and the breeding system reverts to monogamy, because fewer females are in suitable breeding condition (Zabel and Taggart 1989). Such shifts in group size and breeding behaviour are active decisions made by the animals.

Similar indirect influences on mating behaviour will occur if shifts in climate lead to changes in operational sex ratio. A large number of reptiles show temperature dependent sex determination. In painted turtles *Chrysemys picta*, for example, the sex ratio is linked to mean July temperature, with the sex ratio becoming increasingly female biased as temperatures increase (Janzen 1994). Mitchell et al. (2008) speculate that some of the sex bias in tuatara *Sphenodon guntheri* could be offset by behavioural decisions to nest in shaded areas or laying deeper in the soil. The Australian water dragon *Physignathus lesueurii*, for example, uses both techniques to compensate for higher temperatures in the warms regions of its distribution (Doody et al. 2006). It is unknown if this is an active decision making process or an example of local adaptation. It is likely that females are choosing nest sites based on soil temperature and are therefore making active decisions where to nest. To confirm this, one would have to follow the nesting behaviour of individuals over time to show that they respond in a plastic way.

To summarize, there has been a big emphasis on describing phenological responses to climate change, but the vast majority of these responses are simple physiological responses to environmental cues. Whilst it is clear that these changes must also be associated with the corresponding emergence of novel behavioural patterns or innovations, such observations are rare in the primary literature. We can, however, resort to previous examples where animals have been observed responding to natural environmental perturbations (e.g. extended drought periods) and thereby make some predictions about how animals will respond to future climate change scenarios. In future, closer observations of behavioural responses to climate change and accounts of behavioural innovations will come to light.

4.8 Conclusions

While much of the modelling and theoretical evidence suggests that learning will play an important role in how animals cope with shifts in environmental variables induced by human activities, there is surprisingly little empirical evidence of this. In both case studies explored here, urbanization and climate change, there are a very large number of observations that show that animals do indeed adapt their behaviour to suit novel environmental circumstances, but we are still very much in the descriptive stage of the scientific process. Very few experimental or manipulative studies have been conducted. Perhaps the only exceptions being the documentation of birds responding in real time to background levels of noise (Bermudez-Cuamatzin et al. 2011) and evidence of foraging innovation in urban birds. In most cases, we simply can't say if the changes in behaviour are brought about by natural selection or via phenotypic plasticity. Modern statistical approaches will enable us to disentangle the relative contributions of genes and phenotypic plasticity in shaping animal behaviour (e.g. mixed modelling and reaction norms).

The potential relationship between brain size, cognitive ability and behavioural plasticity is certainly worthy of further investigation, in a broader range of species, especially given the link with the propensity for behavioural innovation. It is likely that those animals that have these traits will show the greatest resilience to environmental changes induced by human behaviour. The potential for innovation also seems to be associated with other factors such as a lack of neophobia. Investigations of the traits linked with invasive potential will also likely shed further light on those characteristics that enable animals to respond positively to environmental change.

Acknowledgements

I would like to thank the referees for their helpful comments. This work was supported by an Australian Research Fellowship from the Australian Research Council.

References

Beckmann, J.P. and Berger, J. (2004). Rapid ecological and behavioural changes in carnivores: the responses of black bears (*Ursus americanus*) to altered food. *Journal of Zoology*, 261, 207–12.

Bermudez-Cuamatzin, E., Rios-Chelen, A.A Gil, D., and Garcia C.M. (2011). Experimental evidence for real-time song frequency shift in response to urban noise in a passerine bird. *Biology Letters*, 7, 36–8.

Blair, R.B. (1996). Land use and avian species diversity along an urban gradient. *Ecological Applications*, 6, 506–19.

Blumstein, D.T. (2002). Moving to suburbia: ontogenetic and evolutionary consequences of life on predator-free islands. *Journal of Biogeography*, 29, 685–92.

Both, C., Van Asch, M., Bijlsma, R. G., Van Den Burg, A. B., and Visser, M. E. (2009). Climate change and unequal phenological changes across four trophic levels: constraints or adaptations? *Journal of Animal Ecology*, 78, 73–83.

Boyd, R. and P. J. Richerson. (1985) *Culture and the Evolutionary Process*, pp. 199–202. University of Chicago Press, Chicago.

Bradshaw, C. J. A., Hindell, M. A., Sumner, M. D., and Michael, K. J. (2004). Loyalty pays: potential life history consequences of fidelity to marine foraging regions by southern elephant seals. *Animal Behaviour*, 68, 1349–60.

Bradshaw, W.E. and Holzapfel, C.M. (2006). Climate change—Evolutionary response to rapid climate change. *Science*, 312, 1477–8.

Brown, C. and Braithwaite, V.A. (2005). Effects of predation pressure on the cognitive ability of the poeciliid *Brachyraphis episcopi*. *Behavioural Ecology*, 16, 482–97.

Brown, C., Davidson, T., and Laland, K. (2003). Environmental enrichment and prior experience improve foraging behaviour in hatchery-reared Atlantic salmon. *Journal of Fish Biology*, 63 (s1), 187–96.

Brown, C. and Laland, K. (2003). Social learning in fishes: A review. *Fish and Fisheries*, 4, 280–8.

Cassel-Lundhagen, A., Kanuch, P., Low, M., and Berggren, A. (2011). Limited gene flow may enhance adaptation to local optima in isolated populations of the Roesel's bush cricket (*Metrioptera roeselli*). *Journal of Evolutionary Biology*, 24, 381–90.

Charmantier, A., McCleery, R.H., Cole, L.R., Perrins, C., Kruuk, L.E.B., and Sheldon, B.C. (2008). Adaptive phenotypic plasticity in response to climate change in a wild bird population. *Science*, 320, 800–3.

Clayton, N.S. and Krebs, J.R. (1994). Hippocampal growth and attrition in birds affected by experience. *Proceedings of the National Academy of Sciences of the USA*, 91, 7410–14.

Clergeau, P., Savard, J.P.L., Mennechez, G., and Falardeau, G. (1998). Bird abundance and diversity along an urban—rural gradient: a comparative study between two cities on different continents. *Condor*, 100, 413–25.

Courtney, P.A. and Fenton, M.B. (1976). The effects of a small rural garbage dump on populations of Peromyscus leucopus Rafinesque and other small mammals. *Journal of Applied Ecology*, 13, 413–22.

Crooks, K.R. and Soulé, M.E. (1999). Mesopredator release and avifaunal extinctions in a fragmented system. *Nature*, 400, 563–6.

Day, R. L., MacDonald, T., Brown, C., Laland, K. N., and Reader, S. M. (2001). Interactions between shoal size and conformity in guppy social foraging. *Animal Behaviour*, 62, 917–25.

Diamond, J.M. (1986). Rapid evolution of urban birds. *Nature*, 324, 107–8.

Doody, J.S., Guarino, F., Georges, A., Corey, B., Murray, G., and Ewert, M.W. (2006). Nest site choice compensates for climate effects on sex ratios in a lizard with environmental sex determination. *Evolutionary Ecology*, 20, 307–30.

Dubois, A. and Martens J. (1984). A case of possible vocal convergence between frogs and a bird in Himalayan torrents, *Journal für Ornithologie*, 125, 455–63.

Dukas, R. (1998). Evolutionary ecology of learning. In: Dukas R, (ed.) *Cognitive Ecology: The Evolutionary Ecology of Information Processing and Decision Making*, pp. 129–74. Chicago, University Chicago Press.

Flyger, V. (1970). Urban gray squirrels-problems, management, and comparisons with forest populations. *Transactions of the North Eastern Fisheries and Wildlife Conference*, 27, 107–13.

Ferno, A, Huse, G, Jakobsen, P.J., and Kristiansen, T.S. (2006) The role of fish learning skills in fisheries and aquaculture. In Brown, C, Laland, K and Krause, J. (eds) *Fish Cognition and Behavior*. Cambridge, Blackwell.

Foley, C., Pettorelli, N, and Foley, L. (2008). Severe drought and calf survival in elephants. *Biology Letters*, 4, 541–4.

Fuller, R.A., Warren, P.H., and Gaston, K.J. (2007). Daytime noise predicts nocturnal singing in urban robins. *Biology Letters*, 3, 368–70.

Garcia, J., Hankins, W.G., and Rusiniak, K.W. (1974). Behavioral regulation of the milieu interne in man and rat. *Science*, 185, 824–31.

Hinde, R.A and Fisher, J. (1951). Further observations on the opening of milk bottles by birds. *British Birds*, 44, 393–6.

Inouye, D.W., Barr, B., Armitage, K.B., and Inouye, B.D. (2000). Climate change is affecting altitudinal migrants and hibernating species. *Proceedings of the National Academy of Sciences of the USA*, 97, 1630–3.

Janzen, F.J. (1994). Climate-change and temperature-dependent sex determination in reptiles. *Proceedings of the National Academy of Sciences of the USA*, 91, 7484–90.

Jarman, P.H and Jarman, M.V. (1973). Daily activity of impala. *East Africa Wildlife Journal*, 11, 75.

Jung, K., and Kalko, E.K.V. (2009). Where forest meets urbanization: foraging plasticity of aerial insectivorous bats in an anthropogenically altered environment. *Journal of Mammalogy*, 91, 144–53.

Kark, S., Iwaniuk A., Schalimtzek A., and Banker E. (2007). Living in the city: can anyone become an 'urban exploiter'? *Journal of Biogeography*, 34, 638–51.

Krebs, J.R. and Davies, N.B. (1991). *An Introduction to Behavioural Ecology*. Oxford, Blackwell Scientific.

Lande, R. (2009). Adaptation to an extraordinary environment by evolution of phenotypic plasticity and genetic assimilation. *Journal of Evolutionary Biology*, 22, 1435–46.

Laverty, T.M. and Plowright, R.C. (1988). Flower handling by bumblebees: a comparison of specialists and generalists. *Animal Behaviour*, 36, 733–40.

Leduc, A.O.H.C., Roh, E., Harvey, M.C., and Brown, G.E. (2006). Impaired detection of chemical alarm cues by juvenile wild Atlantic salmon (*Salmo salar*) in a weakly acidic environment. *Canadian Journal of Fisheries and Aquatic Science*, 63, 2356–63.

Lefebvre, L., Whittle, P., Lascaris, E., and Finkelstein, A. (1997). Feeding innovations and forebrain size in birds. *Animal Behaviour*, 53, 549–60.

Licht, T. (1989). Discriminating between hungry and satiated predators: The response of guppies (Poecilia reticulata) from high and low predation sites. *Ethology*, 82, 238–43.

Lima, S. L. and Dill, L.M. (1990). Behavioral decisions made under the risk of predation: a review and prospectus. *Canadian Journal of Zoology*, 68, 619–40.

Lusseau, D., Williams, R., Wilson, B., Grellier, K., Barton, T.R., Hammond, P.S., and Thompson P.M. (2004). Parallel influence of climate on the behaviour of Pacific killer whales and Atlantic bottlenose dolphins. *Ecology Letters*, 7, 1068–76.

Luther, D. and Baptista, L. (2010). Urban noise and the cultural evolution of bird songs. *Proceedings of the Royal Society of London B, Biological Sciences*, 277, 469–73.

Madin, E.M.P, Gaines, S.D., and Werner, R.R. (2010). Field evidence for pervasive indirect effects of fishing on prey foraging behavior. *Ecology*, 91, 3563–71.

Marples, N.M., Roper, T.J., and Harper, D.G.C. (1998). Responses of wild birds to novel preys: evidence of dietary conservatism. *Oikos* 83, 161–5

McCleery, R. (2009). Changes in fox squirrel anti-predator behaviors across the urban–rural gradient. *Landscape Ecology*, 24, 483–93.

Mennechez, G. and Clergeau, P. (2006). Effect of urbanisation on habitat generalists: starlings not so flexible? *Acta Oecologica*, 30, 182–91.

Mitchell, N.J., Kearney, M.R., Nelson, N.J., and Porter, W.P. (2008). Predicting the fate of a living fossil: how will global warming affect sex determination and hatching phenology in tuatara? *Proceedings of the Royal Society of London B, Biological Sciences*, 275, 2181–3.

Møller, A.P. (2008). Flight distance of urban birds, predation and selection for urban life. *Behavioural Ecology and Sociobiology*, 63, 63–75.

Møller, A.P., Saino, N., Adamik, P., Ambrosini, R., Antonov, A., Campobello, D., Stokke, B.G., Fossoy, F., Lehikoinen, E., Martin-Vivaldi, M., Moksnes, A., Moskat, C., Roskaft, E., Rubolini, D., Schulze-Hagen, K., Soler, M. and Shykoff, J. A. (2011). Rapid change in host use of the common cuckoo *Cuculus canorus* linked to climate change. *Proceedings of the Royal Society of London B, Biological Sciences*, 278, 733–8.

Munday, P. L., Dixson, D.L., McCormick, M.I., Meekan, M., Ferrari, M.C.O., and Chivers, D.P. (2010). Replenishment of fish populations is threatened by ocean acidification. *Proceedings of the National Academy of Sciences of the USA*, 107, 12930–4.

Parks, S. E., Johnson, M., Nowacek, D., and Tyack, P. L. (2011). Individual right whales call louder in increased environmental noise. *Biology Letters*, 7, 33–5.

Patricelli, L. and Blickley, J. (2006). Avian communication in urban noise: causes and consequences of vocal adjustment. *Auk*, 123, 639–49.

Petren, K. and Case, T.J. (1996). An experimental demonstration of exploitation competition in an ongoing invasion. *Ecology*, 77, 118–32.

Polovina, J. J. (1996). Decadal variation in the trans-Pacific migration of northern bluefin tuna (*Thunnus thynnus*) coherent with climate-induced change in prey abundance. *Fisheries Oceanography*, 5, 114–19.

Prange, S., Gehrt, S.D., and Wiggers, E.P. (2004). Influences of anthropogenic resources on raccoon (*Procyon lotor*) movements and spatial distribution. *Journal of Mammalogy*, 85, 483–90.

Price, T.D., Qvarnström, A., and Irwin, D.E. (2003). The role of phenotypic plasticity in driving genetic evolution. *Proceedings of the Royal Society of London B, Biological Sciences*, 270, 1433–40.

Reader, S. (2003). Innovation and social learning: individual variation and brain evolution. *Animal Biology*, 53, 147–58.

Reale, D., McAdam, A.G., Boutin, S., and Berteaux, D. (2003). Genetic and plastic responses of a northern mammal to climate change. *Proceedings of the Royal Society of London B, Biological Sciences*, 270, 591–6.

Rodd, F.H., Hughes, K.A., Grether, G.F., and Baril, C.T. (2002). A possible non-sexual origin of mate preference: are male guppies mimicking fruit? *Proceedings of the Royal Society of London B, Biological Sciences*, 269, 475–81.

Ryan, D. A. and Larson, J. S. (1976). Chipmunks in residential environments. *Urban Ecology*, 2, 173–8.

Sasvari, L. (1985). Keypeck conditioning with reinforcements in two different locations in thrush, tit and sparrow species. *Behavioural Processes*, 11, 245–52.

Sherry, D.F., Jacobs, L.F., and Gaulin, S.J.C. (1992). Spatial memory and adaptive specialization of the hippocampus. *Trends Neuroscience*, 15, 298–303.

Shettleworth S.J. (1998). *Cognition, Evolution and Behaviour.* New York, Oxford University Press.

Snell-Rood, E.C., Papaj, D.R., and Gronenberg, W. (2009). Brain size: a global or induced cost of learning? *Brain, Behavior and Evolution*, 73, 111–28.

Sol, D., Timmermans, S., and Lefebvre, L. (2002). Behavioural flexibility and invasion success in birds. *Animal Behaviour*, 63, 495–502.

Stirling, I., Lunn, N.J., and Iacozza, V. (1999). Long-term trends in the population ecology of polar bears in n Hudson Bay in relation to climate change. *Arctic*, 52, 294–306.

Sun, J.W.C. and Narins, P.M. (2005) Anthropogenic sounds differentially affect amphibian call rate. *Biological Conservation*, 121, 419–27.

Thibert-Plante, X. and Hendry, A.P. (2011). The consequences of phenotypic plasticity for ecological speciation, *Journal of Evolutionary Biology*, 24, 326–42.

van Bergen, Y., Coolen, I., and Laland, K.N. (2004). Nine-spined sticklebacks exploit the most reliable source when public and private information conflict. *Proceedings of the Royal Society of London B, Biological Sciences*, 271, 957–62.

Van Franeker, J. A., Creuwels, J.C.S., van der Veer, W., Cleland, S., and Robertson, G. (2001). Unexpected effects of climate change on the predation of Antarctic petrels. *Antarctic Science*, 13, 430–9.

Walther, G.-R., Post, E., Convey, P., Menzel, A., Parmesan, C., Beebee, T. J. C., Fromentin, J.-M., Hoegh-Guldberg, O., and Bairlein, F. (2002). Ecological responses to recent climate change. *Nature*, 416, 389–95.

Warburton, K. (2003). Learning of foraging skills by fish. *Fish and Fisheries*, 4, 203–15.

Webster, S. J. and Lefebvre, L. (2001). Problem solving and neophobia in a columbiform–passeriform assemblage in Barbados. *Animal Behaviour*, 62, 23–32.

Western, D. and Lindsay, W.K. (1984). Seasonal herd dynamics of a savanna elephant population. *African Journal of Ecology*, 22, 229–44.

Wilson, A.C. (1985). The molecular basis of evolution. *Scientific American*, 253, 164–73.

Zabel, C.J. and Taggart, S.J. (1989). Shift in red fox, *Vulpes vulpes*, mating system associated with El Nino in the Bering Sea. *Animal Behaviour*, 38, 830–8.

Zentall, T.R. (2004). Action imitation in birds. *Learning and Behavior*, 32, 15–23.

PART II
Responses

CHAPTER 5

Dispersal

Alexis S. Chaine and Jean Clobert

> ### ⊃ Overview
>
> Dispersal allows individuals to shift to more favourable breeding areas. As environments change, dispersal becomes critical in allowing individuals to exploit new opportunities and to counteract the adverse effects of habitat alterations. The basic fitness equation that underlies dispersal behaviour is central to understanding this response: the expected fitness of travelling to, and breeding in, a new patch must be higher than that of staying locally. Thus, changes in the quality of habitat patches—and in the costs of travelling through the matrix—will largely determine how dispersal responds to environmental change, assuming sufficient information is available to potential dispersers. Using examples from the empirical literature, we illustrate some of the situations that might be mediated by dispersal, and highlight some of the consequences of dispersing in a changing landscape. In doing so, we identify key gaps in our understanding of how global changes will affect dispersal.

5.1 Introduction

Dispersal describes the movement of individuals from one reproductive patch to another. This movement plays an important role in structuring a variety of ecological and evolutionary processes. The decision to disperse is fundamentally a behavioural process that allows individuals to move from one environment to another in order to find the best possible environmental conditions for reproduction. Dispersal—usually movement by juveniles and, less frequently, by adults—leads to shifts in population sizes in habitat patches and the transfer of genes among patches, thereby altering the genetic make-up of populations across the landscape. Dispersal is quite distinct from migration, which reflects a more regular seasonal movement that does not necessarily lead to individuals changing reproductive patches (see Chapter 6) and, thus, dispersal will have different consequences for local population dynamics. In ecological terms, dispersal leads to connectivity among patches and ties local population dynamics to metapopulation dynamics, thereby contributing to the persistence of linked populations and the potential for species to expand their existing range (Levins 1969; Gilpin and Hanski 1991). In genetic terms, the movement of individuals will potentially introduce new genes into a population, thereby affecting the genetic structure of a population and its ability to adapt to local selection pressures (see reviews in Clobert et al. 2001; Hanski and Gaggiotti 2004; Ronce 2007). As environments change, individual dispersal decisions are likely to shift, which, in turn, will have profound effects on the ecological and genetic structure of populations.

When environments change, species can respond by adjusting to new conditions (i.e. plasticity), adapting to new conditions (i.e. genetic change), or moving to areas with better conditions (i.e. range shift). While dispersal will clearly be a major contributor to range shifts, it will also influence population adjustments and adaptation to new

Behavioural Responses to a Changing World. First Edition. Edited by Ulrika Candolin and Bob B.M. Wong.
© 2012 Oxford University Press. Published 2012 by Oxford University Press.

surroundings. Dispersal can allow populations to take advantage of new opportunities created in a changing environment, and can buffer against the adverse effects of a degrading environment (Brown and Kodric-Brown 1977). This buffering effect plays a role similar to phenotypic plasticity in life history traits in helping to maintain a viable population under sub-optimal conditions and, in so doing, provide time for genetic adaptations. Thus, dispersal plays a critical role in directly and indirectly influencing the success and trajectory of populations responding to new ecological conditions. Arguably, then, dispersal is among the most important factors determining population responses to global change. Thus, understanding the effects of this change requires an understanding of how dispersal behaviours respond to shifting environments.

Our goal in this chapter is to explore the role that dispersal behaviour and dispersal decisions can play under rapidly changing environmental conditions. As such, our focus will be centred on cases of dispersal as a flexible behaviour rather than cases where dispersal is a fixed alternative strategy evolved under longer term fitness benefits (i.e. purely genetically fixed dispersal strategies). This is not to say that a changing environment will not influence fixed alternative dispersal strategies; rather, our discussion will focus on cases where dispersal is a flexible behaviour. We will begin with a brief overview of dispersal, with particular emphasis on the view that dispersal is a plastic and condition-dependent behaviour, since this plasticity will play a fundamental role in dispersal ecology under changing environmental regimes. We then focus our discussion on dispersal in a changing world and consider the effects that changes in the landscape will have on dispersal dynamics, as well as the consequences that dispersal can have for populations impacted by anthropogenic modifications of the environment. We have organized this discussion into four major sections. The first two focus on major environmental impacts affecting dispersal: the modification of habitat quality and habitat fragmentation. The remaining sections will concentrate on some of the consequences of dispersal in a changing landscape, including invasion and colonization of new habitats and evolutionary traps.

5.2 Dispersal: a balance of costs and benefits

The basic principles that govern dispersal decisions are fairly simple and provide a useful framework for understanding how changing environmental conditions will affect dispersal. Dispersal will essentially occur if the fitness benefits of finding a new and 'better' habitat patch where reproduction can occur outweigh the costs associated with leaving the natal patch and travelling across an unknown landscape, or uninhabitable 'matrix', to find a new location (Fig. 5.1). Since evolution shapes dispersal behaviour, fitness in this case refers to the long-term average for the individual. In situations where kin competition or cooperation occurs, we must focus on inclusive fitness. Essentially, then, there must be a net (inclusive) fitness benefit of moving relative to the expected fitness of staying at home. This may seem rather obvious since it is a base condition for the evolution of any behaviour or trait. However, a changing environment will essentially modify factors that influence this cost–benefit balance. If we keep these basic conditions in mind, our ability to understand the effects of anthropogenic impacts will be greatly enhanced. For example, if human impacts reduce food availability in a specific population, we might expect an increase in the number of individuals who disperse away from that population in search of better conditions. This should continue until an equilibrium is reached among populations. The real challenge is to identify and measure the costs and benefits of dispersal in terms of expected fitness, and to understand how multiple environmental impacts will combine to affect dispersal decisions. Recent reviews provide nice discussions of the problems and current challenges confronting dispersal research. These problems and challenges centre on identifying the key factors that affect the cost–benefit balance, the extent of variation in dispersal strategies within populations, and the evolution of dispersal itself (Clobert et al. 2001; Hanski and Gaggiotti 2004; Ronce 2007; Benard and McCauley 2008; Bowler and Benton 2009; Clobert et al. 2009).

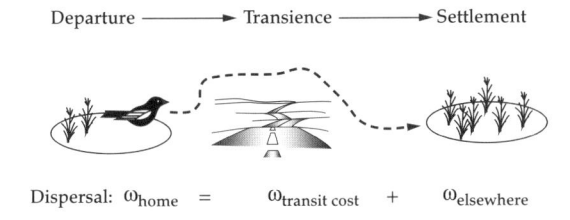

Figure 5.1 Dispersal is the movement from one breeding patch to another. It involves three linked steps: departure, transit, and settlement. Dispersal depends on the expected fitness benefit (ω) of staying in each patch relative to the fitness cost of moving. At equilibrium, the expected fitness of dispersers equals that of non-dispersers. Environmental change can affect the costs and benefits of all three phases of the dispersal process.

$$\omega_{home} = \omega_{transit\ cost} + \omega_{elsewhere}$$

Dispersal can be viewed as a process carried out during three connected phases—departure, movement, colonization—which are independently affected by changing environments (Fig. 5.1; Ims and Yoccoz 1997; Nathan et al. 2008). While an environmental shift can affect all three phases in similar ways, different impacts on one phase relative to others can have different effects on population dynamics. For example, when an environmental shift affects all three phases similarly, a reduction in the motivation to depart can be associated with a reduction in survival during transit and low colonization success. In contrast, environmental shifts that greatly reduce survival during transit between patches, but do not affect the decision to depart, could potentially result in dramatic population declines. Indeed, a given environmental shift is likely to have different consequences for habitat patches where reproduction occurs relative to areas that are already uninhabitable and simply used in transit. Changes in selection during any of these phases will have consequences for selection at other phases since they must operate in synergy for dispersal to be a successful life history strategy (Travis et al., submitted). In the scenario where only transit is impacted, higher mortality rates during this phase will reduce the advantages of dispersal and therefore select for lower departure rates. Consequently, in trying to understand the effects of a changing environment on dispersal behaviour, we need to be cognisant of the short-term implications these changes may have on each phase of the dispersal process, how changes in one phase might affect the others, and how this, in turn, will affect dispersal as an evolutionary strategy.

5.3 Dispersal is a plastic behaviour

Recent work in dispersal ecology has underscored the fact that dispersal behaviour is often highly plastic and condition dependent (Murren et al. 2001; Bowler and Benton 2009; Clobert et al. 2009), as has been suggested for movement behaviour in general (Nathan et al. 2008). The decision to disperse and where an individual should settle will depend not only on evolved imperatives, but also on internal (e.g. condition) and external (e.g. social and ecological) factors (Fig. 5.2a). This view implies an interaction between an individual's genes and prevailing conditions (Fig. 5.2b). The realization that dispersal is a plastic behaviour is key to understanding the interplay between dispersal behaviour and changing environments, because it implies that dispersal decisions will adjust as the surrounding environment shifts. Plasticity also implies that dispersal behaviour can be modified by conditions unrelated to the evolutionary history under which the plastic dispersal strategy evolved.

A broad range of factors can influence dispersal decisions, each of which can be modified by a changing environment. Abiotic factors, such as temperature or humidity, can affect the decision to stay, partly because they interact with a species' physiology and ability to move and, consequently, influence survival and reproduction. For example, wolf spiders *Pardosa purbeckensis* alter their decision to disperse via passive aerial transport depending on the strength of the winds, which determine their ability to find suitable habitats (Bonte et al. 2007). The biotic environment, including food, predators, and competitors, can also influence the decision to disperse (Fig. 5.3). For instance, food supplementation in Carrion crows *Corvus corone* leads to reduced dispersal because of improved patch quality (Baglione et al. 2006). Social interactions, by affecting an individual's fitness prospects (e.g. competition for mates), can also be important. For example, increased population density leads to higher dispersal rates in many species (Matthysen 2005). Finally, an individual's own condition can affect its decision to disperse

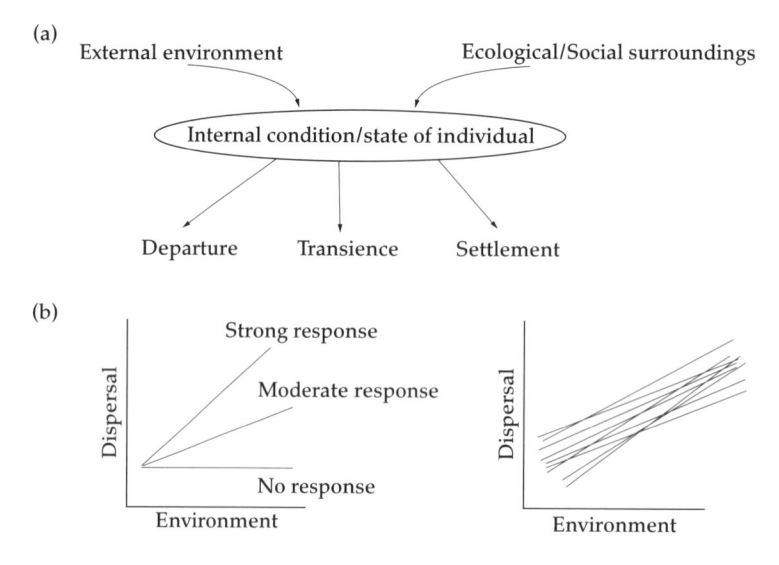

Figure 5.2 Dispersal behaviour is plastic in many organisms. (a) Dispersal decisions depend on both the external and internal environment. Adapted from Clobert et al. (2009). (b) The probability that dispersal will mitigate the effects of environmental change depends on two features of plasticity: the average plasticity of the population (the slope of the reaction norm) and within-population variation in plasticity (the variation in reaction norms).

since condition can often impact potential fitness through its effects on competition, mate choice, and colonization success (Clobert et al. 2009).

In nearly all species, one should expect dispersal behaviour to vary among individuals (Fig. 5.2b; Murren et al. 2001, Clobert et al. 2009). Few studies, however, have explicitly examined among population variation in dispersal behaviour and even fewer have examined within population variation. In a broad scale study of butterflies, Stevens et al. (2010) found that variation in dispersal behaviour among populations within a species was on par with the variation observed among species. While existing data are generally insufficient to determine the underlying cause of this variation, the very existence of variation in dispersal behaviour means that populations will not respond uniformly to new conditions. However, such variation is potentially important in allowing adaptation to take place in response to the new selection pressures introduced by global change. A particularly interesting case is individuals varying extensively in reaction norms, which could both buffer populations against the adverse effects of habitat change as well as provide them with the capacity to colonize new habitats (Fig. 5.2b).

Despite the importance of these factors for dispersal behaviour, empirical data is currently lacking.

There is increasing evidence that dispersive individuals often possess complementary traits—morphological and behavioural—that help them to successfully disperse. For example, dispersive western bluebirds *Sialia mexicana* tend to be larger and more aggressive, which helps them invade new areas (Duckworth and Badyaev 2007). More dispersive individuals may also possess traits that help them to travel or navigate through the landscape more efficiently and, in so doing, more effectively colonize a new habitat (e.g. Phillips et al. 2006; Fjerdingstad et al. 2007). For example, highly dispersive genotypes of *Tetrahymena* ciliates become more streamlined and grow a flagellum for more efficient travel and also have a much higher reproductive rate which increases their colonization success (Fjerdingstad et al. 2007; Schtickzelle et al. 2009). Species or individuals that possess such traits are likely to be affected differently by environmental change compared to those that do not. However, it is worth bearing in mind that certain trait combinations, although beneficial in helping individuals disperse, can also be costly due to, for example, trade-offs with other life history or fitness traits. In both bluebirds and ciliates, individuals that

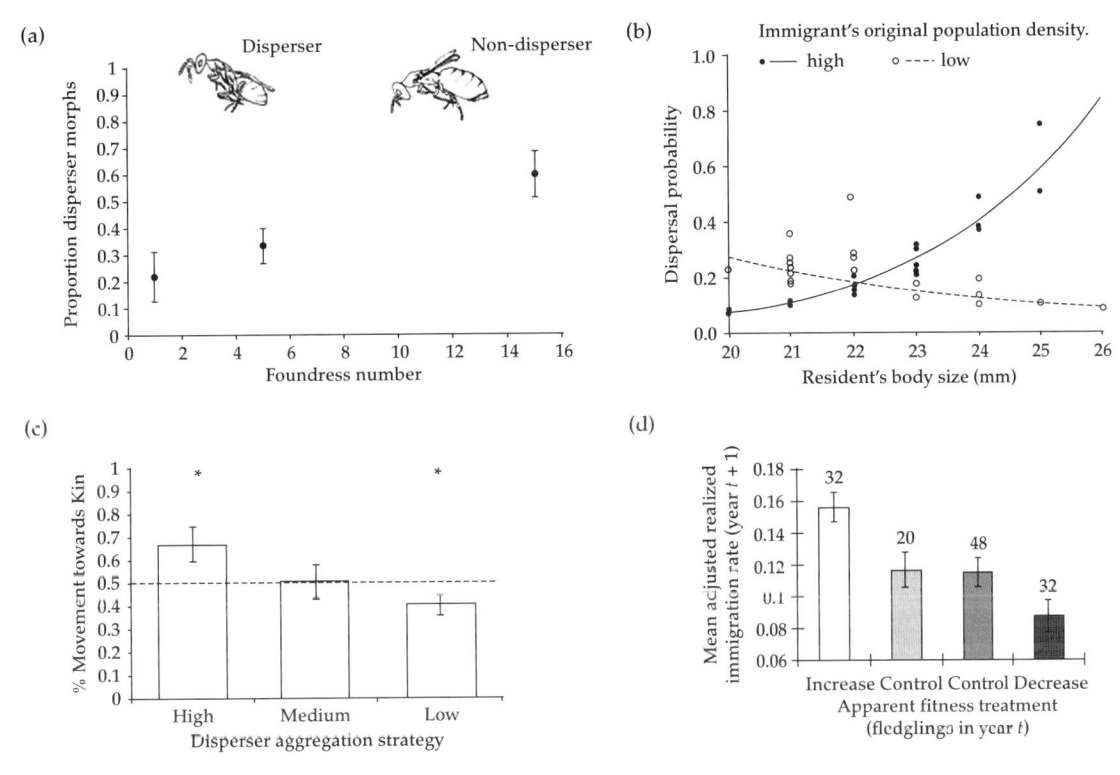

Figure 5.3 Examples of the use of information on environmental conditions at each phase of dispersal. (a) Parasitoid wasps *Melittobia australica* produce more dispersal morphs when local competition is high. Adapted from Innocent et al. (2010). (b) The common lizard gleans information from immigrants to judge the quality of other populations, and modify their dispersal decision accordingly, depending on their own body size. Adapted from Cote and Clobert (2007). (c) *Tetrahymena* cilliates disperse to patches with a compatible kinship group when all patches are occupied, depending on how aggregative they are. Adapted from Chaine et al. (2010). (d) Collared flycatchers estimate the reproductive success of neighbours at the end of the breeding season to judge the quality of the habitats and choose next year's breeding site. Adapted from Doligez et al. (2002).

possess traits associated with improved dispersal and colonization—aggression and movement—come with the cost of being much more asocial and, as such, are more likely to lose out on cooperative interactions with conspecifics (Duckworth and Badyaev 2007; Schtickzelle et al. 2009). Integration or trade-offs among traits and behaviours linked to dispersal has been the subject of very little empirical investigation, but evidence is increasing that they exist and play an important role in colonization and invasion (Clobert et al. 2009; Cote et al. 2010; Fogarty et al. 2011).

Overall, dispersal behaviour is plastic in a broad range of organisms, and the extent of variation among individuals in plasticity, the type of responses to particular environmental cues, and the degree of specialization in dispersal characters are likely to play important roles in determining the effect of environmental changes on the process of dispersal and, thus, on population connectivity and local adaptation.

5.4 Acquisition of information

Dispersal in species with high phenotypic plasticity relies critically on the availability of information and how it is used to optimize the fitness benefits of dispersal. This information can be used to decide whether to disperse, when to disperse, how to navigate, and which patch to choose, so that the costs of dispersal are offset by the benefits of finding a new patch (Fig. 5.4a). Information, in this regard, can come from a range of sources (Fig. 5.3), including individual condition, the local habitat (e.g. density, kin structure, etc.), nearby locations (Doligez et al. 2002), or conspecifics (Cote and Clobert 2007).These sources of information may be used individually or in combination. In the common lizard *Lacerta vivipara* a range of cues was found to influence a juvenile's decision to disperse, including maternal health, sex ratio, kin competition, and indirect social information carried by arriving immigrants (Léna et al. 1998 Massot and Clobert 2000 Le Galliard et al. 2003; Le Galliard et al. 2005; Cote and Clobert 2007). Information use in dispersal is an emerging research field, and the effects of environmental changes on dispersal through modifications of information may be more widespread than is currently known.

The use of cues that predict long-term fitness benefits is a basic tenet of adaptive phenotypic plasticity (including behaviour, see Chapter 11). Its investigation is of critical importance for understanding how dispersal behaviour functions and how it might be altered as the environment changes.

If the effects of environmental change on fitness are reflected in the cues that usually predict fitness, then dispersal behaviour will accurately track and manage these changes. For example, if dispersal decisions are based on habitat cues, then most individuals are likely to leave a degrading habitat as part of their evolved strategy to monitor local habitat quality. However, in some cases, environmental change disconnects the cues from the expected fitness (Fig. 5.4b; Lyon et al. 2008). For instance, environmental change may alter the cue that affects dispersal behaviour even if this does not improve fitness. Alternatively, environmental change may affect fitness in a way that could be ameliorated by shifts in dispersal behaviour, but the cues used in dispersal decisions are not aptly altered (Fig. 5.4b). Good examples come from migratory and breeding behaviour in birds, where global shifts in temperature have disrupted the connection between breeding date and the timing of maximum food availability for offspring (see Chapter 6). Thus, the link between information and expected fitness outcomes will likely determine how dispersal behaviour responds to global change.

(a) Information that could be used in each stage of dispersal

Departure	Do other patches exist? Are those patches better than here?
Transience	Which direction towards good patches? Which path to take? How well can nearby habitats be detected?
Settlement	Stay here or keep going?

(b) Mismatches in dispersal information

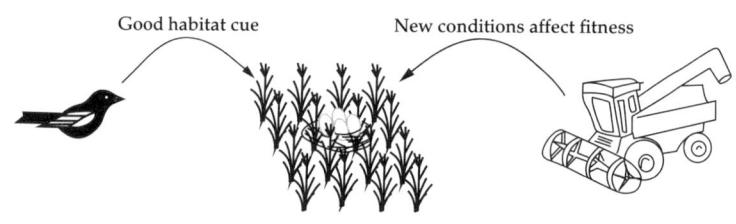

Good habitat cue   New conditions affect fitness

Figure 5.4 (a) Different types of information are important at each phase of dispersal. (b) Cues that predicted fitness in the past but no longer do so accurately can become ecological and/or evolutionary traps. These can cause inappropriate dispersal decisions with detrimental effects on fitness.

5.5 Dispersal in a changing landscape

Human induced impacts will cause a whole host of changes to climate, habitats, populations, and ecosystems as detailed throughout Parts II and III of this book. All of these changes are likely to interact with dispersal. They can alter the quality of the habitats and the uninhabitable matrix, or the reliability of the cues that predict the costs and benefits of dispersal and, in so doing, affect dispersal strategies. In the rest of this chapter, we will focus on the major pathways through which anthropogenic disturbances interact with dispersal, including (1) changed patch quality, (2) habitat fragmentation, (3) species invasion, and (4) ecological traps.

5.5.1 Habitat quality

Human land use and other direct and indirect anthropogenic impacts have a wide range of effects on the quality of habitat patches (Fig. 5.5a). Direct effects include conversion of habitat patches into urbanized areas and farmland. Indirect impacts include a range of effects from climate change to pollution, and the introduction of new species and the extirpation of existing species. A habitat patch that has been altered by humans can often still be used, but the quality of the patch itself may shift. The amount of resources may decline, such as the availability of food, shelter, and nest sites, which increases competition. In some cases, altered sites improve in quality (Bender et al. 1998). For example, orchards provide an increase in food resources for some species. Since dispersal decisions depend on the expected fitness benefits of dispersing in relation to staying, any shifts in patch quality will impact dispersal behaviour. Shifts in the quality of potential destinations, including the addition or elimination of potential patches, are as important as shifts in the quality of the natal patch.

Habitat quality impacts on dispersal
One of the primary effects of habitat alteration is its impact on the fitness of individuals, which in turn will influence their decision to stay or disperse. Research on changes in habitat quality has largely focused on the effects of habitat degradation and reductions in patch size on population dynamics, which often lead to reductions in population sizes

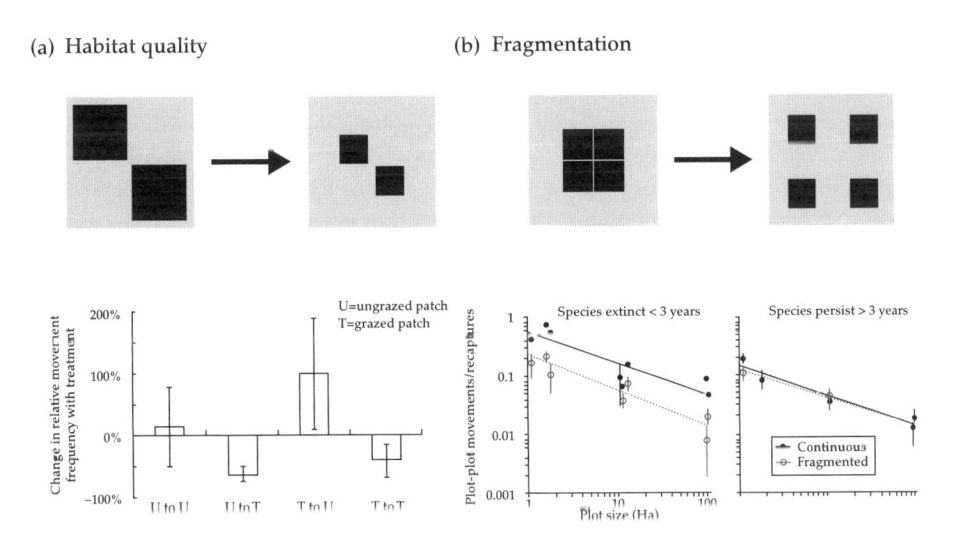

Figure 5.5 Habitat quality and fragmentation are the two major factors that influence dispersal. (a) Alterations of resources can degrade habitat quality and reduce expected fitness gain (illustrated as a reduction in patch size). Experimental grazing of habitat patches of the bog fritillary butterfly reduced the availability of host flowers, which increased dispersal towards intact patches. Adapted from Baguette et al. (2010). (b) Habitat fragmentation results from a reduction in connectivity among patches (illustrated by increased distance among patches). The graphs show that fragmentation lowers dispersal in extinction-prone species but not in species that are able to persist. Adapted from Van Houtan et al. (2007).

and loss of biodiversity (Fahrig 1997; Bender et al. 1998; Fahrig 2003; Watling and Donnelly 2006; Ferraz et al. 2007). Such effects are not universal, and some species—especially those in edge or ecotone habitats—appear less impacted (Bender et al. 1998). However, the effect of dispersal is often ignored. We would expect individuals to disperse away from a patch when habitat quality decreases and the costs of dispersal and the quality of other patches remain the same. Conversely, if patch quality improves, we would expect dispersal from the patch to decrease. Similar predictions can be made for destination patches. If the quality of the patches improves, then increased dispersal towards them is favoured, all else being equal, and *vice versa* for degradation of destination patches. This pattern of movement follows an ideal free distribution, which assumes that individuals have accurate information on habitat qualities across the meta-population (Holt and Barfield 2001). Such information appears to exist in a number of species, as will be discussed below.

There are many habitat characteristics that influence fitness and, thus, dispersal. A common assumption is that resources are limiting such that an increased density of individuals decreases fitness and, accordingly, causes an increase in dispersal. Indeed, there is extensive evidence for density dependent dispersal across a broad range of taxa (Matthysen 2005). Other metrics of habitat quality also influence fitness, such as food quality (Baglione et al. 2006), kin structure (Léna et al. 1998), and social strategies (Sinervo et al. 2006). Decreases in patch quality by experimentally manipulating these factors have led to increases in dispersal, suggesting that individuals gather information on habitat quality across populations. For example, experimental grazing of some habitat patches of the bog fritillary butterfly *Boloria eunomia* reduced the quality of these patches, as vegetation and host plants were consumed. This led to an increase in dispersal out of the grazed patches, as well as a reduction in dispersal out of the patches that had not been grazed (Baguette et al. 2010; Fig. 5.5a). Similarly, in a 50-year study of population dynamics of pike *Esox lucius*, individuals were found to move from a low fitness basin towards a high fitness basin. As expected, an experimental increase in pike density in the high fitness basin reversed this trend and led to a shift in dispersal (Haugen et al. 2006). In these examples, individuals were able to assess the quality of their own patch relative to other patches and, in so doing, make accurate context-dependent dispersal decisions based on the information they had acquired.

Not all species show a clear shift in dispersal behaviour when habitat quality is altered. Indeed, the response is sometimes mixed. For example, in the Rocky Mountain parnassian butterfly *Parnassius smintheus* an experimental manipulation of sex ratio and nectar flowers in prairie patches caused no change in the dispersal behaviour of female butterflies. Males, on the other hand, showed a more nuanced response: immigration rates of males were higher when resource levels (flowers and females) were higher, whereas emigration rates remained unchanged (Matter and Roland 2002). By contrast, common lizards showed the opposite pattern, with females shifting dispersal behaviour more than males. Females reduced their dispersal in male biased populations (probably because females have more mate choice opportunities via sperm choice, despite a lower survival probability), while males showed no change in dispersal rate, even when the local sex ratio was so biased that mating opportunities became dismal (Le Galliard et al. 2005). Neither sex changed their dispersal behaviour as population density increased (Le Galliard et al. 2005), suggesting that the composition of the social group has a bigger impact on fitness than the local density *per se*. Together, these results highlight the difficulties that confront researchers in assessing habitat quality.

Ideal free dispersal?
In essence, dispersal should follow an ideal free distribution according to patch quality (Holt and Barfield 2001). Specifically, dispersal should shift individuals to other locations as patch quality declines. However, as the examples in butterflies and lizards illustrate, such an expectation is not always met for at least four major reasons. First, our measurements of changes in patch quality in terms of an organism's fitness can be erroneous. Obvious changes in patch characteristics may have a negligible impact on the fitness of the species under focus.

Second, when patch quality or density declines, so does the reproductive output of individuals within the patch. A reduction in local density relaxes competition for resources and, thus, improves the quality of a patch, which will reduce dispersal out of the patch. For example, new sites at range margins will be colonized by dispersers, but the low density of individuals in these patches could favour philopatric behaviour and, thus, slow down range expansion. However, if colonists are asocial relative to individuals who arrive later, increased density at range margins could again increase dispersal (Clobert et al. 2009; Cote et al. 2010; Fogarty et al. 2011). Therefore, long-term dispersal dynamics are complex and will shift according to natural changes in habitat quality, reproductive output, and life history strategies. Third, predictions often rest on the assumption that the remaining landscape remains the same. This is seldom the case. Specifically, anthropogenic impacts not only affect the quality of the local patch, but also the matrix landscape between patches that an individual must cross, as well as the quality of the patches that serve as potential destinations. Decreased patch quality throughout the meta-population landscape can accelerate local extinction and favour dispersal as the chances of colonizing an empty patch increase (Heino and Hanski 2001). Furthermore, environmental impacts that introduce greater heterogeneity in quality among patches can affect dispersal rates (Legendre et al. 2008). Finally, changes in dispersal behaviour depend crucially on the information available to individuals. Habitat alterations that differ from those encountered during a species' evolutionary history may not be registered as changes in habitat quality (Fig. 5.4b), or individuals may not have accurate information on the quality of surrounding patches. How fast and accurately a species responds to habitat changes depends largely on the information individuals are able to acquire about the meta-population (patch quality and connectivity), their dispersal capabilities, variation in dispersal strategies among individuals (because of genetic variation and plasticity, Fig. 5.2b), and the reproductive output of the population.

Achievement of an ideal free distribution through dispersal depends crucially on the availability of information with regard to the fitness expectations of different patches including those that are more distant. Evidence that organisms are, indeed, capable of gathering such information is accumulating, although still rare. For example, bog fritillary butterflies reduce their dispersal out of high quality patches when the quality of surrounding patches declines (Baguette et al. 2010, Fig. 5.5a). Collared flycatchers *Ficedula albicollis* monitor the local habitat after breeding, as this allows them to judge the production of fledglings in different patches. Based on this information, they decide whether to disperse to a new breeding patch in the following year (Doligez et al. 2002, Fig. 5.3d). In the common lizard, immigrants bring information about the density of their populations of origin and, thus, of the surrounding habitat patches, which allows residents to make appropriate dispersal decisions (Cote and Clobert 2007, Fig. 5.3b). Thus, in at least some species, individuals are able to acquire accurate information as to the quality of patches throughout the meta-population, and use this information to make decisions about departure and habitat selection.

5.5.2 Habitat fragmentation

Human land use has led to the loss of habitats and the fragmentation of landscapes, which makes dispersal a critical link among populations. Most populations are naturally fragmented and occur in habitat patches with varying degrees of connectivity (i.e. metapopulations). Habitat fragmentation can further influence this connectivity. While there is no single definition of 'habitat fragmentation' (Fahrig 2003), one that is useful in the context of dispersal focuses on how alterations in the matrix affects connectivity of patches and population dynamics, whilst controlling for the effects of habitat loss (Fig. 5.5b). Both direct and indirect modification of the landscape can cause habitat fragmentation, with major impacts on populations as human populations continue to grow.

Connectivity
An important measure for gauging the population-level impacts of human-induced habitat fragmentation is to examine changes in the isolation and

connectivity of patches. The degree of isolation depends on a number of factors that influence the costs of moving between patches. One factor is the degree of spatial or physical isolation of a patch. As the distance between patches increases, the cost of transit is expected to increase, although this widely made assumption remains to be verified empirically. Some species display dispersal polymorphisms, including both long and short distance dispersal behaviours (Roff and Fairbairn 2001; Bonte et al. 2007; Fjerdingstad et al. 2007; Schtickzelle et al. 2009). An increased distance between patches may shift the benefit of the behaviours and, thus, their frequency in the population (phenotypically or genetically). Another factor that affects the costs of transit is the risk of mortality, or of not finding a suitable destination patch. For example, roads create short distance separations between patches, but can incur severe mortality costs for species trying to cross them (Forman and Alexander 1998). Likewise, species with a short perceptual range (relative to the scale of the environmental disturbance) may find it difficult to navigate in modified environments or to locate a suitable destination patch. The third factor that influences the costs of movement is the difficulty of traversing specific matrix habitat types (Ricketts 2001). As humans alter the landscape, the *permeability* of the matrix may change, (i.e. the difficulty of physically moving through specific habitats in the matrix and the behavioural affinity of organisms to specific habitats). A matrix that is less permeable will result in higher costs associated with dispersal (i.e. energy, time, and mortality risk), which will reduce the net benefit of leaving a patch in search of a better one.

Despite its importance, understanding permeability can be challenging; what is permeable for some organisms may pose a barrier to others. For example, a short 100 m cleared pasture in the Brazilian rainforest seems like a trivial barrier for a bird, but could be insurmountable for many forest adapted species that avoid clearings (Sodhi et al. 2004). In contrast, major roads can appear as massive barriers to dispersal, yet some organisms cross them with little difficulty (Forman and Alexander 1998). Permeability can be measured either experimentally, by investigating the behavioural affinity of individuals to specific habitats and their ability to cross them (Stevens et al. 2006, Baguette and Van Dyck 2007), or by determining genetic differentiation, which gives a coarse estimate of permeability (Crooks and Sanjayan 2006). The measurement of permeability is often complicated by extensive variation in habitat affinities both across and within species. As a result, determining the impact of human induced habitat fragmentation on connectivity among patches may be difficult *a priori* and achievable only after fragmentation has already occurred.

Causes and consequences of fragmentation
Habitat fragmentation can potentially result from a range of human activities. Empirical evidence suggests that anthropogenically induced habitat loss and degradation contribute more to the decline and extinction of populations than fragmentation *per se* (reviewed in Fahrig 1997; Fahrig 2003; Watling and Donnelly 2006). However, results have varied considerably among the handful of studies that have sought to examine the relative effects of patch size and connectivity (Fahrig 1997; Debinski and Holt 2000; Fig. 5.5b) and at least some studies show a non-negligible effect of fragmentation. A recent meta-analysis found that patch isolation influenced species distributions (along with patch size), but that the effect was only clear when the matrix habitat between the patches (which influences dispersal behaviour) was taken into account (Prugh et al. 2008). Notably, habitat fragmentation itself can have both positive and negative effects on life history traits, population dynamics, and biodiversity (reviewed in Fahrig 2003).

Fragmentation effects on dispersal behaviour
While community level studies have given variable results on the effects of habitat fragmentation on species persistence and distribution, species specific studies have often detected measurable impacts of habitat fragmentation on dispersal behaviour. Evidence comes from a number of sources, including genetic analyses of movement, capture-mark-recapture (CMR) studies, and tracking of individuals. For example, evidence from genetic analysis shows that the fragmentation of habitats

by roads in northeastern America has reduced the movement of timber rattlesnakes *Crotalus horridus* and led to full separation of their winter dens (Clark et al. 2010). Interestingly, these small country roads, constructed just seven generations ago, have hardly changed the total habitat area. A narrow road would seem like a trivial distance to cross for a rattlesnake, yet the genetic separations are surprisingly strong. Roads could have contributed directly to this separation, through increased mortality during dispersal, but behavioural avoidance of roads could also have reduced connectivity (Forman and Alexander 1998). Similar results have been found in other species (from bighorn sheep to beetles) and, for some, larger roads have a larger impact on dispersal (Keller and Largiadèr 2003; Epps et al. 2005).

Studies that track the movement of individuals rather than their genes can give a much more accurate view of how habitat fragmentation affects dispersal behaviour, since these studies include dispersers that are not integrated into the breeding population but still have an impact on population dynamics (e.g. through competition). In this regard, mark-recapture studies have reported mixed effects of habitat fragmentation on both dispersal (Prugh et al. 2008) and population persistence (Fahrig 2003). An excellent illustration of this comes from a multispecies mark-recapture study in birds. In that study, Van Houtan et al. (2007) found that fragmentation resulted in increased dispersal in some species (n = 8) but reduced dispersal in others (n = 10) (Fig. 5.5b). Interestingly, species that reduced dispersal went extinct in less than three years, whereas those that increased dispersal persisted in the fragmented habitat. Furthermore, dispersal and extinction patterns were not associated with daily movement patterns, suggesting that the decision to disperse depended on external cues specific to dispersal and patch choice.

Mark-recapture studies in butterflies provide some of the best evidence of how habitat fragmentation can impact dispersal. For example, dispersal of the bog fritillary butterfly *Proclossiana eunomia* was found to decrease with increased habitat fragmentation. The species is a habitat specialist naturally found on patchy resources and, as a result, should have been able to cope with fragmented resources (Baguette et al. 2003; Schtickzelle et al. 2006). The decrease was a behavioural decision and was not caused by higher mortality during dispersal, as mortality in the matrix was actually lower in more fragmented landscapes. As this example highlights, the effects of fragmentation are not necessarily straightforward: fragmentation can alter the decision to disperse between patches but have little (if any) effect on actual survival in the matrix. Indeed, increased fragmentation is expected to favour dispersal only in individuals highly adapted to disperse (Roff and Fairbairn 2001; Van Dyck and Baguette 2005).

A small number of studies have used both genetic and demographic methods to determine the effect of fragmentation on dispersal, but virtually none have done so in conjunction with a gradient of habitat fragmentation. One notable exception is a long term study of Florida scrub jays *Aphelocoma coerulescens*. Using both resightings of colour banded individuals and genetic measures of dispersal, Coulon et al. (2010) studied 13 scrub jay populations exposed to differing degrees of habitat fragmentation. The researchers found that increased fragmentation enlarged dispersal distances, as detected by resighting of marked individuals. However, individuals that dispersed longer distances in more fragmented habitats were less successful in joining the breeding population, as revealed by genetic isolation by distance among populations. Together, these results reinforce the notion that the effects of fragmentation on dispersal behaviour are complex. Specifically, adjustment of dispersal behaviour to habitat fragmentation can be overridden by other factors, which reduce the effectiveness of dispersal at the genetic level.

Fragmentation and conservation
Conservation efforts to counter the effects of fragmentation have largely centred on building 'corridors' that provide a safer and more permeable passage than the surrounding environment (Crooks and Sanjayan 2006; Gilbert-Norton et al. 2010). The goal is to increase dispersal among patches to allow for the beneficial effects of dispersal, which include bolstering low population sizes and recolonizing extinct habitat patches. Overall, the

effect of corridors appears positive, as dispersal is, on average, increased by 50%, thereby alleviating some of the costs of fragmentation and small patch size (Gilbert-Norton et al. 2010). Corridors, however, differ in efficiency; natural corridors are more effective than human built corridors (Gilbert-Norton et al. 2010), which reinforces the notion that understanding dispersal behaviour and landscape permeability from the organism's perspective is challenging. Moreover, not all effects of corridors are positive. Recent investigations found that corridors can increase edge effects associated with habitat patches, which can increase predation and/or reduce reproductive success (reviewed in Crooks and Sanjayan 2006). Another potential consequence of corridors, which has so far received very little attention, is whether increased dispersal can counteract the effects of selection and, as a result, hamper local adaptation to a given habitat patch.

5.5.3 Dispersal as a mechanism for invasion and range shifts

As habitats are altered and landscapes become increasingly fragmented, dispersal remains one of the key processes that could allow individuals to colonize both newly created and recently extinct patches. New patches may provide higher fitness benefits than the natal patch and, thus, should favour dispersal and colonization. Colonization of patches where the population has gone extinct has been the focus of meta-population research, from both a theoretical and an empirical perspective. Human land use increases population extinctions in some patches, but also creates new habitats. Habitat restoration and natural recovery of patches after anthropogenic disturbances (e.g. oil spills) rely on dispersal of species to restore the native flora and fauna. In contrast, when land is converted for other uses, such areas may become potentially usable to new species (Phillips et al. 2006; Duckworth and Badyaev 2007).

Research has shown that some species produce distinct dispersal phenotypes, which can aid in their travel to, and colonization of, new habitat patches. For example, in the case of western blue birds, highly dispersive individuals are more aggressive and this, in turn, facilitates their invasion of new habitats by allowing them to outcompete resident species, such as the Mountain bluebird *Sialia currucoides* (Duckworth and Badyaev 2007). However, the type of information used to decide whether to disperse is unclear, as well as how individuals navigate to those habitats. In the common lizard, dispersers who depart due to kin competition in the natal patch have a different phenotype from those who do not disperse. Importantly, these dispersive individuals also have a higher success in colonizing empty patches (Cote et al. 2007).

Natural dispersal is assumed to be the major process underlying range shifts linked to climate change (Parmesan 2006). Short distance dispersal at range margins allows species to expand their ranges when nearby habitats become less hostile, while both dispersal and death due to changing habitats appear responsible for the retraction of range limits. Thus, many of the major shifts in species distributions in response to human-induced environmental change are probably mediated by dispersal behaviour.

5.5.4 Ecological traps

One major consequence of human activities is the creation of conditions that have not been previously encountered by species during their evolutionary history. This can lead to organisms choosing habitats that are detrimental to their fitness, a scenario known as 'ecological traps' (Schlaepfer et al. 2002; Fig. 5.4b). As discussed above, information on the expected fitness returns of settling into different habitats plays an important role in the decision to disperse, as well as the choice of a patch. Anthropogenic disturbances can alter the value of the information acquired by changing the link between a cue and the features of the habitat that determine fitness. For example, Indigo buntings *Passerina cyanea* prefer to nest in natural edge habitats, as these provide excellent nesting and foraging opportunities. Yet, their nesting success in anthropogenically created edge habitat is low because these habitats contain more nest predators (Weldon and Haddad 2005).

Most studies on ecological traps focus on the fitness of individuals within the patches and on the decision to settle into those patches, rather than on the decision to disperse. In the common lizard, increases in temperatures over the past decade have led to a decrease in dispersal. Increased temperature makes some patches unsuitable (too dry) and thereby decreases the potential benefits of dispersal from a currently suitable patch. This could potentially result in an increase in kin competition and prevent the lizards from escaping habitat degradation associated with future increases in climate warming (Massot et al. 2008). Similarly, an individual based model of dispersal showed that the evolution of dispersal distances in one landscape can prevent a population from colonizing new patches if the landscape becomes more fragmented (Travis et al. 2010). For example, reduced dispersal in response to improved habitat quality can decrease connectivity and the persistence of the metapopulation since extinct patches will no longer be recolonized (Poethke et al. 2011). Thus, an evolutionary alteration of dispersal decisions can lead to an 'evolutionary trap'. Similar ecological and evolutionary traps are likely to play a key role in the choice of dispersal paths through a landscape, but this possibility has received very little attention to date. While ecological and evolutionary traps have rarely been discussed in the context of dispersal, the important role that information on habitat quality plays in dispersal behaviour suggests that these traps could be highly influential.

5.6 Conclusions

We have argued that dispersal decisions depend on the fitness benefits of staying in the natal patch versus moving to another patch. The changes humans inflict on landscapes and organisms will shift these benefits and alter dispersal decisions. In this regard, dispersal gives species an opportunity to avoid the negative impacts of environmental change by allowing them to move into more suitable areas. However, it can also affect local adaptation through gene flow. Throughout this chapter, we have sought to draw attention to a range of issues that, we believe, are important for understanding how environmental changes will affect dispersal behaviour and how this, in turn, will alter the ecological and evolutionary dynamics of species.

A first challenge for understanding the impact of environmental change on dispersal is to accurately measure the environment from the organisms' perspective. We still know very little about what species perceive as important resources in a habitat patch and what criteria they use to navigate the matrix towards other patches. This is not a trivial problem. To predict animal movements and subsequent population dynamics, it is imperative that we understand how landscape and habitat features affect fitness and movement decisions.

The rate at which dispersal behaviour can change in response to environmental change depends on the genetic basis of dispersal behaviour. The degree of plasticity in such behaviour largely determines if a species can adjust to profound environmental changes or only to minor changes in the short term, and if this adjustment is sufficient to prevent extinction (Fig. 5.2b). For populations to adapt in the longer term, we must know how much genetic variation in dispersal behaviour exists among individuals. In this regard, the degree of plasticity and variation in genetically determined reaction norms depend on the variability of the environment that the species has experienced during its evolutionary history. Yet, in the context of dispersal, empirical measurement of this variability is virtually non-existent.

We have argued that information plays an important role in dispersal decisions. Unfortunately, we currently know very little about what information individuals use in their decisions about dispersing (Fig. 5.3, 5.4a). Environmental change can potentially affect the acquisition of information on habitat qualities and landscape architecture. Thus, to understand dispersal behaviour in a changing world, we need to know what information is available to individuals and what is actually used, neither of which is currently well known. We also need to know if the information used remains reliable (Fig. 5.4b). If cues that predict the fitness benefits of dispersal are no longer linked to benefits but, instead, become unreliable, then species run the risk of falling into an ecological trap. Moreover, if species use multiple sources of information to make

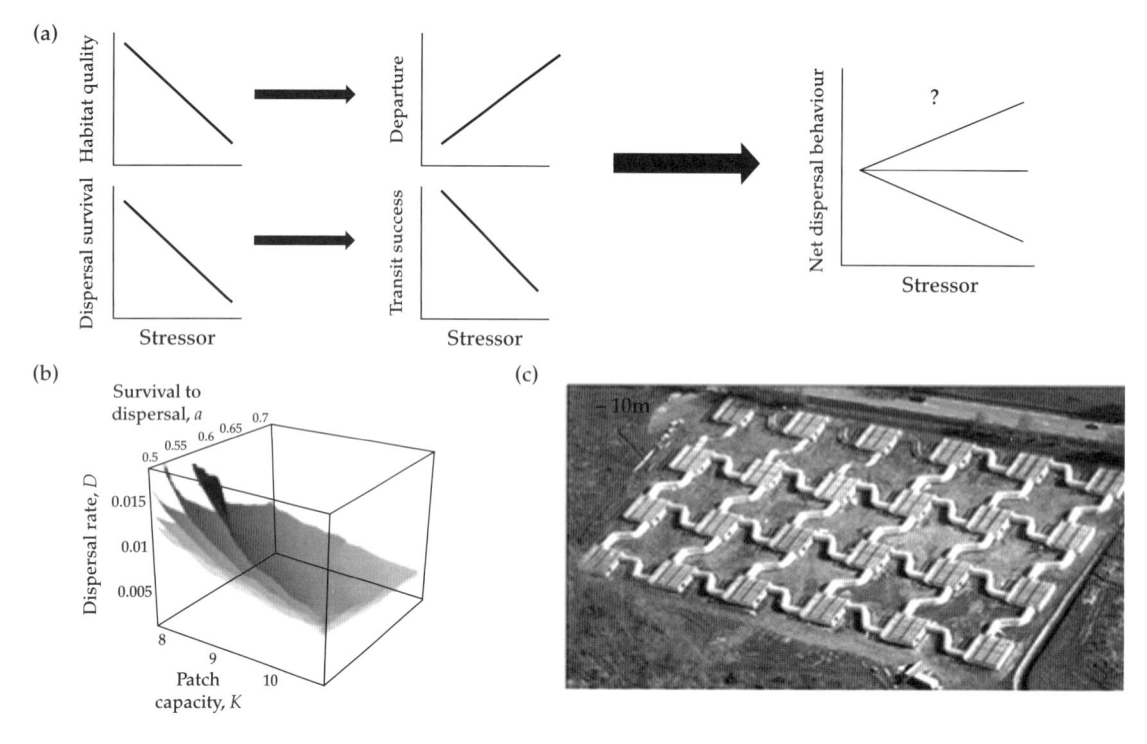

Figure 5.6 Combining the effects of multiple stressors can have nonlinear influences on dispersal and meta-population function. (a) A single stressor can influence different phases of dispersal differently, which in combination, can have unpredictable consequences for dispersal, as illustrated with simple linear relationships (although the effects are more likely to be nonlinear). (b) Multiple impacts on different aspects of habitat quality can influence dispersal. This is illustrated with surfaces for the minimum dispersal rate needed for a meta-population to remain viable when external impacts alter survival, patch quality, and population growth rate (surfaces ranges for growth rates r from 0.025 to 0.045). Populations are viable for values above surfaces. Adapted from Massot et al. (2008). (c) An experimental meta-population, or metatron, at the CNRS in Moulis, designed to test the effects of multiple ecological factors, both within patches and in corridors, on dispersal behaviour and meta-population function.

dispersal decisions, how will they react when information conflicts or varies in reliability?

Dispersal behaviour involves a number of sequential phases—emigration, transience, settlement—that together affect the success of a dispersal decision (Fig. 5.1). Anthropogenic change could affect just one of these phases or a combination of different phases (Fig. 5.6), yet the consequences for dispersal and population viability depend on how impacts at one dispersal phase interact with impacts at others. For example, impacts that reduce emmigration might render degradation of the matrix irrelevant. In contrast, no change in emmigration but increased mortality in a degraded matrix can have negative consequences for fitness and population viability. Impacts on each phase are unlikely to be simply additive and more complex interactions are possible and even likely. We are just beginning to examine how performance in the different phases of dispersal interact, and we are still a long way off from predicting how these relationships will change in altered landscapes.

Environmental change is likely to have multiple, simultaneous impacts on populations. This combination of effects can introduce complex, non-linear dynamics, making dispersal and population trajectories difficult to predict (Fig. 5.6). For example, simultaneous alteration of multiple habitat patches complicates dispersal decisions, especially when species have imperfect information about their environment. Similarly, environmental change that affects both population growth and dispersal dynamics can cause strong fluctuations in population dynamics, which can undermine population persistence. Moreover, environmental change can influence both the condition of individuals and the quality of the habitat, but we currently have little knowledge of how different factors combine to influence dispersal. Understanding the combined effects of multiple impacts at the meta-population level remains a formidable challenge.

Acknowledgements

We thank V. Stevens, U. Candolin, B. Wong, and an anonymous reviewer for valuable comments that improved this chapter.

References

Baglione, V., Canestrari, D., Marcos, J. M., and Ekman, J. (2006). Experimentally increased food resources in the natal territory promote offspring philopatry and helping in cooperatively breeding carrion crows. *Proceedings of the Royal Society of London B, Biological Sciences*, 273, 1529–35.

Baguette, M., Clobert, J., and Schtickzelle, N. (2010). Metapopulation dynamics of the bog fritillary butterfly: experimental changes in habitat quality induced negative density-dependent dispersal. *Ecography*, 34, 170–6.

Baguette, M., Mennechez, G., Petit, S., and Schtickzelle, N. (2003). Effect of habitat fragmentation on dispersal in the butterfly Proclossiana eunomia. *Comptes Rendus Biologies*, 326, 200–9.

Baguette, M. and Van Dyck, H. (2007). Landscape connectivity and animal behavior: functional grain as a key determinant for dispersal. *Landscape Ecology*, 22, 1117–29.

Benard, M. and Mccauley, S. (2008). Integrating across life-history stages: consequences of natal habitat effects on dispersal. *The American Naturalist*, 171, 553–67.

Bender, D. J., Contreras, T. A., and Fahrig, L. (1998). Habitat loss and population decline: a meta-analysis of the patch size effect. *Ecology*, 79, 517–33.

Bonte, D., Bossuyt, B., and Lens, L. (2007). Aerial dispersal plasticity under different wind velocities in a salt marsh wolf spider. *Behavioral Ecology*, 18, 438–43.

Bowler, D. E. and Benton, T. G. (2009). Variation in dispersal mortality and dispersal propensity among individuals: the effects of age, sex and resource availability. *Journal of Animal Ecology*, 78, 1234–41.

Brown, J. H. and Kodric-Brown, A. (1977). Turnover rates in insular biogeography: effect of immigration on extinction. *Ecology*, 58, 445–9.

Chaine, A. S., Schtickzelle, N., Polard, T., Huet, M., and Clobert, J. (2010). Kin-based recognition and social aggregation in a ciliate. *Evolution*, 64, 1290–300.

Clark, R. W., Brown, W. S., Stechert, R., and Zamudio, K. R. (2010). Roads, interrupted dispersal, and genetic diversity in Timber Rattlesnakes. *Conservation Biology*, 24, 1059–69.

Clobert, J., Danchin, E., Dhondt, A. A., and Nichols, J. D. (2001). *Dispersal*, New York, Oxford University Press.

Clobert, J., Galliard, J.-F. L., Cote, J., Meylan, S., and Massot, M. (2009). Informed dispersal, heterogeneity in animal dispersal syndromes and the dynamics of spatially structured populations. *Ecology Letters*, 12, 197–209.

Cote, J. and Clobert, J. (2007). Social information and emigration: lessons from immigrants. *Ecology Letters*, 10, 411–17.

Cote, J., Clobert, J., Brodin, T., Fogarty, S., and Sih, A. (2010). Personality-dependent dispersal: characterization, ontogeny and consequences for spatially structured populations. *Philosophical Transactions of the Royal Society B: Biological Sciences*, 365, 4065–76.

Cote, J., Clobert, J., and Fitze, P. S. (2007). Mother-offspring competition promotes colonization success. *PNAS*, 104, 9703–8.

Coulon, A., Fitzpatrick, J. W., Bowman, R., and Lovette, I. J. (2010). Effects of habitat fragmentation on effective dispersal of Florida Scrub-Jays. *Conservation Biology*, 24, 1080–8.

Crooks, K. R. and Sanjayan, M. (2006). *Connectivity Conservation*, Cambridge, UK, Cambridge University Press.

Debinski, D. M. and Holt, R. D. (2000). A survey and overview of habitat fragmentation experiments. *Conservation Biology*, 14, 342–55.

Doligez, B., Danchin, E., and Clobert, J. (2002). Public information and breeding habitat selection in a wild bird population. *Science*, 297, 1168–70.

Duckworth, R. A. and Badyaev, A. V. (2007). Coupling of dispersal and aggression facilitates the rapid range expansion of a passerine bird. *Proceedings of the National Academy of Sciences of the USA*, 104, 15017–22.

Epps, C. W., Palsbøll, P. J., Wehausen, J. D., Roderick, G. K., Ramey, R. R., and Mccullough, D. R. (2005). Highways block gene flow and cause a rapid decline in genetic diversity of desert bighorn sheep. *Ecology Letters*, 8, 1029–38.

Fahrig, L. (1997). Relative effects of habitat loss and fragmentation on population extinction. *Journal of Wildlife Management*, 61, 603–10.

Fahrig, L. (2003). Effects of habitat fragmentation on biodiversity. *Annual Review of Ecology, Evolution, and Systematics*, 34, 487–515.

Ferraz, G., Nichols, J. D., Hines, J. E., Stouffer, P. C., Bierregaard, R. O., and Lovejoy, T. E. (2007). A large-scale deforestation experiment: effects of patch area and isolation on Amazon birds. *Science*, 315, 238–41.

Fjerdingstad, E., Schtickzelle, N., Manhes, P., Gutierrez, A., and Clobert, J. (2007). Evolution of dispersal and life history strategies in *Tetrahymena* ciliates. *BMC Evolutionary Biology*, 7, 133.

Fogarty, S., Cote, J., and Sih, A. (2011). Social personality polymorphism and the spread of invasive species: a model. *American Naturalist*, 177, 273–87.

Forman, R. T. T. and Alexander, L. E. (1998). Roads and their major ecological effects. *Annual Review of Ecology and Systematics*, 29, 207–31.

Gilbert-Norton, L., Wilson, R., Stevens, J. R., and Beard, K. H. (2010). A meta-analytic review of corridor effectiveness. *Conservation Biology*, 24, 660–8.

Gilpin, M. E. and Hanski, I. (1991). *Metapopulation Dynamics: Empirical and Theoretical Investigations*, London, Academic Press.

Hanski, I. A. and Gaggiotti, O. E. (eds) (2004). *Ecology, Genetics and Evolution of Metapopulations*, Amsterdam: Academic Press.

Haugen, T. O., Winfield, I. J., Vøllestad, L. A., Fletcher, J. M., James, J. B., and Stenseth, N. C. (2006). The ideal free pike: 50 years of fitness-maximizing dispersal in Windermere. *Proceedings of the Royal Society of London B, Biological Sciences*, 273, 2917–24.

Heino, M. and Hanski, I. (2001). Evolution of migration rate in a spatially realistic metapopulation model. *American Naturalist*, 157, 495–511.

Holt, R. D. and Barfield, M. (2001). On the relationship between the ideal free distribution and the evolution of dispersal. In: Clobert, J., Danchin, E., Dhondt, A. A. and Nichols, J. D. (eds) *Dispersal*. Oxford, UK, Oxford University Press.

Ims, R. and Yoccoz, N. (1997). Studying transfer processes in metapopulations: emigration, migration and colonization. In: I, H. and M, G. (eds) *Metapopulation Biology: Ecology, Genetics, and Evolution*. San Diego, Academic Press.

Innocent, T. M., Abe, J., West, S. A. and Reece, S. E. (2010). Competition between relatives and the evolution of dispersal in a parasitoid wasp. *Journal of Evolutionary Biology*, 23, 1374–85.

Keller, I. and Largiadèr, C. R. (2003). Recent habitat fragmentation caused by major roads leads to reduction of gene flow and loss of genetic variability in ground beetles. *Proceedings of the Royal Society of London B, Biological Sciences*, 270, 417–23.

Le Galliard, J. F., Ferriere, R., and Clobert, J. (2003). Mother-offspring interactions affect natal dispersal in a lizard. *Proceedings of the Royal Society of London B, Biological Sciences*, 270, 1163–9.

Le Galliard, J. F., Fitze, P. S., Ferriere, R., and Clobert, J. (2005). Sex ratio bias, male aggression, and population collapse in lizards. *Proceedings of the National Academy of Sciences of the USA*, 102, 18231–6.

Legendre, S., Schoener, T. W., Clobert, J., and Spiller, D. A. (2008). How is extinction risk related to population-size variability over time? A family of models for species with repeated extinction and immigration. *American Naturalist*, 172, 282–98.

Léna, J.-P., Clobert, J., De Fraipont, M., Lecomte, J., and Guyot, G. (1998). The relative influence of density and kinship on dispersal in the common lizard. *Behavioral Ecology*, 9, 500–7.

Levins, R. (1969). Some demographic and genetic consequences of environmental heterogeneity for biological control. *Bulletin of Entomological Society of America*, 15, 237–40.

Lyon, B. E., Chaine, A. S., and Winkler, D. W. (2008). Ecology—A matter of timing. *Science*, 321, 1051–2.

Massot, M. and Clobert, J. (2000) Processes at the origin of similarities in dispersal behaviour among siblings. *Journal of Evolutionary Biology*, 13, 707–19.

Massot, M., Clobert, J., and Ferriere, R. (2008). Climate warming, dispersal inhibition and extinction risk. *Global Change Biology*, 14, 461–9.

Matter, S. F. and Roland, J. (2002). An experimental examination of the effects of habitat quality on the dispersal and local abundance of the butterfly Parnassius smintheus. *Ecological Entomology*, 27, 308–16.

Matthysen, E. (2005). Density-dependent dispersal in birds and mammals. *Ecography*, 28, 403–16.

Murren, C. J., Julliard, R., Schlichting, C. D., and Clobert, J. (2001). Dispersal, individual phenotype, and phenotypic plasticity. In: Clobert, J., Danchin, E., Dhondt, A. A. and Nichols, J. D. (eds) *Dispersal*. Oxford, UK, Oxford University Press.

Nathan, R., Getz, W. M., Revilla, E., Holyoak, M., Kadmon, R., Saltz, D., and Smouse, P. E. (2008). A movement ecology paradigm for unifying organismal movement research. *Proceedings of the National Academy of Sciences of the USA*, 105, 19052–9.

Parmesan, C. (2006). Ecological and evolutionary responses to recent climate change. *Annual Review of Ecology, Evolution, and Systematics*, 37, 637–69.

Phillips, B. L., Brown, G. P., Webb, J. K., and Shine, R. (2006). Invasion and the evolution of speed in toads. *Nature*, 439, 803–803.

Poethke, H. J., Dytham, C., and Hovestadt, T. (2011). A metapopulation paradox: partial improvement of habitat may reduce metapopulation persistence. *American Naturalist*, 177, 792–9.

Prugh, L. R., Hodges, K. E., Sinclair, A. R. E., and Brashares, J. S. (2008). Effect of habitat area and isolation on fragmented animal populations. *Proceedings of the National Academy of Sciences of the USA*, 105, 20770–5.

Ricketts, T. H. (2001). The matrix matters: effective isolation in fragmented landscapes. *American Naturalist*, 158, 87–99.

Roff, D. A. and Fairbairn, D. J. (2001). The genetic basis of migration and its consequences for the evolution of correlated traits. In: Clobert, J., Danchin, E., Dhondt, A. A. and Nichols, J. D. (eds) *Dispersal*. Oxford UK, Oxford University Press.

Ronce, O. (2007). How does it feel to be like a rolling stone? Ten questions about dispersal evolution. *Annual Review of Ecology, Evolution, and Systematics*, 38, 231–53.

Schlaepfer, M. A., Runge, M. C., and Sherman, P. W. (2002). Ecological and evolutionary traps. *Trends in Ecology & Evolution*, 17, 474–80.

Schtickzelle, N., Fjerdingstad, E., Chaine, A. S., and Clobert, J. (2009). Cooperative social clusters are not destroyed by dispersal in a ciliate. *BMC Evol Biol.*, 9, 251.

Schtickzelle, N., Mennechez, G., and Baguette, M. (2006). Dispersal depression with habitat framgentation in the Bog fritillary butterfly. *Ecology*, 87, 1057–65.

Sinervo, B., Chaine, A., Clobert, J., Calsbeek, R., Hazard, L., Lancaster, L., Mcadam, A. G., Alonzo, S., Corrigan, G., and Hochberg, M. E. (2006). Self-recognition, color signals, and cycles of greenbeard mutualism and altruism. *Proceedings of the National Academy of Sciences of the USA*, 103, 7372–7.

Sodhi, N. S., Liow, L. H., and Bazzaz, F. A. (2004). Avian extinctions from tropical and subtropical forests. *Annual Review of Ecology, Evolution, and Systematics*, 35, 323–45.

Stevens, V., Leboulengé, É., Wesselingh, R., and Baguette, M. (2006). Quantifying functional connectivity: experimental assessment of boundary permeability for the natterjack toad (*Bufo calamita*). *Oecologia*, 150, 161–71.

Stevens, V. M., Pavoine, S., and Baguette, M. (2010). Variation within and between closely related species uncovers high intra-specific variability in dispersal. *PLOS ONE*, 5, e11123.

Travis, J. M. J., Smith, H. S., and Ranwala, S. M. W. (2010). Towards a mechanistic understanding of dispersal evolution in plants: conservation implications. *Diversity and Distributions*, 16, 690–702.

Van Dyck, H. and Baguette, M. (2005). Dispersal behaviour in fragmented landscapes: Routine or special movements? *Basic and Applied Ecology*, 6, 535–45.

Van Houtan, K. S., Pimm, S. L., Halley, J. M., Bierregaard, R. O., and Lovejoy, T. E. (2007). Dispersal of Amazonian birds in continuous and fragmented forest. *Ecology Letters*, 10, 219–29.

Watling, J. I. and Donnelly, M. A. (2006). Fragments as Islands: a Synthesis of Faunal Responses to Habitat Patchiness. *Conservation Biology*, 20, 1016–25.

Weldon, A. J. and Haddad, N. M. (2005). The effects of patch shape on Indigo buntings: evidence for an ecological trap. *Ecology*, 86, 1422–31.

CHAPTER 6

Migration

Phillip Gienapp

> ○ **Overview**
>
> Human-induced environmental change has important implications for migratory species. In this chapter, I explore how such changes affect migration by altering conditions along the migration route, as well as the cues that are used in the timing of migration. The chapter focuses on climate change and, in particular, research carried out on the effects of rising temperatures on migratory birds and salmon. Research in these taxa underscore the role of phenotypic plasticity in the timing of migration, with most species migrating earlier to their breeding areas in warmer years. Migration time is linked to fitness, which implies that climate change could affect migration time through both phenotypic plasticity and by imposing selection on arrival time at breeding sites—with potentially dire consequences for the persistence of populations.

6.1 What is migration?

Animal migration is one of the most obvious, spectacular, and widespread phenomena seen in nature. Most people will be familiar with the migration of birds between their breeding and wintering grounds. In the case of the arctic tern *Sterna paradisea*, a round trip pilgrimage can span a distance of up to 80,000 km as birds make their way between their breeding grounds in North America and Europe to their wintering grounds in the Antarctic (Egevang et al. 2010). During their travels, long-distance migrants often have to negotiate challenging environmental terrain as seen, for example, in many songbirds that breed in the temperate latitudes of Europe, but must cross the Mediterranean Sea and Sahara to reach their wintering grounds in sub-Saharan Africa. And of course, migration is not just confined to birds. Arguably one of the most iconic examples of animal migrations is seen in wildebeest *Connochaetes taurinus* as they track the availability of 'green forage' across the African savannah (Boone et al. 2006). In fish, migration often involves transition between fresh and saltwater environments: some species, such as eels, spend most of their lives in freshwater but migrate out to sea to breed (catadromous), while others, such as salmon, exhibit the opposite life history pattern (anadromous). Among the insects, one of the most striking examples of migration is seen in Monarch butterflies *Danaus plexippus* as they journey back and forth across the northeastern United States and Mexico, each leg of the trip completed by a different generation.

These well known examples of animal migration underscore its diversity. Migratory routes can be simple or complex. They may involve vast horizontal distances or movement between elevations. In some species, only some populations are migratory. And depending on a species' life history, migration can involve multiple, repeated journeys (e.g. wildebeest) or a terminal, once-in-a-lifetime event (e.g. salmon). All these examples are commonly referred to as migration. So, what are the common properties categorizing these movements as such? Obviously, all include some movement between different areas (and often also different habitats). The movements are seasonal and the covered distances are often large. Furthermore, reproduction typically takes place at distinct locations. Defining migration becomes less straightforward, however, when migration distances are comparable to those travelled during 'normal' movements, as in insects, or when migration is virtually indistinguishable from foraging (albatrosses, for example, can circle the globe twice during a single foraging trip).

If one looks closely at examples of migration, a common property seems to be movement between distinct breeding and 'non-breeding' areas that relate to seasonally changing environmental conditions. Migration does not have to involve the same individual returning to the non-breeding areas but could involve the offspring of those that made the initial journey (in semelparous species). Individuals may also stay for more than one year in the non-breeding areas before returning to breed. This can occur in birds, especially in waders, when individuals skip their first breeding season and do not migrate to the breeding grounds but, instead, 'oversummer' in the wintering areas (Alerstam 1990).

6.2 Environmental change and migration

Environmental change can affect migration by altering conditions along the migration route or at the breeding and non-breeding areas, and by influencing the cues that are used in the timing of migration. In particular, habitat changes can create physical barriers for migrating individuals. A classic example is seen in the building of dams and weirs that impede the migration of chinook salmon *Oncorhynchus tshawytscha* (Kareiva et al. 2000). Some dams prevent the passage of salmon, with dire consequences for spawning. Other dams allow the passage of fish but increase mortality during migration. These factors have led to drastic declines in salmon populations (Kareiva et al. 2000). Similarly, the construction of roads, railroads, settlements, and the enlargement of agricultural areas have strongly affected migration in ungulates and have led to severe effects on populations, including local extinctions (reviewed in Bolger et al. 2008). Even birds that are potentially capable of flying over destroyed or altered habitats can be affected by habitat modification since very few species migrate in one single trip but must, instead, rest and feed along the way. This is especially important for 'capital breeders', which mainly rely on body resources for egg production (that are built up in the wintering areas or staging sites along the migration route) (Drent et al. 2007). Wind farms can also pose a direct barrier for migratory birds comparable to the effects of roads or fences on ungulates, but the evidence here is inconclusive (reviewed in Drewitt and Langston 2006).

Another environmental change with perhaps more subtle effects on migration is climate change. In many systems, the timing of migration depends on climatic conditions (e.g. patterns of rainfall or rising spring temperature). Climate change has generally advanced the phenology of traits in many taxa, such as flowering in plants, breeding and migration time in birds, or hibernation in mammals (Parmesan 2006). It is important to bear in mind, however, that such changes do not necessarily occur at the same rate as the changes that are taking place in the surrounding environment. For example, the phenology of a predator may advance but the phenology of its prey may occur at a different rate, leading to a mismatch between trophic levels (Visser and Both 2005).

Migrating individuals also need to be able to reliably navigate to their desired destination. In birds, a number of different directional cues may be used to obtain the correct bearing. Such cues include the sun (Schmidt-Koenig 1990), stars (Emlen 1970), and the Earth's magnetic field (Wiltschko and Wiltschko 1995). In addition to this 'clock and compass' strategy, birds may also rely on a 'map' of learned topographic features (Thorup et al. 2007). In insects, true long distance migration is rare, but Monarch butterflies are known to use a suncompass (Reppert et al. 2010). In many marine species, migrating individuals appear to rely on ocean currents, or orient using the Earth's magnetic field, which not only provides information on direction but also on location through its intensity and inclination ('magnetic map') (Lohmann et al. 2008).

Certain navigational cues can potentially be undermined by human-induced changes to the environment. Light pollution, for example, can impair the perception of astronomical cues (stars) at night. Strong light sources, in particular, are known to attract flying species and can lead to deadly collusions (e.g. bird strikes at lighthouses), although the general effects on navigation are largely unknown (Longcore and Rich 2004). Similarly, the efficacy of ocean currents as a cue for migratory marine species is likely to be affected by changes to ocean currents as a result of global warming (Lohmann et al. 2008). Warming temperatures and aquatic pollution can also influence olfactory cues and, in so doing, affect migration in taxa such as fish, which rely on chemical cues to navigate towards their natal rivers (Dittman and Quinn 1996). Lastly, drastic habitat changes, such as large scale clear-cut logging, can undermine the efficacy of topographical 'cues' that are often used by migratory birds (Thorup et al. 2010).

This chapter focuses on the effects of climate change on migration time in birds and salmon. This is not because other human-induced environmental changes would be unimportant but because the effects of climate change on migration time in these taxa have been studied for several decades, which gives the necessary data to address the relevant questions. Furthermore it should be instructive to contrast endothermic (birds) and ectothermic taxa (fish) which are differently affected by ambient temperatures.

6.3 Migration time and fitness

Environmental conditions change during the season and consequently there will be an optimal 'time window' for life cycle events such as migration or breeding. For example, individuals that arrive too early at the breeding area could be confronted with harsh environmental conditions and, as a result, may not survive. However, if they arrive too late, they may miss out on gaining access to the best breeding sites and/or mates, or the optimal time window for breeding may close (i.e. environmental conditions deteriorate too soon for breeding to be successful). Apart from conditions at the breeding grounds, conditions at important staging or resting grounds may also change seasonally, which can restrict passage. Variation in the optimal timing of migration should select for phenotypic plasticity in migration time (see Chapter 11 on phenotypic plasticity). In support of this, migrating individuals generally arrive earlier in warmer years. For exam-

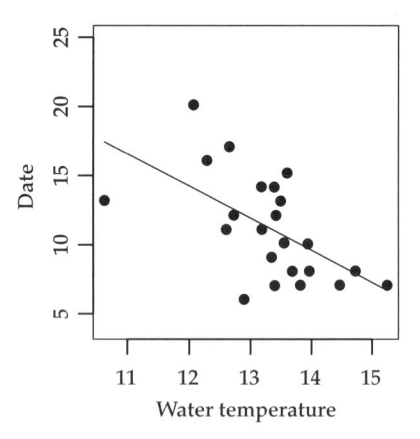

Figure 6.1 Phenotypic plasticity of upstream migration in sockeye salmon. Median passage date at Baker Lake is plotted against mean July temperature of Skagit River. Redrawn from Hodgson et al. (2006) with permission from John Wiley & Sons.

ple, sockeye salmon *Oncorhynchus nerka* migrate earlier under warmer river temperatures (Fig. 6.1). Similarly, arrival time in pied flycatchers *Ficedula hypoleuca* is related to temperatures along the migration route (Fig. 6.2a) and in the breeding area (Fig. 6.2b). Since temperatures along the migration route and in the breeding area are correlated (Fig. 6.2c) and related to prey phenology (Both et al. 2006), temperatures along the migration route can serve as a cue for optimal arrival time.

6.3.1 Migration time and fitness in birds

In migratory birds, mass mortality events due to unusually harsh weather conditions in spring have been recorded (reviewed in Newton 2007). However, most of these cases are anecdotal and, although such events are certainly capable of reducing population numbers (at least in the shorter term), it is unclear how strong an effect such mortality may have on selection on the timing of migration. Good data on the fitness consequences of arrival time at the breeding grounds in birds are scarce, in contrast to what is known about the fitness consequences of breeding time. This is mostly because the exact arrival dates of individuals are often difficult to ascertain by direct observation alone. In two studies on barn swallows *Hirundo rustica* and American redstarts *Setophaga ruticilla*, that captured and marked individuals upon arrival and followed them through the breeding season, early arriving individuals were, indeed, found to breed earlier and produce more nestlings (Smith and Moore 2005; Møller et al. 2009). A similar set of findings has been reported in great cormorants *Phalacrocorax carbo*. Importantly, in that system, selection for early arrival was found to increase over time (Fig. 6.3), possibly as a result of climate change (Gienapp and Bregnballe, unpublished material). Indirect evidence for a benefit of early arrival at the breeding grounds comes from the fact that territory quality is important for reproductive success (e.g. Alatalo et al. 1984; Hasselquist 1998), and early arriving individuals are able to secure better territories (Sergio et al. 2007).

6.3.2 Migration time and fitness in salmon

Most salmon species spawn in autumn or early winter with the eggs hatching in the following spring. Since fish are ectothermic, water temperatures are closely linked to fitness because temperature can directly affect survival at all life history stages (i.e. adults during upstream migration, as well as the developing eggs and juveniles). For example, it has been shown that water temperatures exceeding 18 °C increase the metabolic costs of migration in salmonids, thereby reducing fecundity and increasing mortality (reviewed in McCullough 1999). Individuals should therefore avoid migrating in the summer when water temperatures are high. On the other hand, individuals cannot migrate too late in the season if they are to allow eggs enough time to develop at optimal temperatures. Furthermore, there is an optimal emergence time (i.e. when small fish hatch and start to move actively), as early or late emerging fish have reduced survival (e.g. Einum and Fleming 2000; Letcher et al. 2004). Consequently, there appears to be an optimal window for both migration and spawning.

Hence, the timing of migration has fitness consequences in both birds and salmon. With the exception of severe weather events, survival and reproduction in birds is not directly affected by ambient temperature but, rather, indirectly through temperature effects on abundance and phenology of food. By contrast, as ectotherms, survival and reproductive success of fish are much more directly dependent on temperature. Consequently, ambient temperature is more likely to be a 'cue' in birds

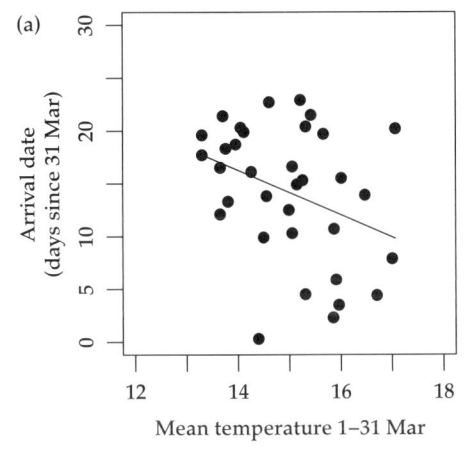

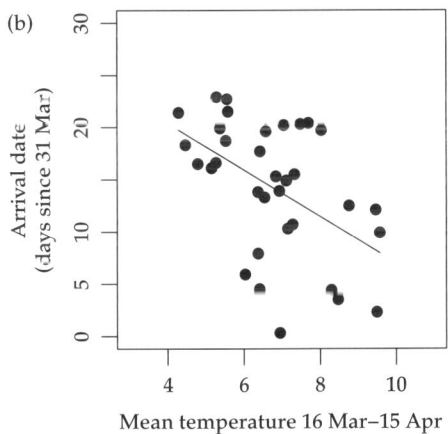

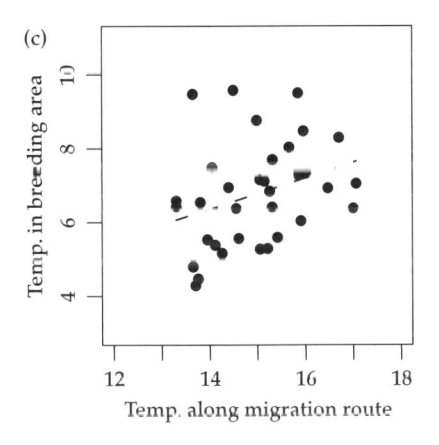

Figure 6.2 Phenotypic plasticity of arrival time in relation to temperature. Mean arrival time of male pied flycatchers *Ficedula hypoleuca* in the Netherlands is plotted against mean temperature along the migration route (a) and against mean temperature in the breeding area (b). (c) Temperatures along the migration route (North Africa) correlate ($r = 0.32$, $t = 1.89$, $p = 0.07$) with temperatures in the breeding areas (Netherlands) and can hence be used as a 'cue' for migration timing. Redrawn and reanalysed from Both et al. (2005) with permission from John Wiley & Sons.

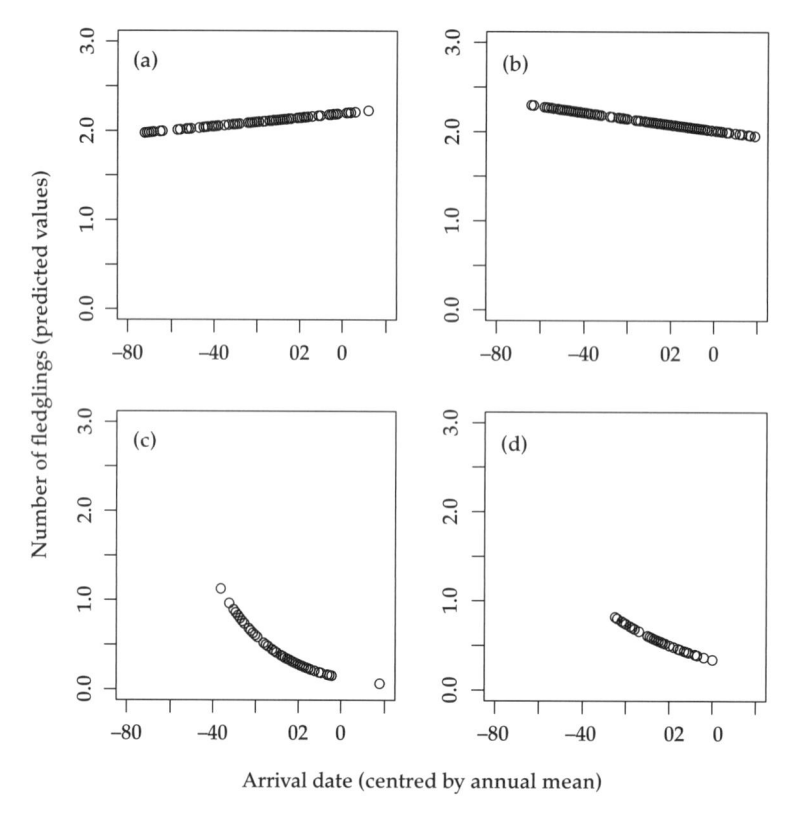

Figure 6.3 Reproductive success in relation to arrival time at the colony in cormorants. The number of fledglings raised by a pair depends on (female) age and arrival time. The relationship between arrival time and reproductive success changed over the years as indicated by a significant interaction between arrival time and year. For illustrative purposes single years (a 1984, b 1990, c 1997, d 2004) have been plotted to show that reproductive success declines with arrival time and that this relationship has become steeper over the years.

while it can act as both a 'cue' and as a 'selective factor' in fish (Hodgson et al. 2006). This difference may have consequences for how climate change affects migration time and fitness in these two taxa.

6.4 Effects of climate change on migration time

6.4.1 Birds

Migration time in birds is a phenotypically plastic trait and covaries with weather and climate. Generally, birds arrive earlier at their breeding grounds after milder winters and in warmer springs. Arrival time, however, has also advanced consistently over the last decades (reviewed in Knudsen et al. 2011). This advancement could be due to several reasons. First, climate change may have altered environmental conditions at the wintering grounds so that individuals are able to depart sooner due to, for example, improved food conditions (i.e. individuals are able to 'fuel up' earlier). Second, conditions *en route* could have been altered allowing a faster passage and/or faster refuelling at stop over sites along the way. And finally, climate change could have induced selection for earlier migration due to phenological mismatches (Both et al. 2006).

Unfortunately, for long-distance migrants, good data on departure time from wintering quarters are even more difficult to obtain than data on arrival time at the breeding grounds. Consequently, the former is often inferred from arrival dates at the breeding areas. Barn swallows and other long-distance migrants wintering south of the Sahara arrive earlier in Europe when ecological conditions in their wintering areas are more favourable (Saino et al. 2004; Gordo et al. 2005; Gordo and Sanz 2006), which indicates that better body condition allows for earlier departure. There is also direct evidence for local weather influencing departure time. Pink-footed geese *Anser brachyrhynchus*, for example, leave their wintering areas in Denmark earlier in warmer winters (Bauer et al. 2008). In this herbivorous species, local temperature could potentially affect departure time directly by affecting plant growth and, hence, body condition. Alternatively, temperature could act as a 'cue' for favourable conditions along the migration route or the breeding grounds on the Svalbard. Disentangling the effects of these two non-mutually exclusive mechanisms on migration time is, however, difficult as it would require manipulating temperature (in the laboratory) or condition (either in the laboratory or in the field).

Environmental conditions along the migration route can affect migration time in several ways. The flying speed of birds depends on wind direction, with tail winds accelerating and head winds slowing down or even inhibiting migration (Alerstam 1990). For instance, the time when migrating song thrushes *Turdus philomelos* pass the Rybachy ringing station in the southeast Baltic depends on both the frequency and strength of tail winds over Europe. Intriguingly, an increase in these tail winds in spring over the last few decades has seen an advancement in migration time in this species (Sinelschikova et al. 2007).

While there is good (correlative) evidence that changed environmental conditions *en route* and, to a lesser extent, in the wintering quarters, have advanced migration time (see Knudsen et al. 2011, for a review), there is no direct evidence for an evolutionary change through selection on arrival time. Demonstrating such an evolutionary change is not straightforward as it requires: (1) quantifying selection in the field, (2) showing that this selection has changed due to climate change, and (3) has led to a genetic change in the population. Hence, there are very few studies that have conclusively demonstrated evolution in response to climate change (for a review see Gienapp et al. 2008).

As already pointed out above, few studies on avian migration time have been able to estimate individual fitness and, hence, selection on migration time or its genetic variation. We know from a number of laboratory studies that various aspects of migration have a genetic basis (reviewed in Pulido 2007) and there is evidence that the observed changes in migratory activity and wintering areas of blackcaps *Sylvia atricapilla* are genetic (Berthold et al. 1992; Pulido and Berthold 2010). In the context of migration time, however, the evidence is much more limited. A few studies have tried to quantify the heritability of migration time but the results have been equivocal. Specifically, while Potti (1998) and Rees (1989) found no evidence for genetic variation in arrival time in pied flycatchers and Bewick's swans *Cygnus bewickii*, Møller (2001) found a moderate heritability of arrival time in barn swallows. Drawing on long-term data sets on arrival time in cormorants, Gienapp and Bregnballe (unpubl.) analysed phenotypic plasticity and showed that individual reaction norms varied in intercept and slope and that this variation is partly genetic.

Based on the examples above, there is indication that evolutionary change in migration time is possible but so far direct evidence is lacking, mainly due to methodological constraints or insufficient data. Without detailed knowledge about phenotypic plasticity and genetic changes over time, it is very difficult to disentangle these two processes, and we have to rely, instead, on indirect evidence. For example, in an avian study of long-distant migrants, Jonzén et al. (2006) argued that an observed advancement of migration time in Europe had to be an evolutionary change since departure time from Africa is not phenotypically plastic. Their reasoning was based on the general assumption that departure time of long-distance migrants from sub-Saharan Africa is determined by internal rhythms and photoperiod (Gwinner 1996) because the weather in sub-Saharan Africa is too weakly correlated with European weather to be deemed useful as a reliable 'cue' for the timing of migration. This view, however, has been criticized for several reasons. For example, departure time could still be phenotypically plastic since it is potentially affected by body condition which, in turn, is affected by environmental conditions that may change due to climate change (e.g. Saino et al. 2004; Gordo et al. 2005).

Another indirect approach to disentangling evolutionary change and phenotypic plasticity relies on the notion of 'sustainable evolution'. Based on this concept, any selection will reduce population mean fitness (selection load) and too strong a selection imposed by environmental change is expected to push a population to the brink of extinction. Theoretical work has shown that a sustainable rate of evolution under which extinction risk is negligible is typically not larger than a few per cent of the phenotypic standard deviation per generation (Lynch and Lande 1993). Gienapp *et al.* (2007) tested whether the advancement of migration time in several bird species would be consistent with sustainable evolution, and concluded that the observed advancements of migration time are most likely due to a phenotypically plastic response to warmer climates.

In general, migration time in many birds has advanced over the last decades and there is very good evidence that this advancement is related to climate change. In this regard, two different—but not mutually exclusive—mechanisms appear to be involved. First, changed environmental conditions in the wintering areas or along the migration route could have advanced migration time through phenotypic plasticity. Second, the observed advancement could be an evolutionary response to selection for earlier arrival time due to changed conditions in the breeding areas. However, it is currently unclear how much of the observed advancement in migration time can be atributed to one or the other of these two mechanisms.

6.4.2 Salmon

As pointed out above, the timing of migration in salmon varies among years and is related to sea and river temperatures (Fig. 6.1.) (Dahl et al. 2004; Hodgson et al. 2006; Crozier et al. 2008). Due to increasing temperatures, the migration of Chinook *Oncorhynchus tshawytscha* and sockeye salmon *O. nerka* have advanced by about two to three days per decade (Quinn and Adams 1996; Crozier et al. 2008).

Surprisingly there is no published evidence for an advancement of migration time in Atlantic salmon *Salmo salar* or brown trout *S. trutta* (Jonsson and Jonsson 2009) although, as I pointed out earlier, migration time in these species is related to water temperatures, which have increased due to climate change.

Salmon generally enter rivers in spring although they may not spawn until autumn. High water temperatures in summer increase migration costs and mortality (Martins et al. 2011). Optimal migration speed is reached at 16 °C (Salinger and Anderson 2006); temperatures above 18 °C reduce survival and fitness, while temperatures above 20 °C can almost completely suppress migration (Fig. 6.4). Consequently, fish try to avoid migrating in summer by either reaching spawning grounds before the arrival of peak summer temperatures or by seeking refuge in areas where water temperatures are lower (e.g. deep pools or lakes; Økland et al. 2001). Again, this advancement in migration time could be due to phenotypic plasticity or microevolution (or both) but—as in birds—disentangling these two processes is difficult. Crozier et al. (2008)

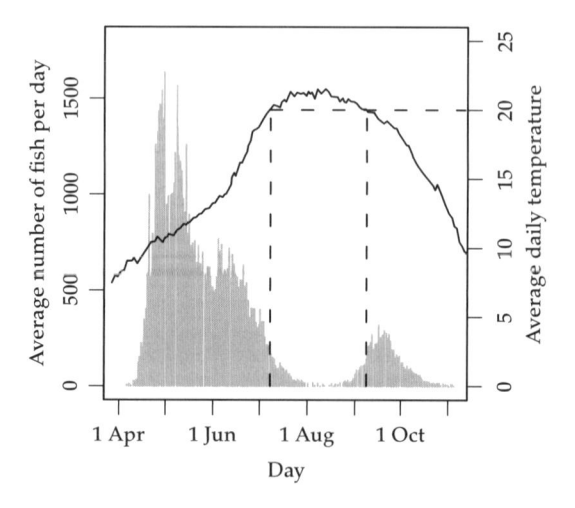

Figure 6.4 Migration time of Chinook salmon in relation to water temperature in Snake River from 1995 to 2006. The vertical bars indicate average daily counts and the solid line average daily water temperature, both at Lower Granite Dam. When water temperatures exceed 20 °C spawning migration is strongly reduced as indicated by the vertical dashed lines. Redrawn from Crozier et al. (2008) with permission from John Wiley & Sons.

argued that the advancement of migration time in Columbia basin sockeye salmon is an evolutionary adaptation rather than due to phenotypic plasticity. First, the correlation between sea surface temperature (which is very likely a 'cue' for optimal timing of migration) and migration time is weak (cf. Hodgson et al. 2006). However, the correlation between April water temperatures and migration time in salmon species in Sweden was, by contrast, high (r = 0.69 Dahl et al. 2004). Secondly, the observed advancement in migration time is in line with an evolutionary change predicted by a model based on survival selection for earlier migration through increased summer water temperatures (Crozier et al. 2008). Further evidence comes from the strong differentiation in migration time among populations related to water temperatures (e.g. Doctor et al. 2010) and generally moderate to high heritabilities of timing-related traits in these species (Carlson and Seamons 2008).

Based on the advancement in migration time, one might expect a corresponding change in spawning time, but so far this has not been reported. Furthermore, as salmon generally migrate early and 'wait' in the river system prior to spawning in the autumn, it may be unsurprising that we do not see a direct link between migration and spawning time. Furthermore, spawning time occurs later under warmer temperatures (Heggberget 1988) and increasing river temperatures should therefore lead to earlier migration but later spawning. Consequently, it does not seem to be the case that spawning time in salmon would be constrained by migration time. This is in contrast to birds, which generally arrive at the breeding grounds only shortly before breeding commences.

6.5 Climate change and migration—consequences for populations

An obvious concern in the context of climate change is how its effects on migration time might impact the persistence of populations and species. For example, there is evidence showing that warming temperatures have moved the distribution of species and populations polewards and towards higher altitudes (Parmesan 2006). Such changes in the distribution of species pose a major threat to biodiversity because, in many cases, very little (if any) suitable habitat may be available in these new areas (Thomas et al. 2004).

Climate change has already disrupted the synchrony between breeding and food supply for several species; the general pattern being that the phenology of the food has advanced faster than the phenology of those that rely on it (Visser and Both 2005). Suppressed reproductive success, in this regard, could ultimately reduce population numbers. Caterpillars, for example, are a main food source for pied flycatchers during chick rearing. Here, there is evidence to suggest that flycatcher populations in deciduous woodlands are being adversely affected by a stronger mismatch with caterpillar phenology and, as a consequence, have declined more strongly than populations inhabiting coniferous woodlands (Both et al. 2006). As pointed out above, we currently do not know—at least for the vast majority of species—whether the observed advancements in arrival time are sufficient to track the likely advancement of the optimal breeding time since very little data on food phenology exists. Growth and development of plants and insects is tightly linked to ambient temperature. The phenology of these lower trophic levels describes the general progression of spring and, hence, may be a suitable proxy for the optimal breeding period for birds. Consequently, populations whose arrival time has lagged behind the advancement of spring as measured by temperature sums or whose arrival time has not advanced at all are currently likely mistimed. Using this rationale, several studies have shown that populations with supposedly insufficient adjustment in migration time have been declining (Møller et al. 2008; Saino et al. 2011). If we follow this line of reasoning, it seems that some populations are currently so strongly mistimed that they may already be experiencing a serious reduction in reproductive success. The question now becomes whether migration time can sufficiently evolve to reduce the current mistiming, and whether this evolution can happen fast enough. As mentioned, there is some evidence that migratory behaviour is heritable (Pulido 2007), and there are examples that it has evolved in response to environmental change (Berthold et al. 1992; Pulido 2007). Consequently, we might expect that migration time in birds is also heritable and hence could respond to selection. Unfortunately, however, we do not have enough data of sufficient detail to answer the question of whether this genetic response can keep pace with the rapid environmental changes that are taking place. The mistiming of migration already apparent in certain species suggests that the pace of genetic change may not be occurring fast enough.

While in birds heterogeneous climate change seems to have led to selection on migration and may already have been responsible for population declines, the picture looks quite different in salmon. Whilst evidence in salmonids suggests that migration time has also advanced in response to climate change (e.g. Quinn and Adams 1996), so far no population declines in the context of migration time and climate change have been reported. In salmon, both migration and spawning itself seem to harbour moderate levels of genetic variation, which would allow evolutionary adaptation to climate change. For example, there is indirect evidence that the observed advanced migration time in sockeye salmon is partly an evolutionary response to selection (Crozier et al. 2008). Furthermore, it seems likely that in salmon the link between 'cue' and 'selective factors' (see Box 6.1) for migration time is closer than in birds since

Box 6.1 Phenotypic plasticity and migration time

Phenotypic plasticity of phenological traits, such as migration time, differs in some aspects from the more 'classic' examples of phenotypic plasticity as seen, for instance, in the adjustment of plant growth to soil nutrients, or prey morphology to predation risk (Pigliucci 2001). In these classic examples, the 'selective factor' that is enforcing selection on the trait (e.g. predation risk) is also the cue that prompts the adjustment of the trait. This, however, is not the case for phenological traits which are adjusted, not in response to the selective factor *per se*, but on some other factor that is correlated with the selective factor.

Assume that food availability in a breeding area increases and decreases during the season in a hump-shaped fashion, and that reproductive success depends on this seasonal shift. Food availability would be the ultimate 'selective factor' and individuals arriving at its peak (or shortly before—in order to prepare for breeding and to find a mate) would obviously enjoy the highest fitness. Assume now that food availability is related to spring temperatures (i.e. the growth of the food—plants and insects—is related to temperature), and that this temperature varies among years. Consequently, there would be an optimal relationship between arrival time and spring temperature. This relationship can be regarded as a 'reaction norm' and the optimal relationship the 'optimal reaction norm'.

If spring temperature is closely correlated with food abundance, migrating individuals should adjust their arrival at the breeding grounds according to temperature and their reaction norm should follow closely the optimal reaction norm. However, it is impossible for migrating individuals to 'measure' local temperatures at the breeding area while they are still *en route*. Consequently, they need to rely on certain 'cues' that are predictive of the ultimate selective factor. Climate and weather tend to be spatially and temporally autocorrelated, and this autocorrelation makes it possible to use other climate or weather variables as 'cues' for an optimal migration time.

Climate change is however affecting different seasons and regions differently, with a generally stronger warming trend in winter and at higher latitudes (Luterbacher et al. 2004) but also variation at smaller spatial scales (e.g. a stronger predicted warming in the Baltic region compared to adjacent Arctic regions; Høgda et al. 2001). These spatially varying effects will likely alter the spatial and temporal autocorrelation between the various climate and weather variables. This, in turn, is expected to reduce the reliability of the 'cues' used for timing of migration. For example, geese *Anser sp.* tracking the onset of vegetation growth in spring (i.e. the 'green wave') in the Baltic could potentially arrive too early in their Arctic breeding areas (Drent et al. 2007). Altered relationships between 'cues' used for the timing of migration and the ultimate selective factors that act on migration time means that individuals should respond differently to the same 'cues' or use other 'cues' to optimally time their migration. Climate change is hence expected to lead to selection on the use of such cues.

water temperature is both a main selection pressure and the 'cue' the individuals are responding to when timing their migration. As also pointed out above, climate change should cause salmon to migrate earlier but spawn later. Hence, unlike birds, migration and breeding times are not closely linked and migration timing is therefore not expected to constrain breeding time.

However, the increasingly early arrival expected under continued global warming could, nevertheless, still affect survival and reproductive success in salmon. Specifically, if individuals migrate earlier into freshwater and have to 'wait' for longer until spawning, this could mean lost foraging opportunities due to the reduced time that individuals are able to spend feeding out at sea. Moreover, depending on the availability of cold water refugia, which may be reduced by rising temperatures and changed water flow patterns, fish may have to 'wait' under higher than optimal water temperatures, which is also expected to have an impact on survival and reproductive success (reviewed in McCullough 1999).

6.6 Conclusions

It is clear that environmental changes can have severe consequences for migration. It is well known that human activities, such as the expansion of settlements and the construction of roads and dams, can obstruct migration routes and, in so doing, threaten the persistence of local populations. Yet, when it comes to more subtle effects of habitat change on migration, our knowledge remains rather limited. Degradation in the quality of 'stop over' (i.e. feeding and resting) sites along the migration route may, for example, affect survival or reproduction but very little is actually known as to where, exactly, populations may be overwintering or stopping-over along their journeys. Many long-distance migratory bird species are declining (Sanderson et al. 2006) and although the effects of climate change have been implicated (Both et al. 2006; Møller et al. 2008) the exact reasons are still unclear. Further development of tracking technology (e.g. miniaturizing of GPS transmitters) may allow us to track individuals along their complete migration route and thereby proffer insights into whether (and how) habitat change in the wintering areas or at stopover sites affect migrating individuals.

Warming climate has advanced the phenology of many traits, including migration time. Migration time of birds and salmon is better studied than in many other species, but our knowledge is far from complete. There are strong indications, for instance, that migration time is related to survival and reproductive success but, in many bird species, good data are still lacking. Migration time shows phenotypic plasticity at the population level (Figs. 6.1 and 6.2), but we know very little about whether and how this plasticity varies among individuals. There is evidence to suggest that migration time is heritable in barn swallows, cormorants, and salmon, but our general understanding of the genetic architecture underlying migration time remains limited. Climate change has already strongly affected migration time in birds and salmon but the consequences for population growth (and the persistence of populations *per se*) are less clear and, interestingly, could differ between the two taxonomic groups. Later migration times in birds are likely to have a negative effect on populations. In salmon, however, the link between migration time and breeding time—and hence reproductive success—is less closely coupled. Further studies on both the fitness consequences and genetics of migration time in birds and salmon would therefore be desirable. More broadly, recent methodological advances (e.g. in tracking technology) also provide an excellent opportunity to study migration in a range of species for which very little is currently known (e.g. bats and whales). Doing so will be vital if we are to fully understand the challenges migratory species face in an increasingly human-dominated world.

Acknowledgements

I would like to thank two anonymous reviewers, Ulrika Candolin, and Bob Wong for helpful comments on the manuscript. Many thanks are also due to Bob Wong for his effort to improve the writing. The Academy of Finland provided funding during the writing of this chapter.

References

Alatalo, R. V., Lundberg, A., and Ståhlbrandt, K. (1984). Female mate choice in the pied flycatcher *Ficedula hypoleuca*. *Behavioral Ecology and Sociobiology*, 14, 253–61.

Alerstam, T. (1990). *Bird Migration*. Cambridge, Cambridge University Press.

Bauer, S., Gienapp, P., and Madsen, J. (2008). The relevance of local environmental conditions for departure decisions changes en route in migrating geese. *Ecology*, 89, 1953–60.

Berthold, P., Helbig, A. J., Mohr, G., and Querner, U. (1992). Rapid microevolution of migratory behaviour in a wild bird species. *Nature*, 360, 668–70.

Bolger, D. T., Newmark, W. D., Morrison, T. A., and Doak, D. F. (2008). The need for integrative approaches to understand and conserve migratory ungulates. *Ecology Letters*, 11, 63–77.

Boone, R. B., Thirgood, S. J., and Hopcraft, J. G. C. (2006). Serengeti wildebeest migratory patterns modelled from rainfall and new vegetation growth. *Ecology*, 87, 1987–94.

Both, C., Bijlsma, R., and Visser, M. E. (2005). Climatic effects on timing of spring migration and breeding in a long-distance migrant, the pied flycatcher *Ficedula hypoleuca*. *Journal of Avian Biology*, 36, 368–73.

Both, C., Bouwhuis, S., Lessells, C. M., and Visser, M. E. (2006). Climate change and population declines in a long-distance migratory bird. *Nature*, 441, 81–3.

Carlson, S. M. and Seamons, T. R. (2008). A review of quantitative genetic components of fitness in salmonids: implications for adaptation to future change. *Evolutionary Applications*, 1, 222–38.

Crozier, L. G., Hendry, A. P., Lawson, P. W., Quinn, T. P., Mantua, N. J., Battin, J., et al. (2008). Potential responses to climate change in organisms with complex life histories: evolution and plasticity in Pacific salmon. *Evolutionary Applications*, 1, 252–70.

Dahl, J., Dannewitz, J., Karlsson, L., Petersson, E., Löf, A., and Ragnarsson, B. (2004). The timing of spawning migration: implications of environmental variation, life history, and sex. *Canadian Journal of Zoology*, 82, 1864–70.

Dittman, A. H. and Quinn, T. P. (1996). Homing in Pacific salmon: mechanisms and ecological basis. *Journal of Experimental Biology*, 199, 83–91.

Doctor, K. K., Hilborn, R., Rowse, M., and Quinn, T. (2010). Spatial and temporal patterns of upriver migration by sockeye salmon populations in the Wood River system, Bristol Bay, Alaska. *Transactions of the American Fisheries Society*, 139, 80–91.

Drent, R. H., Eichhorn, G., Flagstad, A., Van Der Graaf, A. J., Litvin, K. E., and Stahl, J. (2007). Migratory connectivity in Arctic geese: spring stopovers are the weak links in meeting targets for breeding. *Journal of Ornithology*, 148, S501–14.

Drewitt, A. L. and Langston, R. H. W. (2006). Assessing the impacts of wind farms on birds. *Ibis*, 148, 29–42.

Egevang, C., Stenhouse, I. J., Phillips, R. A., Petersen, A., Fox, J. W., and Silk, J. R. D. (2010). Tracking of Arctic terns *Sterna paradisaea* reveals longest animal migration. *Proceedings of the National Academy of Sciences of the USA*, 107, 2079–81.

Einum, S. and Fleming, I. A. (2000). Selection against late emergence and small offspring in Atlantic salmon (*Salmo salar*). *Evolution*, 54, 628–39.

Emlen, S. T. (1970). Celestial rotation: its importance in development of migratory orientation. *Science*, 170, 1198–201.

Gienapp, P., Leimu, R., and Merilä, J. (2007). Responses to climate change in avian migration time—microevolution versus phenotypic plasticity. *Climate Research*, 35, 25–35.

Gienapp, P., Teplitsky, C., Alho, J. S., Mills, J. A., and Merilä, J. (2008). Climate change and evolution: disentangling environmental and genetic responses. *Molecular Ecology*, 17, 167–78.

Gordo, O., Brotons, L., Ferrer, X., and Comas, P. (2005). Do changes in climate patterns in wintering areas affect the timing of the spring arrival of trans-Saharan migrant birds? *Global Change Biology*, 11, 12–21.

Gordo, O. and Sanz, J. J. (2006). Climate change and bird phenology: a long-term study in the Iberian Peninsula. *Global Change Biology*, 12, 1993–2004.

Gwinner, E. (1996). Circannual clocks in avian reproduction and migration. *Ibis*, 138, 47–63.

Hasselquist, D. (1998). Polygyny in great reed warblers: a long-term study of factors contributing to male fitness. *Ecology*, 79, 2376–90.

Heggberget, T. G. (1988). Time of spawning of Norwegian Atlantic salmon (*Salmo salar*). *Canadian Journal of Fisheries and Aquatic Sciences*, 45, 845–9.

Hodgson, S., Quinn, T. P., Hilborn, R., Francis, R. C., and Rogers, D. E. (2006). Marine and freshwater climatic factors affecting interannual variation in the timing of return migration to fresh water of sockeye salmon (*Oncorhynchus nerka*). *Fisheries Oceanography*, 15, 1–24.

Høgda, K. A., Karlsen, S. R., and Solheim, I. (2001) Climate change impact on growing season in Fennoscandia studied by a time series of NOAA AVHRR NDVI data. In *IGARSS 2001 Scanning the Present and Resolving the Future IEEE 2001 International Geoscience and Remote*

Sensing Symposium, pp 1338–40, Vol. 3. Institute of Electrical and Electronics Engineers, Piscataway.

Jonsson, B. and Jonsson, N. (2009). A review of the likely effects of climate change on anadromous Atlantic salmon *Salmo salar* and brown trout *Salmo trutta*, with particular reference to water temperature and flow. *Journal of Fish Biology*, 75, 2381–447.

Jonzén, N., Lindén, A., Ergon, T., Knudsen, E., Vik, J. O., Rubolini, D., et al. (2006). Rapid advance of spring arrival dates in long-distance migratory birds. *Science*, 312, 1959–61.

Kareiva, P., Marvier, M., and Mcclure, M. (2000). Recovery and management options for spring/summer Chinook salmon in the Columbia river basin. *Science*, 290, 977–9.

Knudsen, E., Lindén, A., Both, C., Jonzén, N., Pulido, F., Saino, N., et al. (2011). Challenging claims in the study of migratory birds and climate change. *Biological Reviews of the Cambridge Philosophical Society*, 86, 928–946.

Letcher, B. H., Dubreuil, T., O'Donnell, M. J., Obedzinski, M., Griswold, K., and Nislow, K. H. (2004). Long-term consequences of variation in timing and manner of fry introduction on juvenile Atlantic salmon (*Salmo salar*) growth, survival, and life-history expression. *Canadian Journal of Fisheries and Aquatic Sciences*, 61, 2288–301.

Lohmann, K. J., Lohmann, C. M. F., and Endres, C. S. (2008). The sensory ecology of ocean navigation. *Journal of Experimental Biology*, 211, 1719–28.

Longcore, T. and Rich, C. (2004). Ecological light pollution. *Frontiers in Ecology and the Environment*, 2, 191–8.

Luterbacher, J., Dietrich, D., Xoplaki, E., Grosjean, M., and Wanner, H. (2004). European seasonal and annual temperature variability, trends, and extremes since 1500. *Science*, 303, 1499–503.

Lynch, M. and Lande, R. (1993). Evolution and extinction in response to environmental change. In P. M. Kareiva, J. G. Kingsolver, and R. B. Huey, (eds) *Biotic Interactions and Global Change*. pp. 234–50. Sunderland, MA, Sinauer Ass

Martins, E. G., Hinch, S. G., Patterson, D. A., Hague, M. J., Cooke, S. J., Miller, K. M., et al. (2011). Effects of river temperature and climate warming on stock-specific survival of adult migrating Fraser River sockeye salmon (*Oncorhynchus nerka*). *Global Change Biology*, 17, 99–114.

Mccullough, D. A. (1999) A review and synthesis of effects of alterations to the water temperature regime on freshwater life stages of salmonids, with special reference to Chinook salmon. Seattle, Washington US, Environmental Protection Agency, Region 10.

Møller, A. P. (2001). Heritability of arrival date in a migratory bird. *Proceedings of the Royal Society of London B, Biological Sciences*, 268, 203–6.

Møller, A. P., Balbontin, J., Cuervo, J., Hermosell, I. G., and De Lope, F. (2009). Individual differences in protandry, sexual selection, and fitness. *Behavioral Ecology*, 20, 433–40.

Møller, A. P., Rubolini, D. and Lehikoinen, E. (2008). Populations of migratory bird species that did not show a phenological response to climate change are declining. *Proceedings of the National Academy of Sciences of the USA*, 105, 16195–200.

Newton, I. (2007). Weather-related mass-mortality events in migrants. *Ibis*, 149, 453–67.

Økland, F., Erkinaro, J., Moen, K., Niemelä, E., Fiske, P., Mckinley, R. S., et al. (2001). Return migration of Atlantic salmon in the River Tana: phases of migratory behaviour. *Journal of Fish Biology*, 59, 862–74.

Parmesan, C. (2006). Ecological and evolutionary responses to recent climate change. *Annual Review of Ecology Evolution and Systematics*, 37, 637–69.

Pigliucci, M. (2001). *Phenotypic Plasticity*. Baltimore, John Hopkins University Press.

Potti, J. (1998). Arrival time from spring migration in male pied flycatchers: individual consistency and familial resemblance. *Condor*, 100, 702–8.

Pulido, F. (2007). Phenotypic changes in spring arrival: evolution, phenotypic plasticity, effects of weather and condition. *Climate Research*, 35, 5–23.

Pulido, F. and Berthold, P. (2010). Current selection for lower migratory activity will drive the evolution of residency in a migratory bird population. *Proceedings of the National Academy of Sciences of the USA*, 107, 7341–6.

Quinn, T. P. and Adams, D. J. (1996). Environmental changes affecting the migratory timing of American shad and sockeye salmon. *Ecology*, 77, 1151–62.

Rees, E. C. (1989). Consistency in the timing of migration for individual Bewick's swans. *Animal Behaviour*, 38, 384–93.

Reppert, S. M., Gegear, R. J., and Merlin, C. (2010). Navigational mechanisms of migrating monarch butterflies. *Trends in Neurosciences*, 33, 399–406

Saino, N., Szép, T., Romano, M., Rubolini, D., Spina, F., and Møller, A. P. (2004). Ecological conditions during winter predict arrival date at the breeding quarters in a trans-Saharan migratory bird. *Ecology Letters*, 7, 21–5.

Saino, S., Ambrosini, R., Rubolini, D., Von Hardenberg, J., Provenzale, A., Hüppop, K., et al. (2011). Climate warming, ecological mismatch at arrival and population decline in migratory birds. *Proceedings of the Royal Society of London B, Biological Sciences*, 278, 835–42.

Salinger, D. H. and Anderson, J. J. (2006). Effects of water temperature and flow on adult salmon migration swim speed and delay. *Transactions of the American Fisheries Society*, 135, 188–99.

Sanderson, F. J., Donald, P. F., Pain, D. J., Burfield, I. J., and Van Bommel, F. P. J. (2006). Long-term population declines in Afro-Palearctic migrant birds. *Biological Conservation*, 131, 93–105.

Schmidt-Koenig, K. (1990). The sun compass. *Experientia*, 46, 336–42.

Sergio, F., Blas, J., Forero, M. G., Donazar, J. A., and Hiraldo, F. (2007). Sequential settlement and site dependence in a migratory raptor. *Behavioral Ecology*, 18, 811–21.

Sinelschikova, A., Kosarev, V., Panov, I., and Baushev, A. N. (2007). The influence of wind conditions in Europe on the advance in timing of the spring migration of the song thrush (*Turdus philomelos*) in the south-east Baltic region. *International Journal of Biometeorology*, 51, 431–40.

Smith, R. J. and Moore, F. R. (2005). Arrival timing and seasonal reproductive performance in a long-distance migratory landbird. *Behavioral Ecology and Sociobiology*, 57, 231–9.

Thomas, C. D., Cameron, A., Green, R. E., Bakkenes, M., Beaumont, L. J., Collingham, Y. C., et al. (2004). Extinction risk from climate change. *Nature*, 427, 145–8.

Thorup, K., Bisson, I.-A., Bowlin, M. S., Holland, R. A., Wingfield, J. C., Ramenofsky, M., et al. (2007). Evidence for a navigational map stretching across the continental U.S. in a migratory songbird. *Proceedings of the National Academy of Sciences of the USA*, 104, 18115–19.

Thorup, K., Holland, R. A., Tøttrup, A. P., and Wikelski, M. (2010). Understanding the migratory orientation program of birds: extending laboratory studies to study free-flying migrants in a natural setting. *Integrative and Comparative Biology*, 50, 315–22.

Visser, M. E. and Both, C. (2005). Shifts in phenology due to global climate change: the need for a yardstick. *Proceedings of the Royal Society of London B, Biological Sciences*, 272, 2561–9.

Wiltschko, W. and Wiltschko, R. (1995). *Magnetic Orientation in Animals*. Berlin, Springer.

CHAPTER 7

Foraging

Ronald C. Ydenberg and Herbert H.T. Prins

⊃ Overview

Classical foraging theory considers behavioural modifications in response to changes in the type and array of food available and has, in recent years, also considered how foraging behaviour should adjust to predation danger. In this chapter we use these ideas to consider how well foragers are able to accommodate rapid anthropogenic environmental change. We conclude that foraging behaviour is generally able to mitigate large changes in the amount and type of prey on offer. Predation risk has powerful effects on prey behaviour that can profoundly affect the growth of their populations, even if predators do not actually kill many prey. Moreover, these effects are passed on to other parts of the community. We therefore expect the changing face of predation risk on the planet—broadly speaking, the return of top predators in terrestrial ecosystems and the disappearance of top predators in marine ecosystems—to have important effects.

7.1 Introduction

From atop the new bridge over the River IJssel one has a view back in time over the Dutch landscape. Before the modern office towers that dominate the view to the north were built, the highest structure as far as the eye could see was the fifteenth century 'Peperbus', still visible above the old city of Zwolle just to the east. But much more ancient human occupation has left its imprint on this landscape: indeed, humans built it (Lambert 1985). The river below the bridge is contained within dykes begun hundreds of years ago, and the seasonally-flooded meadows along the river ('uiterwaarden') have been grazed by livestock for a thousand years. The farmlands stretching off to the horizon in all directions were laboriously created by cutting the forests and draining the bogs beginning from the Neolithic, about 7000 years ago during the Swifterbant Culture (5300–3400 BCE; Cappers and Raemaekers 2008) to the Middle Ages (1000–1500 CE). It took perhaps 4000 years for humans to effect the transformation to a completely pastoral landscape (Prins 1998).

The forests and bogs removed by those Neolithic and Iron Age settlers here and elsewhere in northern Europe were merely the most recent in a series of biomes—tundra, steppe, woodlands, coniferous and deciduous forests that succeeded each other after glaciers receded, and the climate warmed. This landscape has changed continuously for the past 12 millenia, and the anthropogenic changes begun seven millenia ago continue today (e.g. Prins 1998). The motivation for this volume is the idea that the current scale and pace of change is unprecedented, and that this human-induced environmental change represents a profound threat to global biodiversity. The aim is to understand the extent to which the behaviour of animals might be able to moderate the impacts of all this change on their populations.

The rate of anthropogenic change has increased since the Industrial Revolution. For example, in the nineteenth century, the American state of Wisconsin was deforested and converted to agricultural lands (see Terborgh 1989), a process almost 100-fold faster than the landscape alteration that occurred around Zwolle. Other agricultural changes are also altering

Behavioural Responses to a Changing World. First Edition. Edited by Ulrika Candolin and Bob B.M. Wong.
© 2012 Oxford University Press. Published 2012 by Oxford University Press.

the foraging environment for many animals. Agricultural abandonment is occurring on a large scale throughout the world as the human population continues to urbanize and farming industrializes, and fields are reverting to savanna or forest. Readily available chemical fertilizers are raising the quality of forage on pastures, and altered harvest practices have made waste or spillage of some widely-grown crops, such as potatoes, corn, and rice easily available. Farming practices such as ditching, draining, mowing, and hedgerow clearing are changing habitats associated with farms. The reduction of phosphates and nitrates has cleaned up waterways in some areas, reducing the nutrient load, increasing water clarity, improving visibility and, thus, the availability of prey to predators. In other areas, such as the Baltic Sea, the nutrient load has increased, and water clarity has diminished, with all kinds of repercussions for visual animals (e.g. Wong et al. 2007).

But, our bridge-top viewpoint suggests that the relation between rapid anthropogenic change and the health of animal populations is not straightforward. For example, we can see a flock of barnacle geese *Branta leucopsis* grazing on the uiterwaarden, while overhead soar several common buzzards *Buteo buteo*, hunting for rodents in the grasslands along the dykes. Both these species are now much more common than when we were students in The Netherlands 35 years ago, before Zwolle's modern skyline appeared. How is it that these species are now thriving in this ultra-anthropogenic landscape? Are they just exceptions to the general trend? Were those species unable to adjust to humans eliminated long ago, leaving those with some adaptability? What characteristics make species such as the red deer *Cervus elephus* and wild boars *Sus scrofus*, present on this landscape since the glaciers retreated, able to hang on in nearby forests?

In this chapter we examine foraging, asking what role, if any, is played by behavioural adjustments to foraging behaviour in accommodating rapid environmental change. Our starting point is the study of behavioural ecology, founded in part by the Dutch ethologist Niko Tinbergen, whose summer camps at Hulshorst just 25 km to the southwest from our viewpoint are famously described in his book *Curious Naturalists* (Tinbergen 1958). In behavioural ecology two major influences are recognized as crucial with regard to individual foraging decisions (Ydenberg et al. 2007). These are: (1) 'economics' (handling time, encounter rate etc.; the stuff of classic 'optimal foraging theory'), and (2) the 'risk of predation'. In the following sections we consider the relevance of these ideas for how rapid environmental change might be accommodated.

7.2 Effects of changes in food on foraging behaviour

The increasingly intensive application of nitrogen fertilizers beginning in the late 1940s is a good example of rapid anthropogenic environmental change that altered the food supply, in this case greatly increasing the protein content of grasses on pastures and, thus, its quality for grazers such as geese. Figure 7.1 plots the protein quality of grasses on natural (saltmarsh) and anthropogenically-created (polder) pastures on the Wadden Sea island of Schiermonnikoog. Until recently, barnacle geese grazed on saltmarshes during the entire non-breeding period, but nowadays they generally graze in pastures during winter where the food quality is higher (Prins and Ydenberg 1985). Prior to the widespread application of nitrogen fertilizers, the annual rhythm in the quality of grazing on pastures resembled that of saltmarshes, with a distinct summer low. The spring migration of geese to the Arctic likely evolved, in part, because the quality of grazing available there in the summer months was higher than in temperate non-breeding areas. Modern pasture management and nitrogen fertilizers have changed that, and since fertilizer usage in northwest Europe intensified, non-migratory populations of geese have established (van der Jeugd et al. 2009; Jonker 2010).

The reduction of hunting by humans has played a major role in the very rapid population growth of barnacle and other temperate goose species in both Europe and North America over recent decades (Fox et al. 2010), but the heightened food quality has undoubtedly also been important. The question asked here, however, is more subtle: what role has flexibility in foraging behaviour played? Would

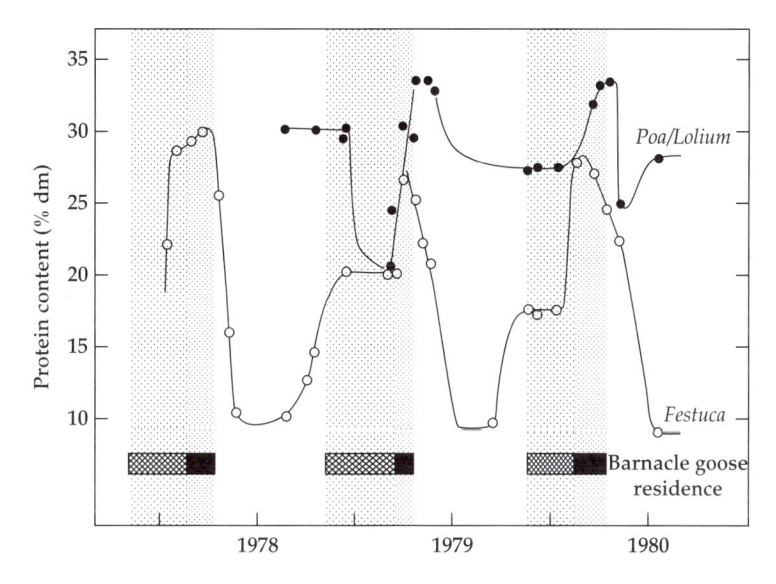

Figure 7.1 Protein content (as percentage of dry mass) of grasses on the polder (*Poa pratensis* and *Lolium perenne*) and saltmarsh (*Festuca rubra*) of Schiermonnikoog over three years. The polder is fertilized with up to 800 kg ha⁻¹ of nitrogen fertilizer, while the saltmarsh is not fertilized artificially. Also indicated is the period of barnacle goose residence on the island. Barnacle geese graze in the polder (light vertical shading) during winter, and abruptly switch to saltmarsh grazing (dark vertical shading) in the early spring. Based on Prins and Ydenberg 1985 (see also Ydenberg and Prins 1981).

barnacle goose populations have grown as much had they not had the flexibility to take up polder grazing, or to cease migration? Or, to put the question the other way round, how much would behavioural flexibility be able to mitigate the effects were the nitrogen fertilization of pastures to cease?

Foraging is an expansive area of investigation (see Stephens et al. 2007), the modern study of which was launched in 1966 by the basic 'diet model' (see Ydenberg 2010), which asks the question: 'When should a predator ignore an encountered prey item in favour of searching for a better one?' This simple question is presumed to underlie decisions about what an organism eats. (In behavioural ecology, the term 'decision' is used whenever one of two or more options is selected. There is no implication that the choice is conscious. See Ydenberg 2010.) The models formalize this situation, and make explicit that this process underlies many foraging situations in the natural world. They identify the parameters to measure and analyse their interaction in an abstract but general way. The review by Sih and Christensen (2001) tallies 134 published studies of this model, showing how broadly this idea applies to the endless variety of foraging processes in the natural world. Sih and Christensen (2001) also identify situations in which the model does not work well. A steady flow of theoretical papers have extended this idea and the other basic models of foraging theory to look at the consequences of changes in basic assumptions (see Ydenberg 2010).

To estimate how much flexibility is able to mitigate the effects of environmental change, we would need to compare the foraging performance of animals with and without 'behavioural flexibility', in a way analogous to Stephens' (1989) comparison of foraging strategies that do and do not use information. Stephens called the difference in the rate of intake between the best strategies that do and do not use information as 'the value of information'. Among other interesting results, he found that information was not always valuable, in the sense that it increased the foraging rate (see also Ydenberg 1998). That behavioural flexibility is advantageous is the very heart of foraging theory, and the ability of foragers to adjust tactics to changing circumstances has often been demonstrated (e.g. Sih and Christiansen 2001), but just how valuable is it?

The following example illustrates as simply as possible how we might conceptualize and estimate the value of behavioural flexibility. Imagine a forager with two prey types, 1 and 2. By definition, prey type 1 is more profitable. There are two states of the environment, A and B. In state A, prey type 1 is encountered at rate λ_{A1} and prey type 2 is encountered at rate λ_{A2}. (The encounter rate is defined as the rate at which a prey type is discovered during search for prey.) When the environment is in state B, prey items of type 1 are encountered at rate λ_{B1} and those of type 2 are encountered at rate λ_{B2}. Rate λ_{A1} is greater than rate λ_{B1} such that the intake-maximizing diet consists of prey type 1 only in state A ('specialist'), but both prey types in state B ('generalist'). State A is therefore better than state B, in the sense that a higher rate of intake is attainable. An inflexible forager would be either a specialist or a generalist, regardless of the state of the environment. But a flexible forager would be able to switch between consuming prey type 1 only and consuming both prey types, depending on whether the environment is in state A or state B. The value of behavioural flexibility can be regarded as the difference it makes, under particular circumstances, to the foraging rate of inflexible versus flexible foragers.

Imagine that the rate of intake of an inflexible specialist is S_A in state A and S_B in state B, while the rate of intake of a flexible forager is F_A ($= S_A$) in state A, and F_B in state B. When $F_B > S_B$ generalism is the better tactic in state B, and flexibility thus mitigates somewhat the reduction in foraging intake when the environment changes to state B. We can say that the value of behavioural flexibility is $F_B - S_B$. For example, if S_A (and thus also F_A) is 10 units, F_B is 7.5 units, and S_B is 5 units, the value of flexibility $F_B - S_B$ is 2.5 units, because without flexibility, intake would have been only 5 units. In percentage terms, the value of behavioural flexibility is 50%, because it makes foraging intake 50% higher (7.5 in place of 5.0 units). The value can be made arbitrarily large by the choice of parameter values.

Exactly this sort of logic led to the creation of the diet model in the first place. David Lack (1955; see Ydenberg 2010, p. 134) anticipated the formal diet model by a decade with his explanation for why swifts *Apus apus* selected only larger insects ('specialize') in fine weather (good conditions = state A) but took both large and small insects ('generalize') in poor weather (= state B): 'in fine weather when larger insects are plentiful, swifts can collect a meal more quickly if they do not go out of their way to catch the smaller kinds. In bad weather, on the other hand, insects are so scarce that the swifts cannot afford to be so selective.'

There are many examples of foraging flexibility in the literature. Pasanen and Sulkava (1971) studied the provisioning of nestlings by rough-legged buzzards *Buteo lagopus* in the Arctic, where there are two states of the environment (high and low vole years). Prey consisted of two types: birds and voles. Birds are desirable prey for transporting to the nest because they are large, and the extensive handling time (plucking, dismemberment; see Ydenberg and Davies 2010) is done by the female at the nest, instead of using up the male's valuable hunting time (who catches most of the prey). Voles, in contrast, are good prey for self-feeding because they can quickly be swallowed whole. Presumably birds are more difficult to catch and occur at lower density than do voles, but they should be taken whenever possible for delivery to the nest.

Overall, buzzards reproduce more poorly in low vole years—but not as poorly as they might if they did not flexibly alter the allocation of prey types between delivery and self-feeding. In high vole years, only 5% of prey items eaten by adults but 29–38% of those delivered are birds. In low vole years, 20% of items consumed by parents, and almost all those delivered, are birds. The change in allocation of birds between self-feeding and delivery to the nest is good evidence that buzzards were flexible in foraging behaviour, rather than just catching what they could. The increase in the use of birds for self-feeding in low vole years shows that they consumed prey valuable for delivery in order to balance their energy budget. Had they not done so, they would have been forced to work at a lower rate and been able to deliver even less food (e.g. Fig. 2 and 3 in Ydenberg 1994).

Studies like these—and hundreds of others on foraging (Stephens et al. 2007)—convincingly show that many animals are very flexible foragers, and able to adjust their foraging behaviour as circumstances change. Of course, the changing circumstances considered in most of these studies are generally encountered naturally (e.g. weather, vole cycles) and, hence, we might expect foragers to have behavioural mechanisms to react accordingly. Rapid anthropogenic changes to the environment may bring foraging situations that have not been encountered in a species' evolutionary history, leaving it helpless, at least in the short term—until the appropriate defences evolve. The best-known examples are the sudden appearance in the environment of novel chemicals, such as DDT, bioaccumulations of which were very injurious to top predators. There are natural analogues. For example, the invasion of the toxic cane toad *Rhinella marina* in Australia has been lethal for the populations of toad-eating snakes along the invasion front. Counter-adaptations quickly evolve (Phillips and Shine 2006), but the behaviour of toad-eating is evidently not behaviourally flexible. It is noteworthy that the extensive reviews of Tuomainen and Candolin (2011) and Sih et al. (2011) on the effects of rapid environmental change both refer to the negative effects of altered food resources, but neither is able to identify many documented instances. In contrast, there are scores of reports of animals adopting new food resources. Sih et al. (2011) do point out that it is not well understood why some species do not use apparently obvious novel food resources (e.g. ornamental plants, bird feeders) while similar species do so. Of course there are instances in which the degree of change, whether anthropogenic or natural, has exceeded the limits of the behavioural flexibility that the foragers possess (e.g. Ronconi and Burger 2008), but it is our contention that most species are well-equipped to deal with an immense range of change in the types, encounter rate, and quality of food.

The literature on food and foraging contains many references to 'flexibility'. A typical example is a paper by Skagen and Oman (1996) entitled 'Dietary flexibility of shorebirds in the Western Hemisphere'. The paper documents that many shorebird species consume a wide range of prey items. However, this is not quite the same as flexibility as we have defined it above. Using our definition, demonstrating behavioural 'flexibility' would require: (1) identifying the various states of the environment (e.g. good vs. poor weather; high vs. low years, etc.); (2) the behavioural tactics available to the forager (e.g. diet choice); and (3) the performance in each environment of the forager using these behavioural tactics. Claims that foragers have, or lack, flexibility need to be evaluated critically because often one or more of the elements necessary to establish the value of flexibility is missing. Though it seems unlikely, the broad diet of shorebirds compiled by Skagen and Oman (1996) could, in theory, arise entirely from environmental variability, rather than from foraging decisions.

The reverse logic also applies. Warren et al. (2009) found that chinstrap penguins *Pygoscelis antarctica* and fur seals *Arctocephalus gazella* showed persistent preferences for particular foraging areas even after a storm reduced the availability of krill *Euphausia superba* in those areas. To know whether this truly demonstrates inflexibility, it would be necessary to document that these marine predators could have done better by moving to other areas.

7.3 Effects of changes in predation danger on foraging behaviour

In the 1980s, ecologists began to realize that predators exert a fundamentally important influence on behaviour (see Lima and Dill 1990; Lima 2002; Prins and Iason 1989) not only by killing prey, but by posing a threat that requires potential prey to adopt countermeasures. And, just as individuals without a functioning immune system quickly succumb to pathogens, animals without functioning antipredator behaviour quickly succumb to predators. The dodo *Raphus cucullatus*, the flightless pigeon of the island of Mauritius, is a metaphor for the importance of antipredator behaviour. Though known to and visited by sailors, Mauritius was uninhabited by humans prior the establishment of a Dutch settlement in 1598. Dodos had no behavioural adaptations to counter the hunting tactics of human predators, and being large, tasty, and flightless, they were completely exterminated in less than a hundred years.

Animal species with functioning antipredator behaviour respond powerfully, as shown by the flight acrobatics of a group of starlings *Sturnus vulgaris* or shorebirds when a peregrine *Falco peregrinus* appears. Ecologists have long understood this of course, and taking things a step further, understood also that most animals almost always behave somewhat cautiously. However, it took much longer to appreciate the profound implications of this caution. The first papers to consider that predators had influences on the ecology and behaviour of prey beyond death and escape behaviour did not appear until the 1970s. The reasons, perhaps, relate to the fact that professional ecology grew up as a discipline during the 1950s and 1960s, working mostly in terrestrial ecosystems of temperate North America and Europe (Ydenberg 1994). Predators, especially raptors, were largely absent at this time, due to persecution and pesticides, and not until DDT was banned and raptor populations began to recover did professional ecologists have the opportunity to experience firsthand how pervasive this influence is. Since then, this literature has documented a very large number of ways that predators affect the behaviour of prey (see reviews by Lima and Dill 1990; Kats and Dill 1998).

Initially, ecologists focused on direct behavioural interactions—the tactics used by predators to stalk and capture prey, and the behaviour of prey to escape, hide from, and avoid predators (e.g. Curio 1974). It took longer to realize that a direct behavioural interaction was not required to create important ecological effects. Prey can adjust their behaviour to the probability that a predator might be present, without ever encountering one directly. One of the first published papers that made this explicit was by Rosenzweig (1974), who by means of a theoretical example showed how the above ground activity of kangaroo rats *Dipodomys spectabilis* balanced benefits (possible encounters with mates) against the costs (possible encounters with predators), which varied with the level of moonlight, it being more dangerous when lighter. This does not require kangaroo rats to experience or even to 'know' about predators: all that is required is that they adjust their 'fearfulness' of light. Their response to light may evolve by natural selection, and may not involve any behavioural modulation. They could, instead, or in addition, learn about the appropriate level of fearfulness by observing conspecifics, or assessing cues about predators themselves (odours, calls etc.).

Fontaine and Martin (2006) experimentally reduced nest predation risk by removing mammalian predators, and showed that parents of 12 songbird species increased investment in young through increased egg size, clutch mass, the rate they fed nestlings, the rate that males fed incubating females at the nest, and decreased the time that females spent incubating. These results demonstrate that birds can assess nest predation risk at large and that nest predation plays a key role in the expression of avian reproductive strategies.

The term 'predation risk', central to this topic, has been used in a variety of ways, some contrary to each other. Lank and Ydenberg (2003) proposed terminology to define its components in a consistent way. In their scheme, organisms employ 'antipredator' behaviour to reduce the chance that a predator captures them. The 'danger' inherent in a situation is the rate (or probability) of mortality that would be observed if no antipredator behaviour were employed. Antipredator behaviour can vary in intensity, a behavioural decision of the prey. Individual prey vary in 'vulnerability', which is the probability they can escape a (standardized) predator attack. Differences in vulnerability may arise due to differences in sex, age, or condition. The rate of mortality that actually occurs in a particular ecological situation emerges from the interaction between danger (a property of the environment), vulnerability (intrinsic to the prey present), and the prey's level of antipredator behaviour (a decision). These parameters could theoretically combine in many ways to give rise to a given level of mortality.

There are a very large number of ways that animals could alter foraging behaviour to increase safety, by adjusting the selection of places, times, techniques, or diet. As just one example, many aquatic creatures undergo 'vertical migration', travelling to the surface at night (or at dawn and dusk) to feed in relative safety (Gliwicz 1986). Most pertinent to this discussion, vertical migration is also adjusted behaviourally in response to cues from predators, increasing in intensity when predators are present. Similarly, long-distance migrant birds adjust their migration schedules, sometimes quite dramatically, to increase safety in response to predation danger (Ydenberg et al. 2004; Jonker et al. 2010).

All over the planet, fauna in places that lacked predators (like dodos on Mauritius) evolved away antipredator behaviour—and were quickly exterminated when humans arrived. If humans did not do the job themselves, the rats, cats, and pigs that accompanied them did so. These cases tell us not only that antipredator behaviour is essential for survival, but that it is costly: else it would not evolve away. Animals evidently must somehow pay for safety. An important decision therefore, central to flexibility in behaviour, concerns the level of investment in antipredator behaviour. Antipredator behaviour is best viewed as a trade-off between benefits (reduced mortality) and costs, which for the purposes of this chapter concern foregone foraging opportunity. If it is ineffective or too expensive, prey may drop antipredator behaviour altogether, and appear nonchalant (e.g. Prins and Iason 1989) or careless (e.g. Ydenberg et al. 2007) even in the face of grave danger. In general, though, a rise in the level of danger makes increased investment in antipredator behaviour worthwhile, though there appear to be interesting and important exceptions.

From a predation danger point of view, two large anthropogenically-driven trends are rapidly altering the planet. The populations of many terrestrial top predators are increasing throughout the northern hemisphere, reversing the patterns of the 1950s and 1960s. Even in densely populated Europe, predators such as wolves, lynxes, and eagles are returning (e.g. Schadt et al. 2002; Helander et al. 2003). There are a variety of reasons, and the factors interacting in any particular case are likely to vary. Certainly important are the banning of DDT in the 1970s, and the elimination of state-sanctioned persecution (in the form of bounties etc.), but the large-scale abandonment by humans of rural areas also plays a role.

In strong contrast, the populations of most top predators in marine systems have, over the past century, been greatly reduced (Myers and Worm 2003, 2009). The ongoing cause is overfishing. Though the reasons for overfishing are complex, the effects are clear, and as the foregoing discussion indicates, we expect prey animals to show major corresponding changes in their behaviour. In the following sections we examine the consequences of these behavioural changes on their populations and communities.

7.4 Consequences for populations

The interactions of predator and prey populations have been studied by ecologists for more than a century. In the classic paradigm, prey population dynamics are created by the opposing forces of predators killing and eating prey, and the reproductive capacity of the prey population. For example, in a book on the ecology and population management of deer *Odocoileus virginianus* entitled *The Science of Overabundance*, McShea et al. (1999) attribute the current very high abundance of these animals in some places to the lack of predators, and the abundance of forage in forests. An analogous case is made for the rapid increase in the populations of some goose species in northwest Europe (Fox et al. 2010). The general assumption is that these recent changes in prey populations are induced directly by changes in the density of food or predators—that is, there is more to eat, and fewer deaths are directly attributable to predators.

Boutin (1990) reviewed studies in which terrestrial vertebrates received supplemental food under field conditions. Most were small scale (less than 50 individuals), short term (<1 year), and the cases are strongly biased toward small-bodied herbivores and birds in north temperate environments. They show that supplemental food led to, among other effects, smaller home ranges, higher body weights, and advanced breeding relative to those on control areas. The typical population response to food supplementation was a two- to three-fold increase in density, but no change in the pattern of population dynamics was noted. A similar literature on the effect of predators is as yet lacking, but Zanette et al. (2006a,b) demonstrated 'synergistic' effects of food and predators on song sparrow demography with field trials in which they manipulated the availability of food in regions of high and low predation danger. They showed that danger greatly amplified the impact of the food treatment (see also Krebs 2011).

In the previous section, we made the case that 'danger management' is as important in the daily lives of animals as is 'energy management', and has driven the evolution of behavioural strategies that weigh carefully the benefits of all their actions against the risk of death. These behavioural strategies come at the expense of foraging proficiency, which suggests that danger management can reduce (or increase!) the reproductive capacity of prey, and hence affect population dynamics. Based on this view, effects of predators on prey populations can be divided into two components. Effects resulting from prey killed by predators are called 'density-mediated' or 'mortality-driven'. This is the component considered (exclusively) in the classic literature. Effects resulting from the behavioural adjustments made by prey are called 'non-lethal', 'non-consumptive', 'trait-mediated', or 'fear-driven'.

The question raised in this volume is to what extent behavioural flexibility might be able to mitigate environmental change for species. To answer this question, we must be able to separate a species' response to changes in both food and danger into density- and trait-mediated components, the idea being that the trait-mediated responses represent flexibility. Boutin (1990) noted that the studies covered in his review (138 studies) do not consider the behaviour of individuals in addressing questions of population regulation, and thus cannot yet partition the response into 'density' and 'trait-mediated' responses.

More recent studies have been able to separate density- and trait-mediated components, and are finding that both are important contributors to the total impact of a predator on a prey species. In one of the first good experimental studies, Nelson et al. (2004) placed damsel bugs *Nabis* spp. into field cages with pea aphids *Acyrthosiphon pisum*, and measured population growth over six days. In one experimental treatment, the damsel bugs were free to attack and kill aphids. In a second experimental treatment, their piercing mouthparts were surgically blunted, so that they could not actually kill and consume aphids, but could still stalk and attack them. Compared to control treatments without damsel bugs, the aphid population growth rate was depressed in both experimental treatments, with the effect in the surgical 'blunting' treatment 39–80% as large as that in the non-surgical treatment. Nelson et al. (2004) did not study how this happened, but pea aphids are known to respond to a variety of predators by interrupting their feeding and walking away or dropping off the plant, which increases mortality and reduces reproductive output. Presumably they reacted in this way to the surgically-altered damsel bugs. The magnitude of the effect measured here appears to be typical. In a review of 166 studies that measured the relative strength of these effects, Priesser et al. (2005) found that trait-mediated effects were at least as strong as density-mediated effects. This surprising discovery is requiring ecologists to think about predator–prey interactions in new ways (Lima 2002).

Other studies have illuminated more clearly how the defensive behaviour of prey creates population growth rate effects. Pangle et al. (2007) measured a large trait-mediated impact of an introduced invertebrate predator *Bythotrephes longimanus* on the growth rates in the field of several native zooplankton species in the Great Lakes, with effect sizes 'on the same order of magnitude as or greater (up to 10-fold)' than density-mediated effects. The effect seemed to arise because zooplankton shifted to deeper portions of the water column as *Bythotrephes* biomass increased, possibly as an avoidance response to predation. This induced migration reduced mortality, but also reduced the birth rate.

Vertical migration occurs in all oceans, and may be the most massive ecological phenomenon on the planet, so the across-the-board reduction of marine top predators is likely to be having an enormous impact. Willis (2007) considers whether the rapid and near-complete elimination of large whales (some 2 million were harvested within 100 years) in the southern oceans may have had effects on the abundance of krill that were their main prey (large whales consumed 175–190 million tonnes annually). Willis (2007) was spurred to ask the question by evidence suggesting that krill populations had not increased at all after the removal of whales, and may even have decreased. How is that possible?

Willis suggests that krill did not migrate vertically when large whales were abundant prior to the advent of industrial whaling. The reason is that due to the size and sophisticated sonar capabilities of baleen whales, deep habitats were not safer for krill, and therefore offered no advantage to offset the absence of food. With the elimination of whales, however, vertical migration became worthwhile for krill because deep habitats do offer safety from their many other predators. Migratory krill grow more slowly and do not become as large as when non-migratory, but live longer and have higher fecundity. The net effects on population productivity are startling: Willis estimates that the krill population delivered over twice the biomass before whaling removed their predators and made their lives safer.

Willis's (2007) results are hypothetical, but analogous effects have been described and measured carefully in other systems. For example, in lakes without predatory bass *Micropterus salmoides*, sunfish *Lepomis macrochirus* populations often lack large individuals, and are composed almost entirely of small fish ('stunted'; see Mittelbach 1986). The reason is that young sunfish normally spend almost all of their time in the relative safety of the littoral zone until they grow large enough to reduce their vulnerability to bass. This takes a long time. Rapid body growth begins once they are large enough to be able to forage with impunity in the pelagic zone during the midsummer zooplankton bloom. In the absence of bass, however, all individuals flood into the pelagic zone during summer, and the resultant competition severely limits the growth that any individual is able to attain.

These examples show that, in contrast to the mitigating effects of behavioural flexibility with changes in food, the behavioural responses of foragers to changes in the danger regime can have large and counterintuitive effects on prey populations (Heithaus et al. 2007). Among many other implications, these studies suggest different types of explanation for the widespread imperilment of marine fishery stocks. The collapse, failure to recover, and population declines recorded for many marine stocks of fish are virtually exclusively interpreted as density-mediated phenomena. It is assumed that too many fish are caught, killed by other predators (e.g. seals) or die as a result of unnatural events (e.g. climate warming) to allow the population to recover. The possibility of trait-mediated contributions is rarely considered.

7.5 Consequences for communities and biodiversity

In the previous sections, we documented some of the evidence suggesting that the flexible behavioural responses of foragers to changes in food and dangers in their environment can have large and sometimes unexpected consequences for their populations. In this section, we consider whether the behavioural adjustments described in the previous section could somehow have effects on entire communities.

Priesser et al.'s (2005) review cited above not only considered 'direct' effects of one species on another (e.g. predators on prey), but also those transmitted to other species (e.g. the resource harvested by the prey). These are called 'indirect' effects. The review found that trait-mediated effects were at least as strong as density-mediated effects, and that in the case of indirect effects, a mean of 85% of the total effect could be attributed to trait-mediated effects. Further, the transmitted effects could be negative or positive for the recipient of the effect.

Top predators, especially, powerfully transmit indirect effects to trophic levels beyond their prey. The prototype example is the well-documented reintroduction of wolves *Canis lupus* to Yellowstone National Park (e.g. Ripple 2003). The threat posed by wolves forced their main prey species, elk *Cervus elephus*, to become much more cautious while foraging, restricting their grazing and browsing to the safest places. The attendant release from herbivory allowed thickets to regenerate, which enabled the return of beavers *Castor canadensis* to the park. The 'ecosystem engineering' effects of beavers in turn affected the park's hydrology. Effects on insect biodiversity and avian breeding success could also be traced to the antipredator behaviour of the elk. Such effects are not unusual. Heithaus et al. (2007) document similar trophic cascades in the marine ecosystems of Shark Bay in northwestern Australia, where the behavioural tactics used by dolphins *Tursiops aduncus* and dugong *Dugong dugong* to evade tiger sharks *Galeocerdo cuvier* structure fish and eelgrass communities, respectively. LeBourdais et al. (2009) surveyed rivers in British Columbia, and found that the densities of fish and breeding harlequin ducks *Histrionicus histrionicus* were inversely related. They concluded that the mechanism was that the presence of fish affected the behaviour of benthic insects in a way that made them less available to harlequins. Their measurements suggested that the trait-mediated (behavioural) effect was much (15 times) stronger than the density-mediated (consumption) effect, a magnitude that agrees well with the review of Priesser et al. (2005). One of the implications of this study is that the widespread introduction of sport fish such as rainbow trout *Oncorhynchus mykiss* into previously fishless rivers and streams has had negative impacts on other biota in these systems.

Ecologists have long appreciated that the presence of some species has especially strong effects on the structure and function of communities. Originating in the 'keystone' species concept of Paine's intertidal work with space-clearing predatory starfish, current research on food webs (e.g Kondoh 2008; Valdovinos et al. 2010) is rapidly expanding and discovering new relationships that are revising our understanding of how food webs function. Central to this is behavioural flexibility in foraging ('adaptive trophic behaviour'), which closely matches the narrow definition of flexibility we developed in Section 7.2. In this still largely theoretical literature, complex food webs are abstracted as composed of several types of interacting subunits, or 'modules'. One type of module, called 'intraguild predation' (IGP), is composed of a predator, a consumer, and a resource species. The predator flexibly preys on both the consumer and the resource, but the consumer is a superior competitor for the resource. IGP modules have properties that seem to give food webs stability and complexity, which are keys to biodiversity. The IGP is itself stable because the predator preys on the consumer when the resource is rare, but consumes the resource (though less efficiently than the consumer) when it is abundant, thus providing resistance to perturbations. This line of research is in its infancy, but is very promising as an avenue to link our good basic knowledge of foraging behaviour to the structure of ecological communities, as envisioned in some of the early literature on optimal foraging.

7.6 Behaviour as a diagnosis tool

The central theme of this chapter is that animals are rather flexible in their behaviour, responding adaptively to changes in their environments. A proper interpretation of behaviour could thus reveal much about an animal's own assessment of the state of its environment. We could learn to interpret foraging decisions to inform us whether the foraging environment was lean or bountiful, or whether individuals perceive a situation as safe or dangerous. This information could be very useful in conservation efforts (Rosenzweig 2007).

Typically, we learn that species are in difficulty from census or monitoring schemes. Rosenzweig (2007) calls these 'trailing indicators'. For example, recent analyses of shorebird census data in North America indicate widespread ongoing declining trends (Morrison et al. 2001), giving ample reason for concern about their population health. Unfortunately, monitoring programmes generally give few solid leads about potential causes of declines. In the case of shorebirds, the data suggest that causes are likely to be general, because so many species seem to be affected. Possible agents of change include climate change, habitat loss in the breeding or wintering ranges, contaminants, reduced quality or loss of critical stopover locations, recovering predator populations, or some combination of these factors. Another possibility is that shorebirds have changed their behaviour (i.e. flexibility) in a way that affects the census (Ydenberg et al. 2004). While some of these causes could be remedied by conservation action, others could not. It is therefore essential to elucidate which of these factors might be contributing to the decline, in order to decide on conservation policy and action that would be effective.

In a paper entitled 'Relationship between stopover site choice of migrating sandpipers, their population status, and environmental stressors', Taylor et al. (2007) gave an example of how this might done. Based on the well-studied western sandpiper *Calidris mauri*, they built an individual-based model of migrating shorebirds moving through a sequence of alternating small and large stopover sites. Larger sites provide more safety, but have poorer feeding than small sites (Pomeroy et al. 2008). They contrasted situations in which there was: (1) a flyway-wide reduction in the amount of food; (2) a flyway-wide increase in predation danger; and (3) an actual lowering of the overall population size. The basic idea is that the mass action of many individuals, each optimizing its migration timing and routing, would lead to the emergence of distinctive patterns of behaviour and site choice under these differing environmental conditions. And, indeed, the model showed different patterns of mass gain, foraging intensity, and usage by migrants of small and large sites under various conditions. The model, of course, is highly abstracted, but it nevertheless suggests that simple behavioural measurements taken at a variety of sites could be very useful in informing us about a species' status. This approach would be easier and quicker than census work, and, importantly, reveals information about the state of the environment. Rosenzweig (2007) calls this a 'leading indicator'.

As just one other example, two main hypotheses are competing to explain recent changes in the migratory behaviour of the barnacle goose in northwestern Europe (Jonker et al. 2010). Jonker (2010) used measurements of the duration of parental care to discriminate between them, reasoning that under the 'food competition' hypothesis parental care should have increased in duration, while under the 'predation danger' hypothesis, it should have decreased. His measures unequivocally support the latter and, in addition, suggest that the shortening parental care has had—as a byproduct—the founding of new breeding populations of barnacle geese, that have shortened or even eliminated migration altogether. A more dramatic trait-mediated effect of a top predator (in this case, the burgeoning population of the white-tailed sea eagle *Haliaeetus albicilla* in the Baltic Sea) could scarcely be imagined.

7.7 Conclusion

In this chapter we have reviewed the role of flexibility in foraging behaviour in allowing animal populations to adjust to rapid environmental change. Our general conclusions are that behavioural flexibility is able to moderate greatly the effects of changes in the amount or quality of food available. Adaptive responses to changes in the predation regime, in contrast, may not only amplify the direct effects of predators on prey populations, but can also be transmitted with even greater effect to other species in a community. These interactions make the outcomes of changes in predation risk less than straightforward and difficult to predict, and our understanding of these ecological interactions is far from complete. Darwin's remark in the famous 'tangled bank' passage closing *The Origin of Species* that species '. . . are dependent upon each other in so complex a manner. . . .' was prescient! It is the sophisticated behavioural adjustments that individual animals are able to make in response to changes in their environment that underlie these complex changes.

Acknowledgements

We thank the editors and three anonymous referees whose comments improved a previous version of this chapter. Herman van Oeveren provided some quick assistance with Figure 7.1. We especially thank the various granting bodies and organizations who have in various ways supported our travels to landscapes in The Netherlands and elsewhere in the world over the past 35 years.

References

Baum, J.K. and Worm, B. (2009). Cascading top-down effects of changing oceanic predator abundances. *Journal of Animal Ecology*, 78, 699–714.

Boutin, S. (1990). Food supplementation experiments with terrestrial vertebrates: patterns, problems, and the future. *Canadian Journal of Zoology*, 68, 203–20.

Cappers, R.T.J. and Raemaekers, D.C.M. (2008). Cereal cultivation at Swifterbant? Neolithic wetland farming on the North European Plain. *Current Anthropology*, 49, 385–402.

Curio, E. (1974). *The Ethology of Predation*. Berlin, Springer-Verlag.

Fontaine, J.J. and Martin, T.E. (2006). Parent birds assess nest predation risk and adjust their reproductive strategies. *Ecology Letters*, 9, 428–34

Fox, A. D., Ebbinge, B.S., Mitchell, C. et al. (2010). Current estimates of goose population sizes in western Europe, a gap analysis and an assessment of trends. *Ornis Svecica*, 20, 115–27.

Gliwicz, M.Z. (1986). Predation and the evolution of vertical migration in zooplankton. *Nature*, 320, 746–8.

Heithaus, M.R., Frid, A., Wirsing, A.J., and Worm, B. (2007). Predicting ecological consequences of marine top predator declines. *Trends in Ecology and Evolution*, 23, 202–10.

Helander, B., Marquiss, M., and Bowerman, W. (2003). *Sea Eagle 2000*. Proceedings from an International Conference at Björkö, Sweden, 13–17 September 2000. Swedish Society for Nature Conservation/SNF & Åta.45 Tryckeri AB, Stockholm.

Jonker, R.M., Eichhorn, G., van Langevelde, F., and Bauer, S. (2010). Predation danger can explain changes in timing of migration: the case of the barnacle goose. *PLOS One*, 5, e11369.

Jonker, R.M., R.H.J.M. Kurvers, A. van de Bilt, M. Faber, S.E. Van Wieren, H.H.T. Prins, and R.C. Ydenberg. (2011). Rapid adaptive adjustment of parental care coincident with altered migratory behaviour. *Evolutionary Ecology*.

Kats, L.B. and Dill, L.M. (1998). The scent of death: Chemosensory assessment of predation risk by prey animals. *Ecoscience*, 3, 361–94.

Kondoh, M. (2008). Building trophic modules into a persistent food web. *Proceedings of the National Academy of Sciences of the USA*, 105, 16631–5.

Krebs, C.J. (2011). Of lemmings and snowshoe hares: the ecology of northern Canada. *Proceedings of the Royal Society of London B, Biological Sciences*, 278, 481–9.

Lack D (1955). *Swifts in a Tower*. London, Chapman and Hall.

Lambert, A.M. (1985). *The Making of the Dutch Landscape: An Historical Geography of the Netherlands*. London, Academic Press.

Lank, D.B. and Ydenberg, R.C. (2003). Death and danger at migratory stopovers: problems with 'predation risk'. *Journal of Avian Biology*, 34, 225–8.

Lebourdais, S.V., Ydenberg, R.C., and Esler, D. (2009). Fish and harlequin ducks compete on breeding streams. *Canadian Journal of Zoology*, 87, 31–40.

Lima, S.L. (2002). Putting predators back into behavioral predator-prey interactions. *Trends in Ecology and Evolution*, 7, 70–5.

Lima, S.L. and Dill, L.M. (1990). Behavioral decisions made under the risk of predation: a review and prospectus. *Canadian Journal of Zoology*, 68, 619–40.

McShea, W.J., Underwood, H.B., and Rappole, J.H. (1999). *The Science of Overabundance: Deer Ecology and Population Management*. Washington D.C., Smithsonian Institution Press.

Mittelbach, G.G. (1986). Predator-mediated habitat use: some consequences for species interactions. *Environmental Biology of Fishes*, 16, 159–69.

Morrison, R.I.G., Abry, Y., Butler, R.W., Beyersbergen, G.W., Donaldson, G.M., Gratto-Trevor, C.L., Hicklin, P.W., Johnston, V.H., and Ross, R.K. (2001). Declines in North American shorebird populations. *Wader Study Group Bulletin*, 94, 39–43.

Myers, R.A. and Worm, B. (2003). Rapid worldwide depletion of predatory fish communities. *Nature*, 423, 280–3.

Nelson, E.H., Matthews, C.E., and Rosenheim, J.A. (2004). Predators reduce prey population growth by inducing changes in prey behaviour. *Ecology*, 85, 1853–8.

Pangle, K.L., Peacor, S.D., and Johannsson, O.E. (2007). Large nonlethal effects of an invasive invertebrate predator on zooplankton population growth rate. *Ecology*, 88, 402–12.

Pasanen, S. and Sulkava, S. (1971). On the nutritional biology of the rough-legged buzzard, *Buteo lagopus lagopus* Brunn, in Finnish Lapland. *Aquilo Series Zoologica*, 12, 53–63.

Phillips, B. L. and Shine, R. (2006). An invasive species induces rapid adaptive change in a native predator: cane toads and black snakes in Australia. *Proceedings of the Royal Society of London B, Biological Sciences*, 273, 1545–50.

Pomeroy, A. C., Acevedo Seaman, D. A., Butler, R. W., Elner, R., Williams, T. D., and Ydenberg, R.C. (2008). Feeding–danger trade-offs underlie stopover site selection by migrants. *Avian Conservation and Ecology*, 3, 7 [online].

Preisser, E.L., Bolnick, D.I., and Benard, M.F. (2005). Scared to death? The effects of intimidation and consumption in predator–prey interactions. *Ecology*, 86, 501–9.

Prins, H.H.T. (1998). The origins of grassland communities in northwestern Europe. In Wallis de Vries, M.F., Bakker, J.P., and van Wieren, S.E. (eds) *Grazing and Conservation Management*, pp. 55–105. Boston, Kluwer Academic Publishers.

Prins, H.H.T. and Iason, G. (1989). Dangerous lions and nonchalant buffalo. *Behaviour*, 108, 262–96.

Prins, H.H.T. and Ydenberg, R.C. (1985). Vegetation growth and a seasonal habitat shift of the barnacle goose. *Oecologia*, 66, 122–5.

Ripple, W.J. (2003). Wolf reintroduction, predation risk, and cottonwood recovery in Yellowstone National Park. *Forest Ecology and Management*, 184, 299–313.

Ronconi, R.A. and Burger, A.E. (2008). Limited foraging flexibility: increased foraging effort by a marine predator does not buffer against scarce prey. *Marine Ecology Progress Series*, 366, 245–58.

Rosenzweig, M. (2007). On foraging theory, humans and the conservation of diversity: a prospectus. In Stephens, D.W., Brown, J.S., and Ydenberg, R.C. (eds) *Foraging: Behavior and Ecology*, pp. 400–11. Chicago, University of Chicago Press.

Rosenzweig, M. L. (1974). On the optimal aboveground activity of bannertail kangaroo rats. *Journal of Mammalogy*, 55, 193–9.

Schadt, S., Revilla, E., Wiegand, T., Knauer, F., Kaczensky, P., Breitenmoser, U., Bufka, L., Cerveny, J., Koubek, P., Huber, T., Staniša, C., and Trepl, L. 2002). Assessing the suitability of central European landscapes for the reintroduction of Eurasian lynx. *Journal of Applied Ecology*, 39, 189–203.

Sih, A. and Christensen, B. (2001). Optimal diet theory: when does it work, and when and why does it fail? *Animal Behaviour*, 61, 379–90.

Sih, A., Ferrari, M.C.O., and Harris, D.J. (2011). Evolution and behavioural responses to human-induced rapid environmental change. *Evolutionary Applications*, 4, 367–87.

Skagen, S. and Oman, H. (1996). Dietary flexibility of shorebirds in the Western Hemisphere. *Canadian Field-Naturalist*, 110, 419–44.

Stephens, D.W. (1989). Variance and the value of information. *The American Naturalist*, 134, 128–40.

Stephens, D.W., Brown, J.S., and Ydenberg, R.C. (2007). *Foraging*. Chicago: University of Chicago Press.

Taylor, C.M., Lank, D.B., Pomeroy, A.C., and Ydenberg, R.C. (2007). Relationship between stopover site choice of migrating sandpipers, their population status, and environmental stressors. *Israel Journal of Ecology and Evolution*, 53, 245–61.

Terborgh, J. (1989). *Where Have All the Birds Gone?* Princeton, Princeton University Press.

Tinbergen, N. (1958). *Curious Naturalists*. London, Country Life Limited.

Tuomainen, U. and Candolin, U. (2011). Behavioural responses to human-induced environmental change. *Biological Reviews of the Cambridge Philosophical Society*, 86, 640–57.

Valdovinos, F.A., Ramos-Jiliberto, R., Garay-Narvaez, L., Urbani, P., and Dunne, J.A. (2010). Consequences of adaptive behaviour for the structure of food webs. *Ecology Letters*, 13, 1546–59.

van der Jeugd, H. P., Eichhorn, G., Litvin, K.E., Stahl, J., Larsson, K., van der Craaf, A.T., and Drent, R.H. (2009). Keeping up with early springs: Rapid range expansion in an avian herbivore incurs a mismatch between reproductive timing and food supply. *Global Change Biology*, 15, 1057–71.

Warren, J.D., Santora, J.A., and Demer, D.A. (2009). Submesoscale distribution of Antarctic krill and its avian and pinniped predators before and after a near gale. *Marine Biology*, 156, 479–91.

Willis, J. (2007). Could whales have maintained a high abundance of krill? *Evolutionary Ecology Research*, 9, 651–62.

Wong, B.B.M., Candolin, U., and Lindström, K. (2007). Environmental deterioration compromises socially enforced signals of male quality in three-spined sticklebacks. *The American Naturalist*, 170, 184–9.

Ydenberg, R.C. (1994). The behavioural ecology of provisioning by birds. *Ecoscience*, 1, 1–14.

Ydenberg, R.C. (1998). Behavioral decisions about foraging and predator avoidance. In Dukas, R. (ed.) *Cognitive Ecology*, pp. 343–78. Chicago, University of Chicago Press.

Ydenberg, R.C. (2007). Provisioning. In Stephens, D.W., Brown, J.S., and Ydenberg, R.C. (eds) *Foraging: Behavior and Ecology*, pp. 273–303. Chicago, University of Chicago Press. Ydenberg, R.C. (2010). Decision theory. In Westneat, D.F. and Fox, C.W. (eds) *Evolutionary Behavioural Ecology*, pp. 131–47. Oxford, Oxford University Press.

Ydenberg, R.C., Butler, R.W., and Lank, D. B. (2007). Effects of predator landscapes on the evolutionary ecology of routing, timing and molt by long-distance migrants. *Journal of Avian Biology*, 38, 523–9.

Ydenberg, R.C., Butler, R.W., Lank, D. B., Smith, B.D., and Ireland, J. (2004). Western sandpipers have altered migration tactics as peregrine falcon populations have recovered. *Proceedings of the Royal Society of London B, Biological Sciences*, 271, 1263–9.

Ydenberg, R.C. and Davies, W.E. (2010). Resource geometry and provisioning routines. *Behavioral Ecology*, 21, 1170–8.

Ydenberg, R.C. and Prins, H.H.T. (1981). Spring grazing and the manipulation of food quality by barnacle geese. *Journal of Applied Ecology*, 18, 443–53.

Ydenberg, R.C., Stephens, D.W., and Brown, J.S. (2007). Foraging: an overview. In Stephens, D.W., Brown, J.S., and Ydenberg, R.C. (eds) *Foraging: Behavior and Ecology*, pp. 1–28. Chicago, University of Chicago Press.

Zanette, L., Clinchy, M., and Smith, J. N. M. (2006a). Food and predators affect egg production in song sparrows. *Ecology*, 87, 2459–67.

Zanette, L., Clinchy, M., and Smith, J. N. M. (2006b). Combined food and predator effects on songbird nest survival and annual reproductive success: results from a bi-factorial experiment. *Oecologia*, 147, 632–40.

CHAPTER 8

Reproductive behaviour

Anders Pape Møller

⊃ Overview

Human-induced changes to the natural environment are experienced by every single species on this planet. Understanding the consequences of such changes, how they affect the reproductive behaviour of animals, and how that ultimately affects populations of wild animals remain an important area of research. Because reproduction and survival constitute the two avenues contributing to fitness, understanding how environmental change affects reproduction and reproductive behaviour is of utmost importance. Three traditions of research are relevant for understanding the direction and the magnitude of changes in reproductive behaviour caused by human-induced environmental change. These are domestication, urbanization, and global change biology. I provide an overview of these three avenues of research and how they relate to reproductive behaviour, providing examples that illustrate the advantages and limitations of these approaches. Finally, I provide an extensive list of research topics of particular importance for understanding these phenomena.

8.1 Introduction

Every single species on earth is subject to direct or indirect effects of humans. These effects range from biochemicals with feminizing effects on animals including humans (e.g. Colborn et al. 1994; Kuiper et al. 1998; Saaristo et al. 2009), to the omnipresent disturbance by humans (e. g. Møller 2008b), and to rapidly changing climatic conditions (Møller et al. 2010b). Human-induced environmental change happens at an unprecedented pace, raising questions about the ability of animals to adapt. Each factor of global change poses its own problems for animals ranging from the ability to sustain the presence of humans to the ability to survive and reproduce in an altered environment. Even worse, the presence of multiple factors changing simultaneously such as climate change, eutrophication, intensification of agriculture, forestry and fisheries, and urbanization poses questions about the ability to adapt when each of the rapidly changing factors on its own already pose great problems for numerous organisms.

A growing number of studies has shown that deterioration of environmental conditions can hamper mate choice by influencing mate encounter rates and the ability to evaluate potential suitors (review in Tuomainen and Candolin 2011; Chapter 15). Similarly the needs of the offspring and the ability of parents to provide offspring care heavily depend on environmental conditions (Clutton-Brock 1991). Alterations to the environment can therefore have a major impact on reproductive success and, hence, the growth of populations. However, it is often not straightforward to demonstrate such effects because they may not be discernible under benign conditions. For example, Møller (2011a) showed recently, based on analyses of long-term time series of behaviour, that behavioural responses to climate change such as spring arrival date, reproductive decisions, and degree of breeding sociality were only present during a few rare, but extreme years, when

Behavioural Responses to a Changing World. First Edition. Edited by Ulrika Candolin and Bob B.M. Wong.
© 2012 Oxford University Press. Published 2012 by Oxford University Press.

behaviour changed completely compared to the situation during normal years. In addition, Moreno and Møller (2011) demonstrated—based on a review of extreme climatic events—that the rate of complete reproductive failure on average increased by a factor of three in adverse years compared to normal years, while simultaneously causing a reduction in adult survival rate. This is important because the frequency and the amplitude of such extreme climatic events are already increasing, and they are predicted to increase further during the next few decades (IPCC 2007).

There are numerous studies of responses of different kinds of organisms to environmental change. While information on short-term responses may be important and useful, it is the long-term trends that are particularly interesting for addressing evolutionary questions and for assessing the possibility of evolutionary adaptation. Estimates of spatial and temporal consistency in behaviour are crucial for identifying the limits to adaptation. An example of a study of consistency in ability to cope with environmental perturbations across evolutionary time related historical mutation rates in more than 30 different species of breeding birds to the effects of radioactive contamination from Chernobyl on the abundance of the same species (Møller et al. 2010a). Bird species that had higher cytochrome *b* mitochondrial DNA base pair substitution rates, and apparently have been most susceptible to factors that have caused high mutation rates, were the exact same species that were poor at coping with current levels of radiation. Likewise, questions about spatial patterns of replication are important because local adaptation to a changing environment will depend on local patterns of selection and genetic variation and on the extent of gene flow from other populations. For example, McKinnon et al. (2010) analysed geographical patterns of migration and nest predation in birds along a latitudinal gradient of 3000 km in Canada, showing decreasing nest predation with increasing latitude, thereby providing a reproductive advantage that could compensate the high costs of long migration distances.

Here I will review three different scientific approaches that may enlighten our understanding of how animals cope with humans and human-induced environmental change in the context of reproduction: domestication, urbanization, and adaptation to global change. These three processes have several commonalities for reproductive behaviour, including adaptation to human proximity, rapid change that may be phenotypic or evolutionary, and qualitative, discontinuous change of the selective environment. In the following paragraphs I briefly describe these approaches and provide examples of how they, in combination, may enlighten current studies of the effects of environmental change on reproductive behaviour. Many of the examples that I use are based on studies of birds, which reflects availability of studies rather than taxonomic bias.

8.2 Domestication and its effects on reproductive behaviour

8.2.1 Domestication and reproductive behaviour

Domestication, and its effects on reproductive behaviour, has played an important role in evolutionary biology since Charles Darwin's (1868) book. Domestication constitutes the process by which specific changes in the behaviour, physiology, and life history of animals allow successful coexistence and reproduction in the proximity of humans. Animals naturally show fear responses to humans because such fear responses provide individuals with survival benefits (Møller et al. 2008). In an increasingly populated world any species that has difficulties coping with omnipresent human beings will see their natural habitat shrinking and the rate at which it is disturbed increasing. For example, human disturbance of tigers *Panthera tigris* affect their ability to capture and bring back prey to their cubs, and this reduces their reproductive success (Kerley et al. 2002). Not surprisingly bird species with long flight distances, when approached by human beings (Blumstein 2006), are facing population declines across Europe, while that is not the case for species with short flight distances that are fully capable of coping with the proximity of humans (Møller 2008b).

Frequent disturbance due to long flight distances of certain species will result in a negative energy balance, but may also cause reproductive failure due to nest desertion in species that fly away from their nest even when only approached by humans at a distance of several hundred metres. Indeed, urbanized bird species even prefer the presence of human beings because it may deter predators that have not adapted to human proximity and hence increase reproductive success considerably, both directly by reducing mortality (Møller 2010c) and indirectly by increasing investment in reproduction (Lima and Dill 1990).

Tonic immobility that reflects the motionless state that some prey individuals assume upon capture, sometimes allowing an apparently lifeless individual to escape, reflects the extent of stress that an individual experiences from the presence of predators including humans (Boissy 1995; Forkman et al. 2007). Tonic immobility also differs between domesticated and wild animals due to microevolutionary change in this heritable trait during the domestication process (Forkman et al. 2007). Species that cope with the proximity of humans through domestication show less tonic immobility. However domesticated animals reared under stressful farming conditions such as those associated with high stocking density, continuous light, or inability to move all show increased levels of tonic immobility. Indeed, tonic immobility differs between males and females and between sexually dichromatic and monochromatic species (Møller et al. 2011), implying that mating and reproductive behaviour is associated with this aspect of antipredator behaviour. However, in a rapidly changing world with human populations reaching perhaps ten billion at the end of this century from seven billion today, there is a high premium on all living organisms to adapt to human proximity. Rapid change in the behaviour of domesticated animals due to the domestication process has shown consistent changes involving reduced aggression, increased docile behaviour, reduced stress responses including reduced corticosterone release, reduced melanin-based coloration, and changes in a range of reproductive behaviours such as earlier start of reproduction, larger litter size, and shorter intervals among reproductive events and consequently a shorter lifespan (Clutton-Brock 1987).

Species that manage to cope with humans through temporal change in flight responses resemble domesticated animals because the latter also show reduced fear reactions and weak stress responses without abandoning their current reproductive event in the presence of humans and their pets such as cats and dogs (e.g. Wirén et al. 2009). Such behavioural adaptation to domestication has been experimentally induced in for example fruit flies (Kohane and Parsons 1987), foxes (Trut et al. 2009), and chickens (Campler et al. 2009; Wirén et al. 2009), implying that there is a direct causal link between selection for domestication, altered reproductive behaviour, and altered fear responses. Domestication is well-known to affect hormone levels, including androgens and corticosteroids that are involved in the expression of reproductive behaviour and fear responses (review in Kohane and Parsons 1988). Thus domestic guinea pigs *Cavia aperea f. porcellus*, domestic ducks *Anas platyrhynchos*, domestic trout *Salmo trutta*, and domestic Arctic foxes *Vulpes lagopus* that were selected for tameness exhibited a decreased reactivity of the stress axis compared to their wild or unselected conspecifics (Künzl and Sachser 1999; Lepage et al. 2000; Gulevich et al. 2004). This reduced stress response allowed successful reproduction even in the presence of humans without causing reproductive failure and desertion of offspring by females (Trut et al. 2009).

Domestication is associated with selection against black eumelanic coloration and for phaeomelanic colours (reviews in Kohane and Parsons 1988; Clutton-Brock 1992). Mutations that produce changes in colour of the integument are normally associated with reduced fitness in wild animal species, but are amplified in domestic animals where phenotypes associated with different colours are established by selective breeding (Klungland and Våge 2003). Selected mutations in domestic animals can also appear subsequent to domestication as documented particularly well for the horse (Fang et al. 2009). These mutations mainly occur in the melanocortin-1-receptor (MC1R) gene in both domestic mammals and birds, and therefore affect the process of melanin synthesis, but also reproductive behaviour because of linkage with hormones (Fang et al. 2009). The change in dark coloration during domestication is closely associated with changes in hormone levels because dark coloration is linked to high levels of testosterone and reactivity of the stress axis. Most tellingly, selection experiments on Arctic foxes that changed the colour of their pelage from black to pale had as a correlated response a reduction in testosterone and corticosterone, a loss of fear, and a change in reproductive behaviour (Trut et al. 2009).

I briefly describe in detail two examples of the impact of domestication on reproductive behaviour: (1) changes in mate choice and mating behaviour, and (2) changes in parental care.

8.2.2 Domestication and changes in mate choice and mating behaviour

Domestication implies proximity of humans, but also of conspecifics and other domesticated animals such as dogs and cats. This switch towards a higher degree of sociality, as reflected by the proximity of many conspecifics, that are usually close kin, will have consequences for sexual selection and mating behaviour (Clutton-Brock 1987). The most obvious example is the domestic cat that is completely solitary in its ancestral state, but highly social when domesticated, and similar patterns can also be seen in other species (Darwin 1868; Clutton-Brock 1987). Proximity of many conspecifics provides easier access to potential mates, and it facilitates comparison of such potential mates for choosy females. Aggregation of large numbers of females provides males with better opportunities for monopolizing females. High mating success by a few males can change the genetic composition of domesticated populations rapidly, especially when the population is inbred. The most extreme examples of increased variance in reproductive success derives from artificial insemination in cattle, where single bulls may sire hundreds of thousands of offspring, sometimes with extremely detrimental consequences in terms of other traits like milk production. A high degree of sociality creates opportunities for infanticide as in domestic cats and male control of reproduction (Natoli 1990). Because aggression is selected against during domestication (Darwin 1868; Clutton-Brock 1987), male–male competition is reduced. This may have consequences for who is mating, but also for other aspects of reproduction influenced by testosterone such as spermatogenesis and sexual behaviour. However, these effects still remain to be quantified.

8.2.3 Domestication and changes in parental effort and parental care

A common consequence of domestication is the production of large and more litters per reproductive season (Darwin 1868; Clutton-Brock 1987). Domesticated animals like pigs have more than doubled litter size compared to ancestral wild boars *Sus scrofa* (Darwin 1868; Clutton-Brock 1987). Many domesticated animals such as sheep, goats, pigs, and cats used to have a single litter per year, but following domestication they are now able to raise two litters per year (Darwin 1868; Clutton-Brock 1987). High rates of reproduction imply reduced investment in each individual offspring with a tendency towards a higher degree of precociality. Domesticated animals generally have much shorter reproductive lifespans than their wild ancestors (Darwin 1868; Clutton-Brock 1987). These changes in life history that have evolved repeatedly in birds such as pigeons, chickens, and turkeys; rodents such as mice, rats, and guinea pigs; hoofed mammals such as pigs, cattle, buffaloes, goats, sheep, horses, and donkeys; and dogs and cats imply a switch from K selected towards r selected reproductive strategies (Darwin 1868; Clutton-Brock 1987).

8.3 Urbanization and its effects on reproductive behaviour

8.3.1 Urbanization and reproductive behaviour

Urban habitats cover increasingly large fractions of land, especially in Europe and North America, with a further predicted dramatic increase during the current century (European Commission 2006). Urbanization represents the transition of animals from living in natural habitats to an ability to live in and subsequently adapt to urban environments (Klausnitzer 1989). Urban environments are characterized by the omnipresence of humans, altered noise and light environments, longer growing seasons and hence changed phenology, and a superabundance of food provided by humans. Urban populations typically differ from rural populations of the same animal species in terms of advanced phenology, altered endocrinology (corticosterone, testosterone), reduced dispersal, and higher population density. There are also consistent changes in migration, life history, and behaviour (e.g. Marzluff et al. 2001). Some urban birds have changed properties of their songs and where they sing (Brumm 2004; Slabbekoorn and Peet 2003). Many urban birds have also changed their phenology (Gilbert 1989), visual sexual signals (Slagsvold and Lifjeld 1985), and their degree of inbreeding (Evans et al. 2010).

I briefly describe three examples of how urbanization has changed reproductive behaviour that permits successful reproduction in urban areas with high densities of humans and their pets that include predators such as cats and dogs: (1) changes in fear responses between rural and urban populations that will allow animals to cope with human presence and hence reproduce successfully, (2) changes in timing and duration of reproductive seasons between rural and urban populations, and (3) changes in life history strategies between rural and urban populations.

8.3.2 Changes in fear responses due to urbanization

Because reproduction depends on surplus energy stores, urban animals should have more energy available for reproduction due to their weaker fear responses. Relative flight distances, measured as the distance at which an animal flees from an approaching human (after adjusting for longer flight distances in species with larger body mass), reflect the risk of being eaten by a real predator such as a sparrowhawk *Accipiter nisus*: shorter flight distances from humans imply greater susceptibility to predation (Møller et al. 2008). In addition, bird species that sing from more exposed sites in the vegetation take greater risks when approached by a potential predator (Møller et al. 2008). Urban animals are tame and show drastic reductions in fear responses compared to rural conspecifics, and this can be considered an adaptive change because reduced fear responses result in reduced energetic costs of disturbance in densely populated urban areas (Cooke 1980; Møller 2008a). Indeed, corticosterone levels in urban blackbirds *Turdus merula* were reduced during winter and spring when compared to those of conspecifics from a rural population reared in the same common environment (Partecke et al. 2006). However, this conclusion should be drawn with care because there was only one replicate of each habitat. There are no similar data on other taxa. Thus reproduction is only feasible in urban environments if frequent and excessive fear reactions do not result in nest desertion or other kinds of behaviour that cause reproductive failure. Indeed, human presence may affect parental care and cause complete reproductive failure (e.g. Fernandez and Azkona 1993). This reduction in fear response in urban birds compared to their rural ancestors seems to reflect microevolutionary adaptation rather than habituation because individuals are highly consistent in their flight distances among observations, and because time since urbanization is correlated with fear response in urban relative to rural populations (Fig. 8.1). If habituation was at work, we should expect changes in fear responses over periods of days or weeks rather than decades or a century. Successful avian invaders of urban environments were less fearful in their ancestral rural habitats than unsuccessful invaders (Møller 2009). In addition, species with short mean flight distance, but also large variances in flight distance, were particularly successful in colonizing urban habitats (Møller 2010a). Variance in fear response can be considered an indicator of the degree of heterogeneity in fear response in a population because individuals are repeatable in their flight distances among observations. Initial colonization resulted in a reduced variance in flight distance compared to the ancestral situation in rural habitats, increasing subsequently as adaptation to the urban environment progressed (Møller 2010a). Thus species that have been urbanized for longer periods of time, and

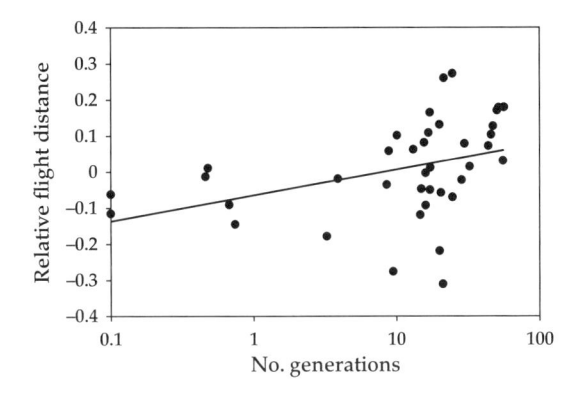

Figure 8.1 Relative flight distance in different species of urban birds in relation to number of generations since urbanization. Relative flight distance was log$_{10}$-transformed mean flight distance of rural birds minus log$_{10}$-transformed mean flight distance of urban birds of the same species. Adapted from Møller (2008a).

that have large population densities in urban compared to rural habitats, show greater increase in variance in flight distance. Therefore, fear responses provide important information about the process of adaptation to urban habitats.

8.3.3 Urbanization and changes in timing and duration of reproductive seasons

Urban areas often differ in climate from nearby rural areas, with important consequences for urban species. Temperatures are typically one or two degrees higher in urban areas, especially in winter at higher latitudes, the so-called heat island effect (Gilbert 1989). Rainfall patterns also differ between urban and rural habitats because pollution causes rain to be concentrated on days late in the week in urban areas, while being distributed evenly in rural habitats, and urban areas often have increased rainfall downwind of the urban centre (e.g. Shepherd et al. 2002). Because of higher temperatures and thresholds for accumulated spring temperatures (so-called degree days) being reached earlier, urban areas have longer growing seasons with spring starting earlier and fall extending for longer. The duration of the breeding season for many species of birds is up to three weeks longer in urban than in nearby rural areas (Klausnitzer 1989). This change in timing of reproduction is also reflected in the annual endocrine and gonadal cycles of urban blackbirds that show earlier onset in spring and, for females, later regression in summer compared to rural birds (Partecke et al. 2004). The latter conclusion is based on data for only one urban and one rural population, so other factors may confound this effect.

8.3.4 Urbanization and changes in life history strategies

Urban areas have higher food abundance (Klausnitzer 1989). Therefore, urban areas generally have much higher population densities of urbanized species compared to ancestral rural habitats, often exceeding the ancestral rural density by two orders of magnitude. Such higher population density implies higher transmission rates for parasites. Thus it is surprising that the prevalence of blood parasites transmitted by invertebrate vectors and ticks was lower in urban compared to rural blackbird populations (Evans et al. 2009). Higher density of urban species should also affect success of predators due to altered functional and numerical responses. Bird species that have become fully urbanized with short flight distances are less susceptible to sparrowhawk predation than less urbanized species with long flight distances (Møller 2008a). Higher population density should imply more intense competition for territories and for access to mates and hence altered reproductive behaviour. Indeed, male birds singing to attract females or repel male competitors sang from higher positions in the vegetation than rural conspecifics (Møller 2011b). This effect was partly modified by susceptibility to predation by cats because species that suffered disproportionately from cat predation sang from much higher positions in urban, than in rural, habitats (Møller 2011b). There was also evidence of a gradual adaptation to the urban environment because the difference in position of song posts chosen by males in urban relative to rural habitats increased with time since urbanization of different species of birds (Møller 2011b).

The life history of urban populations also appears to be altered compared to rural conspecifics.

Elevated adult survival in urban areas due to reduced predation pressure should result in reduced probability of recruitment. Indeed, Evans et al. (2009) found reduced frequency of yearling blackbird breeders in urban compared to rural habitats. Life history theory would predict that higher adult survival coupled with longer reproductive seasons should select for increased frequency of breeding and lower investment in each clutch including reduced clutch size (Roff 2002). Indeed clutch sizes appear to be reduced in urban populations (review in Chamberlain et al. 2009).

8.4 Global change and its effects on reproductive behaviour

8.4.1 Global change and behaviour

As already mentioned, the Earth is rapidly changing due to the massive impact of humans. Such global change has a number of components ranging from climate change to altered farming, forestry, and fisheries. Chemicals including pollutants, food additives and hormones, contamination with radioactive particles, and many others affect all natural environments. The study of such effects is only in its infancy, and we only have the faintest idea of the behavioural and population consequences.

I briefly describe four examples of the effects of such global change on reproductive behaviour: (1) changes in singing behaviour in response to climate change, (2) changes in intensity of sexual selection and climate change, (3) changes in infanticidal behaviour and climate change, and (4) impacts of human harvesting on abundance and composition of animal populations, with respect to sex-specific consequences for morphology and secondary sexual characteristics.

8.4.2 Changes in singing behaviour in response to climate change

Birds and many other animals produce vocal displays from exposed sites, apparently because this facilitates transmission of sound to competing males or females searching for mates. Climate change has advanced spring phenology in many temperate and arctic environments. Thus, leafing and flowering of plants has advanced considerably during the last 20 years (Menzel et al. 2006). The start of singing in birds has become earlier in spring in recent years (Rubolini et al. 2010). Timing of leafing will affect reverberation and sound transmission of bird song with important consequences for choice of song posts. Møller (2010b) recently analysed changes in the position in the vegetation used by male birds for singing during the late 1980s, before recent climate change, and again in 2010 20 years after the start of recent climate change. This recent change resulted in a 20% increase in temperature and a 30% increase in precipitation during spring and summer. Average song post height increased by 18% or 1.2 m during the study. The increase in song post position was particularly pronounced for bird species singing in trees where the effects of leafing would be expected to be most pronounced (Fig. 8.2). Because the increase in song post height should depend on relative costs and benefits of change, Møller predicted that sexually dichromatic species and species with increasing populations, and hence more intense intraspecific competition for mates, should increase their song post height, while high susceptibility to preda-

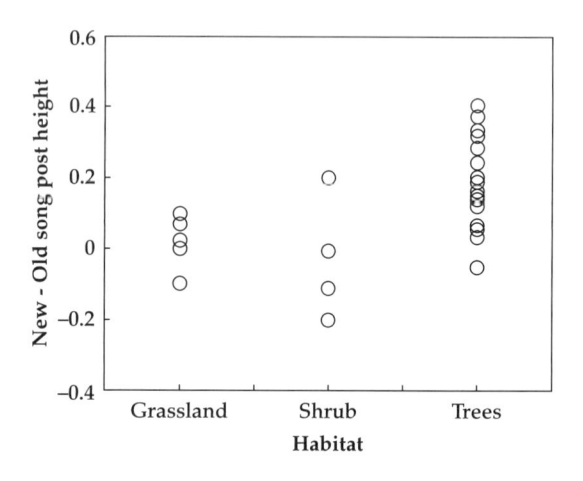

Figure 8.2 Change in song post position between 1986–1989 and 2010 in different species of birds for three main breeding habitats. Thus zero implies no change, while positive values imply that song posts are higher in the vegetation in 2010 than during 1986–1989, as found in species inhabiting trees. Adapted from Møller (2010b).

tion by the sparrowhawk should prevent increases in song post height because sparrowhawks preferentially prey on birds high in the vegetation. That was indeed the case. These results suggest that display sites for singing birds can change rapidly, with potential consequences for optimal design of songs, variance in mating success, and predator–prey interactions. It remains to be tested if these changes in singing behaviour affected female mate choice or territory ownership.

8.4.3 Changes in intensity of sexual selection and climate

Several studies have indicated that climate change affects the sex ratio, and thereby the opportunity for sexual selection, and that the expression of secondary sexual characteristics depends on environmental conditions (see also Chapter 15). There are good reasons to expect that sexual signals would reliably reflect changing environmental conditions, simply because many secondary sexual characteristics are condition-dependent, and individuals in prime condition thus should be better able to respond to changing climatic conditions than individuals in poor condition (Spottiswoode and Saino 2010). Advancement of spring migration by migratory birds under climate change correlates with the strength of sexual selection, because increasing temperatures reduce the intensity of natural selection that opposes sexual selection for early arrival (Spottiswoode et al. 2006). Similarly, the degree of polygyny in grey seals *Halichoerus grypus* depends on local weather conditions, and climate change could therefore affect the intensity of sexual selection (Twiss et al. 2007). Likewise, environmental conditions in the Antarctic winter quarters of the migratory Arctic tern *Sterna paradisaea* influence the extent of black colour on the tip of the beak (a signal of quality) at the breeding grounds in the northern hemisphere more than 25,000 km away (Møller et al. 2007). I briefly discuss two studies that have investigated in detail how climate affects the expression of secondary sexual characteristics, and how this has consequences for sexual selection.

In a first study, Møller and Szép (2005) analysed long-term trends in expression of a secondary sexual characteristic in the barn swallow *Hirundo rustica*, the length of the outermost tail feathers in males. This characteristic is associated with increased male mating success, extra-pair paternity, and quality of mate acquired, independent of age, territory quality, and behaviour (Møller 1994). Tail length of males in a Danish population increased by more than one standard deviation during 1984–2003. Environmental conditions during spring migration in Algeria, after having crossed the Sahara, predicted tail length and change in tail length across generations. When springs were rainy, many males survived, and mean tail length decreased because even males in poor condition survived. In contrast, mainly long-tailed males survived in dry springs because males in poor condition died (Fig. 8.3). Intensity of selection on breeding date and survival declined during the study period, and the response to selection in these traits, as estimated from the product of heritability and total selection, was very similar to the observed response in tail length. These results suggest a rapid microevolutionary change in tail length during a very short period associated with a rapid change in environmental conditions. Expression of two secondary sexual characteristics (white forehead and wing

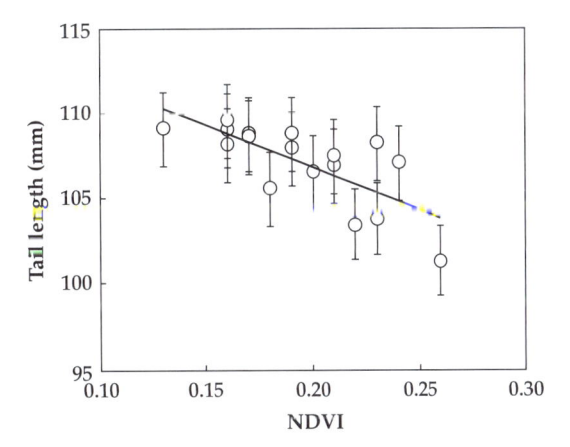

Figure 8.3 Tail length (± SD) of male barn swallows *Hirundo rustica* in relation to environmental conditions during spring migration at the spring staging grounds in Algeria as reflected by satellite imaging based on the Normalized Difference Vegetation Index (NDVI). NDVI is an index of greenness of the vegetation and hence the amount of precipitation during the previous few weeks. The line is the linear regression line. Adapted from Møller and Szép (2005).

patches) in the collared flycatcher *Ficedula albicollis* showed similar evidence of rapid change (Hegyi et al. 2006, 2007). The two characteristics develop during winter in Africa and summer in Europe, respectively, and their size reflects environmental conditions during these two seasons, providing competitor males and choosy females with important information about the ability of males to cope with environmental conditions at different times of the year. Several recent studies have shown similar effects of environmental conditions on secondary sexual characteristics, and even delayed effects reflecting conditions after two annual moults (e.g. Balbontín et al. 2011; Saino et al. 2004). Thus the information content of secondary sexual characteristics can track changes in environmental conditions in disparate parts of the world.

8.4.4 Changes in infanticidal behaviour and climate change

Differential effects of climate change on sex-specific mortality have many consequences. Males (and females) of many species engage in infanticidal behaviour as a means of mate acquisition, if there are no other opportunities for obtaining a mate, or if time till reproduction in this way is shortened (Hausfater and Hrdy 1984). Adult survival of male, but not female, barn swallows during their trans-Saharan spring migration from the South African

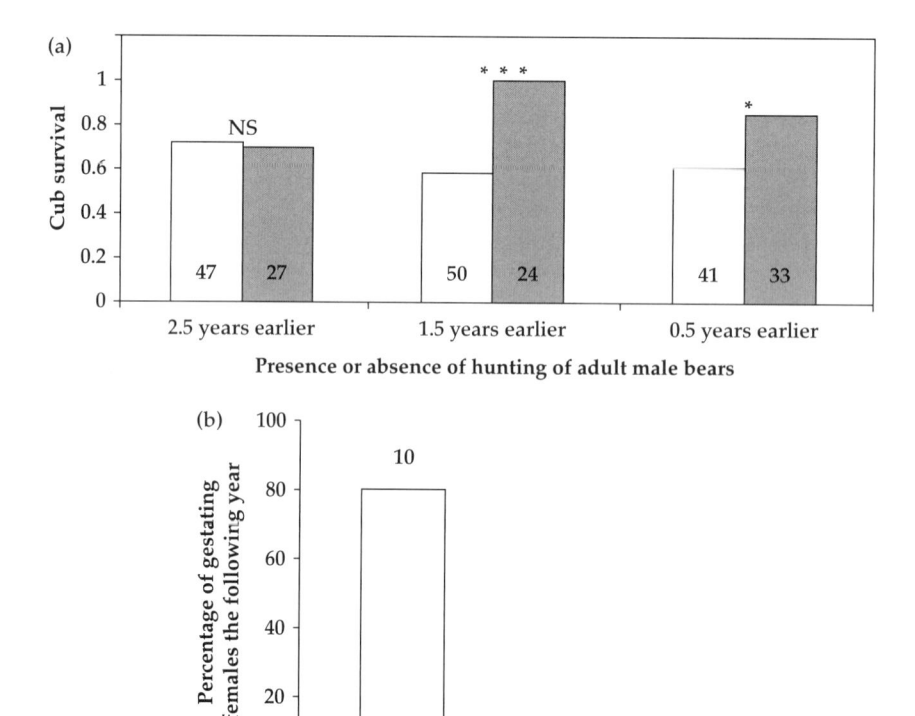

Figure 8.4 (a) Cub survival rate in the presence (open bars) or absence (hatched bars) of hunting of male brown bears *Ursus arctos* 0.5, 1.5, and 2.5 years earlier. Numbers in bars are number of cubs. * implies cub survival differed significantly in the presence and absence of hunting of adult male bears at the $P < 0.05$ level, *** at the $P < 0.001$ level, and NS implies that cub survival did not differ between presence and absence of hunting. (b) Percentage of adult female bears gestating depending on whether or not their cubs were lost the previous year to infanticide. Adapted from Swenson et al. (1997).

winter quarters to their breeding areas in Denmark was significantly depressed in years with low rainfall in northern Africa, where migrants recuperate following the crossing of the Sahara Desert (Møller and Szép 2005). Drier climatic conditions in Northern Africa also reduced population size, and this is important because the frequency of infanticide is elevated at high population densities (Møller 1988). Thus the frequency of infanticide decreased from being a major cause of nestling mortality that accounted for 25% of all mortality to hardly any in recent years (Møller 2004). Furthermore, weather conditions at the breeding sites had an additional effect on the tertiary sex ratio. While the surplus of unmated males was on average 8.2% in climatically normal summers, it fell to 2.9% in warm summers, dramatically reducing the risk of infanticide (Møller 2011a). Thus, climate change can, during very short periods, radically alter important sources of offspring mortality.

8.4.5 Changes in human harvesting and composition of animal populations

Humans may directly interfere with the composition of animal populations through harvesting, as in hunting or fisheries. Swenson et al. (1997) provided a particularly telling example caused by hunting. Sex-specific hunting patterns targeting males increased the frequency of infanticide in brown bears *Ursus arctos*. By shooting males, replacement males will have no incentive to spare the lives of unrelated cub that constitute an obstacle to their own reproductive success by delaying time till next estrous for the mother. While cub survival was 98% in areas without hunting, this was reduced to 72% in areas with hunting (Fig. 8.4a). Infanticide advanced estrous because hardly any females became pregnant the following year if cubs were not lost, while most became pregnant following loss of cubs (Fig. 8.4b). The population effects of hunting due to increased levels of infanticide amounted to a decrease in population growth rate from 1.18 to 1.14, or a net decline in population productivity by 30%. Thus, patterns of harvesting can have dramatic indirect effects on sexual selection and ultimately on population dynamics.

8.5 Synthesis

What are the commonalities and differences among the three approaches to study effects of human-induced environmental change on reproductive behaviour that I have described here?

Domestication and urbanization have important similarities because both reflect adaptation to the proximity of humans, the presence of buffered environmental conditions, and the presence of food in adequate amounts. The two approaches are also similar because the environment changes rapidly and, therefore, the patterns of selection change abruptly. In this respect they may to some extent have common features with the effects of climate change. The history of the Earth is dotted with numerous dramatic environmental changes that have had drastic effects on life. The current situation differs in one respect dramatically from what has happened previously through the diversity of drastic environmental changes, changes that in combination may produce synergistic effects that make adaptation considerably more difficult than when a single major driving factor is changing.

8.6 Future prospects for research

This chapter provides a brief overview of some aspects of effects of environmental change on reproductive behaviour, but cannot possibly constitute a comprehensive review due to space limits. Even so, it is clear from the examples discussed here that several avenues of research are in need of investigation. Here I emphasize five areas of research.

First, studies of changes in reproductive behaviour over time, such as the change in behaviour of urban birds from the rural ancestors being related to time since urbanization (e.g. Møller 2008a, 2011b), may constitute a means by which to investigate whether such changes are gradual and hence most likely evolutionary.

Second, we are currently experiencing dramatic changes in climate, habitat, and pollution, posing multiple problems of adaptation in free-living organisms. We need to know what is the degree of consistency in behavioural responses to different kinds of global change.

Third, we know very little about the temporal and spatial scale at which behavioural responses during reproduction to environmental change are consistent. For example Møller et al. (2010a) showed that responses of different species of birds to an extreme environmental perturbation at an ecological scale (reduction in abundance in response to radioactive contamination from Chernobyl) were strongly positively correlated with responses at an evolutionary scale (mitochondrial mutation rates). Is this example unique, or does it reflect a general pattern?

Fourth, we need to revisit study sites used in the past to assess how reproductive behaviour has changed in response to global change.

Finally, even if we cannot address to what extent changes in a behavioural trait associated with reproduction reflect plastic or evolutionary responses, we may be able to determine whether responses depend on population size and amount of neutral genetic variation (e.g. Møller and Nielsen 2010). In this context, urban populations may be particularly interesting because reduced dispersal, small populations, and high levels of inbreeding may all contribute to genetic divergence, local adaptation, and rapid urbanization.

Acknowledgements

I thank U. Candolin and B. Wong for their invitation to contribute this chapter, and M.D. Jennions and an anonymous reviewer for comments on an earlier draft.

References

Balbontín, J., de Lope, F., Hermosell, I.G., Mousseau, T.A., and Møller, A.P. (2011). Determinants of age-dependent change in a secondary sexual character. *Journal of Evolutionary Biology*, 24, 440–8.

Blumstein, D.T. (2006). Developing an evolutionary ecology of fear: How life history and natural history traits affect disturbance tolerance in birds. *Animal Behaviour*, 71, 389–99.

Boissy, A. (1995). Fear and fearfulness in animals. *Quarterly Review of Biology*, 70, 165–91.

Brumm, H. (2004). The impact of environmental noise on song amplitude in a territorial birds. *Journal of Animal Ecology*, 73, 434–40.

Campler, M., Jongren, M., and Jensen, P. (2009). Fearfulness in red junglefowl and domesticated White Leghorn chickens. *Behavioral Processes*, 81, 39–43.

Chamberlain, D.E., Cannon, A.R., Toms, M.P., Leech, D.I., Hatchwell, B.J., and Gaston, K.J. (2009). Avian productivity in urban landscapes: A review and meta-analysis. *Ibis*, 151, 1–18.

Clutton-Brock, J. (1987). *A Natural History of Domesticated Mammals*. Austin, TX, University of Texas Press.

Clutton-Brock, T.H. (1991). *The Evolution of Parental Care*. Princeton, Princeton University Press.

Clutton-Brock, J. (1992). The process of domestication. *Mammal Review*, 22, 79–85.

Colborn, T., vom Saal, F.S., and Soto, A.M. (1994). Developmental effects of endocrine-disrupting chemicals in wildlife and humans. *Environmental Impact Assessment Review*, 14, 469–89.

Cooke, A.S. (1980). Observations on how close certain passerine species will tolerate an approaching human in rural and suburban areas. *Biological Conservation*, 18, 85–8.

Darwin, C. (1868). *The Variation of Animals and Plants under Domestication*. London, John Murray.

European Commission. (2006). *Urban Sprawl in Europe*. Copenhagen, European Environmental Agency.

Evans, K.L., Gaston, K.J., Sharp, S.P., McGowan, A., Simeoni, M., and Hatchwell, B.J. (2009). Effects of urbanization on disease prevalence and age structure in blackbird *Turdus merula* populations. *Oikos*, 118, 774–82.

Evans, K.L., Hatchwell, B.J., Parnell, M., and Gaston, K.J. (2010). A conceptual framework for the colonisation of urban areas: The blackbird *Turdus merula* as a case study. *Biological Reviews of the Cambridge Philosophical Society*, 85, 643–67.

Fang, M., Larson, G., Soares Ribeiro, H., Li, N., and Andersson, L. (2009). Contrasting mode of evolution at a coat color locus in wild and domestic pigs. *PLOS Genetics*, 5, e1000341.

Fernandez, C. and Azkona, P. (1993). Human disturbance affects parental care of marsh harriers and nutritional status of nestlings. *Journal of Wildlife Management*, 57, 602–8.

Forkman, B., Boissy, A., Meunier-Saluen, M.C., Canali, E., and Jones, R.B. (2007). A critical review of fear tests used on cattle, pigs, sheep, poultry and horses. *Physiology and Behavior*, 92, 340–74.

Gilbert, O.L. (1989). *The Ecology of Urban Habitats*. London, Chapman and Hall.

Gulevich, R.G., Oskina, I.N., Shikhevich, S.G., Fedorova, E.V., and Trut, L.N. (2004). Effect of selection for behavior on pituitary-adrenal axis and proopiomelanocortin gene expression in silver foxes (*Vulpes vulpes*). *Physiology and Behavior*, 82, 513–18.

Hausfater, G. and Hrdy, S.B., (eds) (1984). *Infanticide: Comparative and Evolutionary Perspectives*. New York, Aldine.

Hegyi, G., Török, J., Garamszegi, L.Z., Rosivall, B., Szöllos, E., and Hargitai, R. (2007). Dynamics of multiple sexual signals in relation to climatic conditions. *Evolutionary Ecology Research*, 9, 905–20.

Hegyi, G., Török, J., Tóth, L., Garamszegi, L.Z., and Rosivall, B. (2006). Rapid temporal change in the expression and age-related information content of a sexually selected trait. *Journal of Evolutionary Biology*, 19, 228–38.

IPCC (2007). *Climate Change 2007*. Cambridge, Cambridge University Press.

Kerley, L.L., Goodrich, J.M., Miquelle, D.G., Smirnov, E.N., Quigley, H.B., and Hornocker, N.G. (2002). Effects of roads and human disturbance on Amur tigers. *Conservation Biology*, 16, 97–108.

Klausnitzer, B. (1989). *Verstädterung von Tieren*. Wittenberg Lutherstadt, Neue Brehm-Bücherei.

Klungland, H. and Våge, D.I. (2003). Pigmentary switches in domestic animal species. *Annals of the New York Academy of Science*, 994, 331–8.

Kohane, M.J. and Parsons, P.A. (1987). Mating ability in laboratory-adapted and field-derived *Drosophila melanogaster*: The stress of domestication. *Behavioral Genetics*, 17, 541–58.

Kohane, M.J. and Parsons, P.A. (1988). Domestication: Evolutionary change under stress. *Evolutionary Biology*, 23, 30–48.

Kuiper, G.G.J.M., Josephine G. Lemmen, J.G., Carlsson, B., Corton, J.C., Safe, S.H., van der Saag, P.T., van der Burg, B., and Gustafsson, J.-Å. (1998). Interaction of estrogenic chemicals and phytoestrogens with estrogen receptor ß. *Endocrinology*, 139, 4252–63.

Künzl, C. and Sachser, N. (1999). The behavioral endocrinology of domestication: A comparison between the domestic guinea pig (*Cavia aperea f. porcellus*) and its wild ancestor, the cavy (*Cavia aperea*). *Hormones and Behavior*, 35, 28–37.

Lepage, O., Overli, O., Petersson, E., Järvi, T., and Winberg, S. (2000). Differential stress coping in wild and domesticated sea trout. *Brain, Behavior and Evolution*, 56, 259–68.

Lima, S.L. and Dill, L.M. (1990). Behavioral decisions made under the risk of predation: a review and prospectus. *Canadian Journal of Zoology*, 68, 619–40.

Marzluff, J.M., Bowman, R., and Donnelly, R. (2001). *Avian Ecology and Conservation in an Urbanizing World*. Dordrecht, Kluwer Academic Press.

McKinnon, L., Smith, P.A., Nol, E., Martin, J.L., Doyle, F.I., Abraham, K.F., Gilchrist, H.G., Morrison, R.I.G., and Bêty, J. (2010). Lower predation risk for migratory birds at high latitudes. *Science*, 327, 326–7.

Menzel, A., Sparks, T.H., Estrella, N., et al. (2006). European phenological response to climate change matches the warming pattern. *Global Change Biology*, 12, 1969–76.

Møller, A.P. (1988). Infanticidal and anti-infanticidal strategies in the swallow *Hirundo rustica*. *Behavioral Ecology and Sociobiology*, 22, 365–71.

Møller, A.P. (1994). *Sexual Selection and the Barn Swallow*. Oxford, Oxford University Press.

Møller, A.P. (2004). Rapid temporal change in frequency of infanticide in a passerine bird associated with change in population density and body condition. *Behavioral Ecology*, 15, 462–8.

Møller, A.P. (2008a). Flight distance of urban birds, predation and selection for urban life. *Behavioral Ecology and Sociobiology*, 63, 63–75.

Møller, A.P. (2008b). Flight distance and population trends in European breeding birds. *Behavioral Ecology*, 19, 1095–102.

Møller, A.P. (2009). Successful city dwellers: A comparative study of the ecological characteristics of urban birds in the Western Palearctic. *Oecologia*, 159, 849–58.

Møller, A.P. (2010a). Interspecific variation in fear responses predicts urbanization in birds. *Behavioral Ecology*, 21, 365–71.

Møller, A.P. (2010b). When climate change affects where birds sing. *Behavioral Ecology*, 22, 212–17.

Møller, A.P. (2010c). The fitness benefit of association with humans: Elevated success of birds breeding indoors. *Behavioral Ecology*, 21, 913–18.

Møller, A.P. (2011a). Behavioral and life history responses to extreme climatic conditions. *Current Zoology*, 57, 351–62.

Møller, A.P. (2011b). Song post height in relation to predator diversity and urbanization. *Ethology*, 117, 529–38.

Møller, A.P. and Szép, T. (2005). Rapid evolutionary change in a secondary sexual character linked to climatic change. *Journal of Evolutionary Biology*, 18, 481–95.

Møller, A.P. and Nielsen, J.T. (2010). Fear screams and adaptation to avoid imminent death: Effects of genetic variation and predation. *Ethology Ecology Evolution*, 22, 1–20.

Møller, A.P., Erritzøe, J., Karadas, F. and Mousseau, T.A. (2010a). Historical mutation rates predict susceptibility to radiation in Chernobyl birds. *Journal of Evolutionary Biology*, 23, 2132–42.

Møller, A.P., Fiedler, W. and Berthold, P., ed. (2010b). *Birds and Climate Change*. Oxford, Oxford University Press.

Møller, A.P., Flensted-Jensen, E., and Mardal, W. (2007). Black beak tip coloration as a signal of phenotypic quality in a migratory seabird. *Behavioral Ecology and Sociobiology*, 61, 1561–71.

Møller, A.P., Nielsen, J.T., and Garamszegi, L.Z. (2008). Risk taking by singing males. *Behavioral Ecology*, 19, 41–53.

Møller, A.P., Christiansen, S.S., and Mousseau, T.A. (2011). Sexual signals, risk of predation and escape behavior. *Behavioral Ecology*, **22**, 800–7.

Moreno, J. and Møller, A.P. (2011). Extreme climatic events and life histories. *Current Zoology*, 57, 375–89.

Natoli, E. (1990). Mating strategies in cats: A comparison of the role and importance of infanticide in domestic cats, *Felis catus* L. and lions, *Pathera leo* L. *Animal Behaviour*, 40, 183–6.

Partecke, J., Schwabl, I., and Gwinner, E. (2006). Stress and the city: urbanisation and its effects on the stress physiology in European Blackbirds. *Ecology*, 87, 1945–52.

Partecke, J., Van't Hof, T. and Gwinner, E. (2004). Differences in the timing of reproduction between urban and forest European Blackbirds (*Turdus merula*): result of phenotypic flexibility or genetic differences? *Proceedings of the Royal Society of London B, Biological Sciences*, 271, 1995–2001.

Roff, D.A. (2002). *Life History Evolution*. Sunderland, Sinauer.

Rubolini, N., Saino, N., and Møller, A.P. (2010). Does migratory behaviour constrain the phenological response of birds to climate change? *Climate Research*, 42, 45–55.

Saaristo, M., Craft, J.A., Lehtonen, K.K., Björk, H., and Lindström, K. (2009). Disruption of sexual selection in sand gobies (*Potamoschistus minutus*) by 17alpha-ethinyl estradiol, an endocrine disruptor. *Hormones and Behavior*, 55, 530–7.

Saino, N., Szép, T., Ambrosini, R., Romano, M., and Møller, A.P. (2004). Ecological conditions during winter affect sexual selection and breeding in a migratory bird. *Proceedings of the Royal Society of London B, Biological Sciences*, 271, 681–6.

Shepherd, J.M., Pierce, H., and Negri, A.J. (2002). Rainfall modification by major urban areas: Observations from spaceborne rain rador on the TRMM satellite. *Journal of Applied Meteorology*, 41, 869–701.

Slabbekoorn, H. and Peet, M. (2003). Birds sing at a higher pitch in urban noise: Great tits hit the high notes to ensure that their mating calls are heard above the city's din. *Nature*, 424, 267.

Slagsvold, T. and Lifjeld, J.T. (1985). Variation in plumage color of the great tit *Parus major* in relation to habitat, season and food. *Journal of Zoology*, 206, 321–8.

Spottiswoode, C.N. and Saino, N. (2010). Sexual selection and climate change. In Møller, A.P., Fiedler, W., and Berthold, P., (ed.) *Birds and Climate Change*, pp. 169–90. Oxford, Oxford University Press.

Spottiswoode, C.N., Tøttrup, A.P., and Coppack, T. (2006). Sexual selection predicts advancement of avian spring migration in response to climate change. *Proceedings of the Royal Society of London B, Biological Sciences*, 273, 3023–9.

Swenson, J.E., Sandegren, F., Söderberg, A., Bjärvall, A., Franzén, R. and Wabakken, P. (1997). Infanticide caused by hunting of male bears. *Nature*, 386, 450–1.

Trut, L., Oskina, I., and Kharlamova, A. (2009). Animal evolution during domestication: The domesticated fox as a model. *Bioessays*, 31, 349–60.

Tuomainen, U. and Candolin, U. (2011). Behavioral responses to human induced environmental change. *Biological Reviews of the Cambridge Philosophical Society*, 86, 640–57.

Twiss, S.D., Thomas, C., Poland, V., Graves, J.A., and Pomeroy, P. (2007). The impact of climatic variation on the opportunity for sexual selection. *Biology Letters*, 22, 12–15.

Wirén, A., Gunnarsson, U., Andersson, L., and Jensen, P. (2009). Domestication-related genetic effects on social behavior in chickens: Effects of genotype at a major growth quantitative trait locus. *Poultry Science*, 88, 1162–6.

CHAPTER 9

Social behaviour

Daniel T. Blumstein

⊃ Overview

Social structure and social behaviour are influenced by environmental factors. Hence, human induced environmental changes are likely to have a variety of impacts on sociality and, because sociality often has demographic consequences, on population biology. We can learn a lot by capitalizing on intraspecific and interspecific variation in sociality to identify key environmental drivers of demography. In this chapter, I discuss some of the many ways that social behaviour and social structure are dependent upon the distribution and abundance of resources and other environmental factors. Armed with such knowledge, we can begin to develop individual-based models that will allow us to evaluate the relative importance of these anthropogenically-influenced environmental drivers and, ultimately, better predict the consequences of anthropogenic change on a variety of animals.

9.1 Introduction

Many anthropogenic activities ultimately influence the environment. These environmental changes affect the distribution and abundance of resources that animals use as well as the predators, parasites, and pathogens that they interact with. At the individual level, changes in resources influence the frequency and type of social interactions. Social structure, that includes the number and type of individuals in a group, and the duration and nature of their interactions, emerges through interactions between individuals (Hinde 1976; Whitehead 2008). We often assume that these social interactions are influenced by their benefits and costs, which may vary both temporally and spatially (Krause and Ruxton 2002) and will be influenced by the distribution and abundance of resources and predators, parasites, and pathogens. We should care about identifying the links between resources and other key factors influencing sociality because social structure has a variety of fitness and, hence, demographic consequences (Blumstein 2010; Blumstein and Fernández-Juricic 2010).

How can we learn about these links? In many cases the emergent social systems that describe patterns of space use and grouping, as well as breeding systems, vary both intra- and interspecifically (Lott 1991). It is the intraspecific variation in social systems that can help us understand the link between anthropogenic activities, their demographic consequences, and ultimately will influence whether a population goes extinct or persists in response to anthropogenic activities (Fig. 9.1). Identifying how populations vary may allow us to predict how anthropogenic stressors may cause systematic changes in key demographic parameters like survival and reproduction, and hence a population's persistence.

In this prospective review, I will describe some potential links between the environment and sociality that should influence demography. I will mostly focus on the individual, but clearly the links between anthropogenic activities and population persistence or extinction span multiple ecological levels (Fig. 9.1). Space prevents a comprehensive review of all possible relationships between sociality and envi-

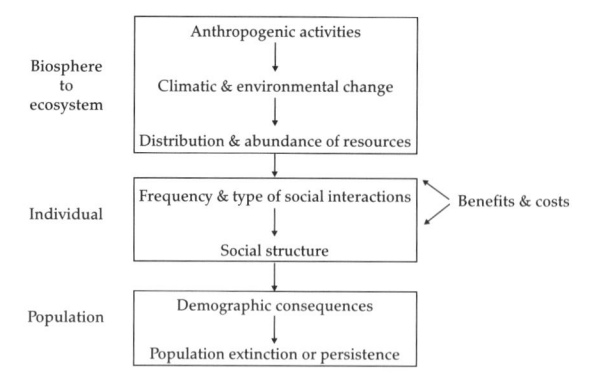

Figure 9.1 The relationship between anthropogentic activities and population extinction or persistence works through individuals interacting with important resources in their environment. The benefits and costs of interacting with others are influenced by the distribution and abundance of resources, and the benefits and costs of these interactions influence the resulting social structure.

ronmental and anthropogenic change. Suffice it to say that most aspects of sociality may be either directly, or indirectly, influenced by the environment. Rather, I selectively discuss some potentially important links between sociality and the environment focusing mostly on birds and mammals, but the ideas apply more generally to other taxa as well. I adopt a pluralistic approach when thinking about animal social groups and will discuss those that are simply ephemeral aggregations not characterized by strong social bonds, and those in which individuals may have more formal relationships, roles (e.g. breeder/non-breeder, territorial member/non-territorial floater), and bonds. Ultimately, any density-dependent activities may have fitness consequences and hence be important for understanding population biology (e.g. Courchamp et al. 2008). I will emphasize the importance of developing a mechanistic understanding of this social variation and will suggest that, once developed, we can use it to construct individual-based models that will allow us to predict the demographic consequences of anthropogenic factors, such as climate change and habitat alteration.

9.2 What environmental factors might influence sociality and how do humans impact them?

Whether directly through habitat destruction or modification, or indirectly through climate change, humans have a profound impact on the distribution and abundance of food and other essential resources that animals may require. Many factors influence where individuals are found and how they may interact in areas where they are found: food and predators are two important ones. For instance, consider two areas, one with predators and one without predators. All else being equal, more prey will be found in the area without predators than with predators. Humans may influence the distribution and abundance of predators both by killing predators and by introducing (or reintroducing them), or by modifying predators' food and creating mesopredator releases (Ritchie and Johnson 2009).

Animals often form groups when under predation threat because, simply by grouping, individuals may reduce their risk of predation. At least three models of predation hazard assessment (detection—Pulliam 1973; dilution—Pitcher and Parrish 1993; predator confusion—Landeau and Terborgh 1986; Krakauer 1995) predict that animals should forage more and allocate less time to antipredator vigilance as group size increases because predation risk declines with the addition of alternative prey. All three models assume a constant attack rate (i.e. that by grouping, individuals are not attracting more predators). This is likely not true in all cases (e.g. Cresswell 1994; Botham and Krause 2005), and if it is not, the form of density-dependence must be identified.

The distribution of predators may change naturally and independently from any changes in resources. For instance, if a predator's range is more closely tied with some environmental driver than its prey, under human induced environmental change, there could suddenly be more predators around, thus increasing predation risk, and the algebraic benefits of grouping shifts towards grouping. Yet the perception of risk may also drive intraspecific variation in grouping. Indeed animals may be more inclined to group in exposed areas to reduce their risk, as seen in coral trout *Plectropomus leopardus* (Goeden 1978), rainbow fish *Melanotaenia eachamensis* (Brown and Warburton 1997), and a variety of ungulates (Eisenberg 1981). Grouping in a specific habitat type influences resource depletion in those locations and may be a potent indirect effect of predation on the larger community (e.g. Laundré et al. 2010). And, if human activities force prey to group in a specific habitat type, there will be an enhanced effect of human activities on overall biodiversity.

Humans also move predators around (Bradshaw and Bekoff 2001). Consider the remarkable experiment that has been conducted in the Greater Yellowstone ecosystem since 1995 with the reintroduction of wolves *Canis lupus* (Smith et al. 2003). Wolves have directly and indirectly changed ungulate behaviour; ungulates flee them and avoid areas around willows *Salix* spp.—because these areas have limited visibility. The consequences of avoiding willows have led to an increased density of both willows and the birds that rely on them. Thus, the introduction of wolves led to an increase in breeding bird density. Grouping dynamics have also been changed by the reintroduction of wolves. Such changes in behaviour are not always expected. For instance, cape buffalo *Syncerus caffer* do not respond to spatial variation in lion *Panthera leo* predation risk, even though lions are a major source of predation (Prins and Iason 1989).

Similarly, humans may influence the distribution and abundance of pathogens and parasites. Parasites and pathogens are known to have a variety of direct and indirect effects by influencing, for example, mortality (e.g. Atkinson et al. 2009) or the behaviour of their hosts to facilitate disease transmission. In the case of the latter, we see direct links between the presence of a parasite or pathogen and host behaviour. Parasites and pathogens may also influence animals by affecting the abundance or distribution of their resources (e.g. changes in plant pathogens may influence the abundance of key plants). Human-mediated changes to the environment, such as climate change, are therefore expected to shift distributions of parasites and pathogens and create situations where animals may be suddenly exposed to novel parasites (Parmesan 2006).

We know from studies of birds (e.g. Brown and Brown 1986), mammals (e.g. Hoogland 1995), and lizards (Godfrey et al. 2009) that individuals in larger groups are more likely to have ectoparasites and other directly transmitted pathogens. Thus, changes in ecological factors that influence interaction rates can influence parasitemia. Such changes might be as simple as reducing available habitat, or as complex as changing the distribution of key resources to which individuals are attracted, or key players in the system, that is, individuals with a disproportionate effect on others (Borgatti 2006). In turn, parasetemia can directly influence mortality (Nunn and Altizer 2006), or indirectly influence mortality through making animals more susceptible to predation (Møller and Nielsen 2007). Parasites may change the adaptive value of sociality, and sociality may itself also select for antiparasite behaviour, such as allogrooming that mediates the effects of parasites (Bordes et al. 2007).

Human-mediated changes in the distribution of food and cover (e.g. through habitat fragmentation, habitat conversion and urbanization) can also be important in a social context. In particular, changes in resource distribution may influence the likelihood that individuals interact. Indeed, classic behavioural ecological models of sociality highlight the importance of resource distribution on group living (e.g. Johnson et al. 2002; but see Revilla 2003). If resources are clumped, individuals using those resources may aggregate to harvest them. This is seen in black-shouldered kites *Elanus caeruleus* (Mendelsohn 1988), golden jackals *Canis aureus* (MacDonald 1979), and spotted hyenas *Crocata crocata* (Kruuk 1972); three (of many) species for whom grouping varies intraspecifically and where more individuals are associated around locally dense food.

Clumping can occur along spatial and temporal domains and humans can create artificial clumps through agriculture, habitat fragmentation, or through our concentration of waste at dumps. Consider a fruiting tree that attracts frugivores. One benefit to group-living frugivorous primates (or bats) is that by living socially, individuals in a group are more likely to find these dispersed but important resources (Garber 1987). If resources have a more homogeneous distribution both in space and time, defence costs may exceed any benefits associated from defending them. A well-studied example of just how dynamic these decisions may be comes from a study of pied wagtails *Motacilla alba*. When food (aquatic insects that wash up on the shore) is scarce, individuals defend a territory on both sides of the bank. As food becomes more abundant, territory holders may share their territories with an associate without suffering any costs (Davies 1976). And, if food is superabundant, territoriality breaks down entirely. Humans, through our habitat modifications may thus affect territoriality.

Recent work (López-Sepulcre et al. 2010) illustrates that the relationship between resources, territoriality, and demography is not necessarily simple or straightforward. Seychelles magpie robins *Copsychus sechellarum* live on variable quality territories that are distributed in space. Individual robins compete more for access to the best territories. This territorial competition is 'bad' for the population as a whole because territorial replacements interfere with effective reproduction. Thus, dispersed resources—which are increasingly likely under human-induced environmental change—may be likely to enhance competition, and ultimately reduce population productivity.

Mating systems also illustrate the link between resources and distribution nicely. Here, female dispersion is often influenced by resource distribution. Males, in turn, track females. Thus, if resources are clumped, females may clump around those resources and males may compete for access to females. The underlying logic of this economic defensibility argument (Orians 1969; Bradbury and Vehrencamp 1977; Emlen and Oring 1977) is that we often assume that females, because they produce relatively fewer gametes than males, are a somewhat limiting resource. Importantly, female reproductive success can only increase by enhancing the survival of their young while males could conceivably mate with additional females. Thus, female fitness is strongly linked to resources whereas male fitness is linked to female distribution and abundance.

A nice example of this comes from a comparison of pinniped mating systems. On sea ice, females are widely distributed and males are unable to defend more than one or a few females. By contrast, on beaches, breeding colonies may have hundreds (or thousands) of females and dominant males can (and do) defend large harems (Le Boeuf 1978). Reproduction on these harems is highly skewed (Le Boeuf and Ritter 1988). As year-round sea ice melts out because of climate change (Markus et al. 2009), breeding females will be forced to have their pups on beaches and, if many of them aggregate on the same beaches, we may see greater reproductive skew.

Reproductive skew has genetic and hence potentially demographic consequences, particularly through its effects on a population's genetic heterozygosity (Anthony and Blumstein 2000). In a given year, with a fraction of males reproducing, the population will be more homozygous than in a mating system where all males have an equal probability of breeding. This reduced heterozygosity could enhance the likelihood of population extinction should a new parasite or pathogen infect the population. Indeed, if anything, we need more genetic heterozygosity to enable populations to respond to a variety of anthropogenic assaults, which will create variable environments, than is needed during periods of stasis.

Yet, the opportunity to have multiple mates need not be all bad. Because clumped females may attract more males, and because females may benefit from mating with more than a single male (Gowaty et al. 2010), it is conceivable that changes in resource distribution could influence the frequency of polygyny by females, and this could influence female fitness and hence population size. At this point, such effects are somewhat speculative and need to be studied in more systems.

9.3 Adaptive social behaviour has demographic consequences

I suggest that anything that influences demography—via its impacts on survival or reproduction—

is worthy of study if one wants to understand how to manage populations under human-induced environmental change (Anthony and Blumstein 2000; Blumstein and Fernández-Juricic 2010). Individuals may engage in social behaviour in an attempt to increase their fitness and it is largely through this enhanced survival and reproduction that social behaviour has its demographic consequences. Knowledge of social behaviour can be applied to wildlife management problems to either increase a threatened or endangered population, or to decrease a 'problem' population (Blumstein and Fernández-Juricic 2010).

However, social behaviour is complex and resources may drive aggregation but aggregation may not necessarily be beneficial to the population. Individuals react to other individuals in the environment and by doing so they also may affect the fitness of other individuals. It is important to realize that what is good for an individual may not be good for the population as a whole. Such 'Tragedy of the Commons' (Hardin 1968) may be common—particularly with respect to mating behaviour, where individual decisions may have negative population consequences (e.g. Blumstein 1998) and sexual conflict, where males may reduce female fitness (Rankin et al. 2011). However, as Rankin et al. (2007) point out, such conflicts are not restricted to reproductive behaviour. Indeed, any situation where individuals compete for depletable resources could lead to suboptimal outcomes for a population. For instance, the added value to a solitary animal joining a group might be positive, but beyond the optimal group size, individuals in the group may have their fitness reduced by the addition of extra individuals (Giraldeau 1988). And, when social cooperation has evolved, cheaters reduce the fitness of others (e.g. Rainey and Rainey 2003) and may drive a population extinct.

Animals living together often compete for depletable or patchy resources, and key resources such as nesting sites or burrows may be in short supply (Krause and Ruxton 2002). In many cases, individuals who have an option, opt out of social living. Such systems illustrate the often-facultative nature of sociality and group living. Indeed, observations like this suggest that sociality is sometimes environmentally forced because of resource limitations. This is most notable when we see subordinate animals failing to breed when living socially, but reproducing quite well when not, as seen in a variety of taxa (Brown 1987; Koenig et al. 1992; Brockmann 1997; Solomon and French 1997).

A defining characteristic of many species living in long-term social groups (*contra* those in social foraging aggregations) is the potential for reproductive suppression. Reproductive suppression is seen when potentially fertile females do not breed (Solomon and French 1997), although it also may occur in males, as seen in bluehead wrasse *Thalossoma bifasciatum* (Warner and Swearer 1991). It often emerges when key resources are limited and dominant (often older) individuals monopolize them (e.g. Komdeur 1992). Such resources may include burrow, nest, or shelter sites, or food. Importantly, reproductive suppression, which may reduce genetic variability, is another example of something that may be good for the dominant breeder, but may not be good for the population as a whole.

In an elegant and now classic experiment with Seychelles warblers *Acrocephalus sechellensis*, Komdeur (1992) demonstrated that habitat saturation explained variation in warbler nesting success. When individuals who were reproductively mature but failed to breed were translocated to an empty island, they bred. Until this habitat became saturated, breeding continued. This demonstrates a profound social cost to living in groups.

When anthropogenic disturbances influence the distribution or abundance of resources, the benefits of ability to aggregate will shift—as may the proportion of individuals that breed. Such effects may not be absolute, but socially induced suppression can also delay the onset of reproduction, as has been suggested in yellow-bellied marmots *Marmota flaviventris* (Armitage 2003). The age of first reproduction has profound impacts on individual fitness (Oli and Armitage 2003), and ultimately demography.

Kinship may also influence the adaptive value of sociality and the benefits of engaging in potentially cooperative behaviour that may enhance both individual fitness as well as group productivity. Human-induced habitat changes that affect where animals forage, or how animals group, may directly influence the likelihood that individuals interact with relatives. Consider a hypothetical species that lives in kin groups and defends a group territory. Imagine that suitable habitat that houses multiple groups is drastically reduced by habitat destruction (development, deforestation, desertification, etc.). As pressure to live in the remaining suitable habitat increases, social structure may break down and individuals are more likely to interact with non-kin. If cooperative behaviour is biased towards relatives, interacting with more non-relatives could reduce the benefits obtained by grouping, and thus reduce individual and perhaps group fitness.

9.4 Individual based models link environmental drivers with demographic outcomes

I suggest that the key to understanding how anthropogenic stressors may influence sociality and ultimately demography and population persistence is to build mechanistic models that link environmental drivers of sociality to individual social 'decisions' and then to demographic outcomes of these decisions. To do so, one must first identify the sorts of ecological drivers of sociality. Then, one could build individual based models and conduct sensitivity analyses to better understand the importance of climatic factors on emergent sociality.

Individual based models link behavioural decisions with demographic outcomes (Huston et al. 1988). To build an individual based model one must start with a clear question. These can be varied—but for now let's focus on models that have examined population persistence (e.g. Grimm et al. 2003; Rossmanith et al. 2006) and population regulation (e.g. Ridley et al. 2003).

Given a focused question, it's then important to identify key behavioural decisions that could influence this outcome as well as ecological drivers of these decisions. Such drivers could be specific sources of mortality (e.g. winter mortality, as shown in individual-based models in marmots), and the behavioural decisions could include dispersal (in marmots and Seychelles warblers), the frequency of polyandry (in woodpeckers), or the incidence of reproductive suppression (in Seychelles warblers).

With a set of decisions, it's important to then find a range of parameter values that will drive dynamics in a way that can be studied. Finding these may require a bit of computational work, but is an essential step. When possible, these parameters may be based on real parameter values from real systems (e.g. Grimm et al. 2003; Rossmanith et al. 2006).

Ultimately, the goal of building such a model is to look for emergent dynamics, but the relative importance of environmental drivers on demography can be identified using this approach. Let's review an example to better illustrate the method.

Lesser-spotted woodpeckers *Picoides minor* are typically monogamous, but sometimes mate polyandrously (Rossmanith et al. 2006). Such facultative polyandry is seen when a female has two separate nests, each with a male who helps care for chicks. However, the reproductive successes of the primary and secondary males are different; primary males have higher reproductive success. This polyandry is more likely in years when there are relatively more males in the population and is likely explained because secondary males are making the best of a bad job. At the population level, however, such polyandry is good in that it increases the population's growth rate. Thus, populations are expected to be more likely to persist with polyandry.

The individual-based model that Rossmanith et al. (2006) constructed to study behavioural flexibility of mating systems assumed that there were a series of annual time steps that followed each individual from birth to death. What happened in each time step was described by a series of rules. Such rules quantified the likelihood of a pair persisting, the probability of eggs not hatching, the probability of nest predation, the probability of an individual being killed, the probability of when an individual is killed its nestlings all die, and so on. Demographic noise was introduced by randomly distributed predation. The model was formally set up as a population viability analysis (Beissinger and McCullough 2002) in that the authors were interested in the estimated time to extinction. A sensitivity analysis was conducted where key parameters (such as juvenile survival rate and rate of polyandry) were individually varied and the resultant times to extinction estimated. The key result was that polyandry rate was negatively associated with time to extinction; polyandrous populations persisted longer.

To be particularly useful in predicting responses to anthropogenic change, models must parameterize and study the effects of environmental drivers on sociality. Such mechanistic, individual-based models may be best able to capture this variation and help us understand the links between the environment, sociality, and their consequences.

9.5 Possible consequences in the Anthropocene

Paul Crutzen, the Nobel Prize-winning atmospheric chemist who helped discover the effects of Freon on ozone, along with colleague Eugene F. Stoermer have described our current geological epoch as the Anthropocene because of humanity's profound impact on the environment (Crutzen and Stoermer 2000). Two of the drivers of this impact are anthropogenic climate change and anthropogenic habitat alteration.

Climate change will influence the distribution and abundance of plants and animals, and there will likely be 'winners' and 'losers' in both the shorter term and longer term. For instance, in the short term, we will have more (and more toxic) poison ivy *Toxicodendron radicans* (Mohan et al. 2006), and yellow-bellied marmots (in at least some locations) than before (Ozgul et al. 2010). Yet we also know that global warming is responsible for the local (and ultimately potentially global) extinction of many species and we currently see widespread evidence of changes in evolved phenologies (Parmesan 2006).

To better predict these winner and losers, we must understand the link between environmental variation and demographic success. In many cases the link goes through the social structure or breeding system. For these species, mechanistic models may be useful. Thus, when animals clump over a resource patch, they may be more likely to become diseased, but when living in groups with more males, females may have greater reproductive success through polyandrous matings. The devil of predicting a species' response is in the details of the complex algebra of the costs and benefits of sociality. However, behavioural ecologists are supremely well positioned to study these costs and benefits because of our well-developed toolkit studying the adaptive basis of behaviour.

With care, bioclimatic modelling may be a useful tool in helping to predict range changes for species with well-known thermal and moisture needs (Heikkinen et al. 2006; Jeschke and Strayer 2008). A bioclimatic model helps identify ecological drivers by mapping the current distribution and determining what potential ecological drivers best explain the current distribution. With some assumptions, it is possible to simulate into the future based on anticipated changes in environmental factors like projected rainfall and projected temperature. By creating bioclimatic models for both plants and animals we may find situations where one organism will face a thermal or moisture limit and thus its distribution will be limited, but the climate is nevertheless suitable for another organism. Species that are out of phase with their resources may be especially vulnerable (Parmesan 2006). However, the real utility may come from modelling the distribution of parasites and pathogens and projecting the vulnerability of populations to relatively unknown diseases. Such exposure may change the benefits of living socially, and in extreme cases potentially select for solitary living as an antiparasitic strategy.

The links between parasites and their social hosts is a fascinating one with some economic value. Studies of bovine tuberculosis *Mycobacterium bovis* in both territorial brush-tailed possums *Trichosurus vulpecula* and social (but not cooperative—Dugdale et al. 2010) European badgers *Meles meles* has shown that killing territorial residents in one location (whether they are infected or not) increases the movement of floaters and has the undesired consequence of increased movement of the disease (Smith 2001; Ramsey et al. 2002; Jenkins et al. 2007). Indeed, allowing infected residents to remain may be a superior strategy to widespread lethal control because of the tendency for non-territorial floaters to move into areas with other residents. In this case, it is the social behaviour (territorial defence) that reduces parasite transmission by reducing animal movement.

While there is some uncertainty over the rate of climate change, there is less uncertainty and more control over the rate of deforestation and other anthropogenic habitat modifications. With the possible exception of beavers, *Castor* spp., humans are the only animal that has had such a profound impact on the physical structure of the environment. This has led to homogenous stands of tarmac, trees, and vast monocrops of corn, soy, and potatoes. This extreme lack of habitat heterogeneity has either removed suitable habitat for many species, or created a place where pest species are favoured. Both of these influence the resultant social behaviour of species that relied on the original inhabitants.

9.6 Prospectus

Ecological models of sociality give us some predictive value assuming we know a sufficient amount about the resource needs of a given species. The challenge is to acquire this information. Additionally, while we can develop a laundry list of social factors that influence survival and reproductive success, we are far from developing the sort of broad conceptual understanding and detailed models that allow us to make more general predictions about how changes in the environment will influence a given species. Nor, I suspect, will we ever have many broad insights.

Much of conservation behaviour requires, by necessity, a single-species approach (Blumstein and Fernández-Juricic 2010). This should not be viewed negatively. Rather, by having a detailed and mechanistic understanding of the adaptive basis and value of sociality, we can predict how species will respond to anthropogenic assaults. I suggest that by adopting an individual-based modelling approach, we should be able to study one species at a time and better understand the links between the environment and demography.

Over time, we may be able to use empirical findings to make more general conclusions about the relationship between sociality and vulnerability. Nonetheless, our need for such approaches has never been greater. Behavioural ecologists will have an important role in these developing predictive models, and indeed in helping prevent extinctions (Schroeder et al. 2011).

Acknowledgements

Many thanks to Bob Wong, Ulrika Candolin, and two anonymous reviewers for astute, persistent, and very constructive comments that helped improve this chapter.

References

Anthony, L. L. and Blumstein, D. T. (2000). Integrating behaviour into wildlife conservation: the multiple ways that behaviour can reduce N_e. *Biological Conservation*, 95, 303–15.

Armitage, K. B. (2003). Reproductive competition in female yellow-bellied marmots. In Ramousse, R., Allaine, D., and Le Berre, M. *Adaptive Strategies and Diversity in Marmots*, pp. 133–42. Lyon, International Marmot Network.

Atkinson, C. T. and LaPointe, D. A. (2009). Introduced avian diseases, climate change, and the future of Hawaiian honeycreepers. *Journal of Avian Medical Surgery*, 23, 53–63.

Beissinger, S. R. and McCullough, D. R. eds (2002). *Population Viability Analysis*. Chicago, IL, University of Chicago Press.

Blumstein, D.T. (1998). Female preferences and effective population size. *Animal Conservation*, 1, 173–7.

Blumstein, D. T. (2010). Social behaviour in conservation. In Moore, A. J., Szekely, T., and Komdeur, J. (eds) *Social Behaviour: Genes, Ecology and Evolution*, pp. 654–72. Cambridge University Press, Cambridge.

Blumstein, D. T. and Fernández-Juricic, E. (2010). *A Primer of Conservation Behavior*. Sunderland, MA, Sinauer Associates.

Bordes, F., Blumstein, D. T., and Morand, S. (2007). Rodent sociality and parasite diversity. *Biology Letters*, 3, 692–4.

Borgatti, S. P. (2006). Identifying sets of key players in a social network. *Computational Mathematics and Organizational Theory*, 12, 21–34.

Botham, M. S. and Krause, J. (2005). Shoals receive more attacks from the wolf-fish (*Hoplias malabaricus* Bloch 1794). *Ethology*, 111, 881–90.

Bradbury, J. W. and Vehrencamp, S. L. (1977). Social organization and foraging in emballonurid bats. III. Mating systems. *Behavioral Ecology and Sociobiology*, 2, 19–29.

Bradshaw, G. A. and Bekoff, M. (2001). Ecology and social responsibility: the re-embodiment of science. *Trends in Ecology and Evolution*, 16, 460–5.

Brockmann, H. J. (1997). Cooperative breeding in wasps and vertebrates: the role of ecological constraints. In Choe, J.C. and Crespi, B.J. (eds) *The Evolution of Social*

Behavior in Insects and Arachnids, pp. 347–71. Cambridge, Cambridge University Press.

Brown, C. and Warburton, K. (1997). Predator recognition and anti-predator responses in the rainbowfish, *Melanotaenia eachamensis*. *Behavioral Ecology and Sociobiology*, 41, 61–8.

Brown, C. R. and Brown, M. B. (1986). Ectoparasitism as a cost of coloniality in cliff swallows (*Hirundo pyrrhonota*). *Ecology*, 67, 1206–18.

Brown, J. L. (1987). *Helping and Communal Breeding in Birds*. Princeton, Princeton University Press.

Courchamp, F., Berec, L., and Gascoigne, J. (2008). *Allee Effects in Ecology and Conservation*. Oxford, Oxford University Press.

Cresswell, W. (1994). Flocking is an effective anti-predation strategy in redshanks, *Tringa totanus*. *Animal Behaviour*, 47, 433–42.

Crutzen, P. J. and Stoermer, E. F. (2000). The 'Anthropocene'. *Global Change Newsletter*, 41, 17–18.

Davies, N. B. (1976). Food, flocking and territorial behaviour of the Pied Wagtail (*Motacilla alba yarellii* Gould) in winter. *Journal of Animal Ecology*, 45, 235–53.

Dugdale, H. L., Ellwood, S. A., and Macdonald, D. W. (2010). Alloparental behaviour and long-term costs of mothers tolerating other members of the group in a plurally breeding mammal. *Animal Behaviour*, 80, 721–35.

Eisenberg, J. F. (1981). *The Mammalian Radiations*. Chicago, University of Chicago Press.

Emlen, S. T. and Oring, L. W. (1977). Ecology, sexual selection, and the evolution of mating systems. *Science* 197, 215–22.

Garber, P. A. (1987) Foraging strategies among living primates. *Annual Review of Anthropology*, 16, 339–64.

Giraldeau, L.-A. (1988). The stable group and the determinants of foraging group size. In Slobodchikoff, C. N. (ed.) *The Ecology of Social Behavior*, pp. 33–53. San Diego, Academic Press, Inc.

Godfrey, S. S., Bull, C. M., James, R., and Murray, K. (2009). Network structure and parasite transmission in a group living lizard, the gidgee skink, *Egernia stokesii*. *Behavioral Ecology and Sociobiology*, 63, 1045–56.

Goeden, G. B. (1978). *A Monograph of the Coral Trout*, Plectropomus leopardus (Lacépède). Queensland Fisheries Service, Research Bulletin, No. 1, Brisbane.

Gowaty, P. A., Kim, Y.-K., Rawlings, J., and Anderson, W. W. (2010). Polyandry increases offspring viability and mother productivity but does not decrease mother survival in *Drosophila pseudoobscura*. *Proceedings of the National Academy of Sciences of the USA*, 107, 13771–6.

Grimm, V., Dorndorf, N., Frey-Roos, F., WIssel, C., Wyszomirski, T., and Arnold, W. (2003). Modelling the role of social behavior in the persistence of the alpine marmot *Marmota marmota*. *Oikos*, 102, 124–36.

Hardin, G. (1968). The tragedy of the commons. *Science* 162, 1243–8

Heikkinen, R. K., Luoto, M., Araújo, M. B., Virkkala, R., Thuiller, W., and Sykes, M. T. (2006). Methods and uncertainties in bioclimatic envelope modelling under climate change. *Progress in Physical Geography*, 30, 751–77.

Hinde, R. A. (1976). Interactions, relationships and social structure. *Man*, 11, 1–17.

Hoogland, J. L. (1995) *The Black-tailed Prairie Dog: Social Life of a Burrowing Mammal*. Chicago, University of Chicago Press.

Huston, M., DeAngelis, D., and Post, W. (1988). New computer models unify ecological theory. *BioScience*, 38, 682–91.

Jenkins, H. E., Woodroffe, R., Donnelly, C. A., Cox, D. R., Johnston, W. T., Bourne, F. J., Cheeseman, C. L., Clifton-Hadley, R. S., Gettinby, G., Gilks, P., Hewinson, R. G., McInerey, J. P. and Morrison, W. I. (2007). Effects of culling on spatial associations of *Mycobacterium bovis* infections in badgers and cattle. *Journal of Applied Ecology*, 44, 897–908.

Jeschke, J. M. and Strayer, D. L. (2008). Usefulness of bioclimatic models for studying climate change and invasive species. *Annals of the New York Academy of Science*, 1134, 1–24.

Johnson, D. D. P., Kayes, R., Blackwell, P. G., and Macdonald, D. W. (2002). Does the resource dispersion hypothesis explain group living? *Trends in Ecology and Evolution*, 17, 563–70.

Koenig, W. D., Pitelka, F. A., Carmen, W. J., Mumme, R. L., and Stanback, M. T. (1992). The evolution of delayed dispersal in cooperative breeders. *Quarterly Review of Biology*, 67, 111–50.

Komdeur, J. (1992). Importance of habitat saturation and territory quality for evolution of cooperative breeding in the Seychelles warbler. *Nature*, 358, 493–5.

Krakauer, D. C. (1995) Groups confuse predators by exploiting perceptual bottlenecks: a connectionist model of the confusion effect. *Behavioral Ecology and Sociobiology*, 36, 421–9.

Krause, J. and Ruxton, G. D. (2002). *Living in Groups*. Oxford: Oxford University Press.

Kruuk, H. (1972). *The Spotted Hyena*. Chicago, University of Chicago Press.

Landeau, L. and Terborgh, J. (1986). Oddity and the confusion effect in predation. *Animal Behaviour*, 34, 1372–80.

Laundré, J. W., Hernández, L., and Ripple, W. J. (2010). The landscape of fear: ecological implications of being afraid. *Open Ecology Journal*, 3, 1–7.

Le Boeuf, B. (1978). Social behavior in some marine and terrestrial carnivores. In Reese, E.D. and Lighter, F.J., *Contrasts in Behavior*, pp. 251–79. New York, John Wiley & Sons.

Le Boeuf, B. J. and Reiter, J. (1988). Lifetime reproductive success in northern elephant seals. In Clutton-Brock, T.H. (ed.) *Reproductive Success: Studies of Individual Variation in Contrasting Breeding Systems*, pp. 344–62. Chicago, University of Chicago Press.

López-Sepulcre, A., Kokko, H., and Norris, K. (2010). Evolutionary conservation advice for despotic populations: habitat heterogeneity favours conflict and reduces productivity in Seychelles magpie robins. *Proceedings of the Royal Society of London B, Biological Sciences*, 277, 3477–82.

Lott, D. F. (1991). *Intraspecific Variation in the Social Systems of Wild Vertebrates*. Cambridge, Cambridge University Press.

Macdonald, D. W. (1979). The flexible social system of the golden jackal, *Canis aureus*. *Behavioral Ecology and Sociobiology*, 5, 17–38.

Markus, T., Stroeve, J. C., and Miller, J. (2009). Recent changes in Arctic sea ice melt onset, freezeup, and melt season length. *Journal of Geophysical Research*, 114, C12024, doi:10.1029/2009JC005436.

Mendelsohn, J. (1988). Communal roosting and feeding conditions in blackshoulder kites. *Ostrich*, 59, 73–5.

Mohan, J. E., Ziska, L. H., Schlesinger, W. H., Thomas, R. B., Sicher, R. C., George, K., and Clark, J. S. (2006). Biomass and toxicity responses of poison ivy (*Toxicodendron radicans*) to elevated atmospheric CO_2. *Proceedings of the National Academy of Sciences of the USA*, 103, 9086–9.

Møller, A. P. and Nielsen, J. T. (2007). Malaria and risk of predation: a comparative study of birds. *Ecology*, 88, 871–81.

Nunn, C. L. and Altizer, S. (2006). *Infectious Diseases in Primates: Behavior, Ecology and Evolution*. Oxford, Oxford University Press.

Oli, M. K. and Armitage, K. B. (2003). Sociality and individual fitness in yellow-bellied marmots: insights from a long-term study (1962–2001). *Oecologia*, 136, 543–50.

Orians, G. H. (1969). On the evolution of mating systems in birds and mammals. *American Naturalist*, 103, 589–603.

Ozgul, A., Childs, D. Z., Oli, M. K., Armitage, K. B., Blumstein, D. T., Olson, L. E., Tuljapurkar, S., and Coulson, T. (2010). Coupled dynamics of body mass and population growth in response to environmental change. *Nature*, 466, 482–5.

Parmesan, C. (2006). Ecological and evolutionary responses to recent climate change. *Annual Review of Ecology, Evolution, and Systematics*, 37, 637–69.

Pitcher, T. J. and Parrish, J. K. (1993). Functions of shoaling behaviour in teleosts. In Pitcher, T. J. (ed.) *Behaviour of Teleost Fishes*, pp. 363–439. London, Chapman & Hall.

Prins, H. H. T. and Iason, G. R. (1989). Dangerous lions and nonchalant buffalo. *Behaviour*, 108, 262–96.

Pulliam, H. R. (1973). On the advantages of flocking. *Journal of Theoretical Biology*, 38, 419–22.

Rainey, P. B. and Rainey, K. (2003). Evolution of cooperation and conflict in experimental bacterial populations. *Nature*, 425, 72–4.

Rankin, D. J., Bargum, K., and Kokko, H. (2007). The tragedy of the commons in evolutionary biology. *Trends in Ecology and Evolution*, 22, 643–51.

Rankin, D. J., Dieckmann, U., and Kokko, H. (2011). Sexual conflict and the tragedy of the commons. *American Naturalist*, 177, 780–91.

Ramsey, D., Spencer, N., Caley, P., Efford, M., Hansen, K., Lam, M., and Cooper, D. (2002). The effects of reducing population density on contact rates between brushtail possums: implications for transmission of bovine tuberculosis. *Journal of Applied Ecology*, 39, 806–18.

Revilla, E. (2003). Moving beyond the resource dispersion hypothesis. *Trends in Ecology and Evolution*, 18, 380.

Ridley, J., Komdeur, J., and Sutherland, W. J. (2003). Population regulation in group-living birds: predictive models of the Seychelles warbler. *Journal of Animal Ecology*, 72, 588–98.

Ritchie, E. G. and Johnson, C. N. (2009). Predator interactions, mesopredator release and biodiversity conservation. *Ecology Letters*, 12, 982–98.

Rossmanith, E., Grimm, V., Blaum, N., and Jeltsch, F. (2006). Behavioural flexibility in the mating system buffers population extinction: lessons from the lesser spotted woodpecker *Picoides minor*. *Journal of Animal Ecology*, 75, 540–8.

Schroeder, J., Nakagawa, S., and Hinsch, M. (2011). Behavioural ecology is not an endangered discipline. *Trends in Ecology and Evolution*, 26, 320.

Smith, D. W., Peterson, R. O., and Houston, D. B. (2003). Yellowstone after wolves. *BioScience*, 53, 330–40.

Smith, G. C. (2001). Models of *Mycobacterium bovis* in wildlife and cattle. *Tuberulosis*, 81, 51–64.

Solomon, N. G. and French, J. A. (eds) (1997). *Cooperative Breeding in Mammals*. Cambridge, Cambridge University Press.

Warner, R. R. and Swearer, S. E. 1991. Social control of sex change in the bluehead wrasse, *Thalassoma bifasciatum* (Pisces: Labridae). *Biological Bulletin*, 181, 199–204.

Whitehead, H. (2008). *Analyzing Animal Societies: Quantitative Methods for Vertebrate Social Analysis*. Chicago, University of Chicago Press.

CHAPTER 10

Species interactions

Shelley E.R. Hoover and Jason M. Tylianakis

Overview

The behavioural responses of species to environmental change are affected directly by interactions with other species, and indirectly via linkages across ecological networks. In this chapter we discuss the general mechanisms through which global change drivers can alter interactions among species, then outline the observed effects of environmental change on different types of species interactions. We discuss recent advances in our understanding of the consequences of network architecture for the effects of environmental changes on species interactions, and the interactive effects of multiple drivers acting in concert in natural systems. We conclude that the effects of global change on species interactions are complex and extensive, and there is an urgent need to examine the effects of multiple global change drivers on interactions among species and throughout ecological networks.

10.1 Introduction

Ecosystems worldwide are undergoing rapid and dramatic change (Millennium Ecosystem Assessment 2005). The main drivers of this global environmental change (GEC) are predicted to have increasing impacts as human exploitation of the environment intensifies (Sala et al. 2000). Numerous studies to date have focused on the effects of GEC on species ranges, population abundances, phenology, and physiology (Sala et al. 2000). However, we are only beginning to understand the less-obvious effects of global change on animal behaviour and, in particular, how these behavioural changes affect interactions among species (Tylianakis et al. 2008). Interspecific interactions may be particularly sensitive to environmental change, as they are the product of the behaviour, physiology, phenology, and ranges of multiple species (Suttle et al. 2007; Tylianakis et al. 2008). Importantly, all species are connected indirectly to others via linkages within an ecological network (Tylianakis et al. 2007; Walther 2010), such that behavioural responses of some species to environmental change may have ecosystem-wide effects (see Chapter 13). The structure of complex networks of interactions plays an important role in the maintenance of biodiversity (Bascompte et al. 2006), mediation of community and ecosystem responses to GEC (Brooker 2006; Suttle et al. 2007), and the stability of ecosystem services on which human well-being depends. Yet, changes to this crucial network structure may not be predictable from changes to individual species or pairwise interactions, even in cases where the latter are well known.

We begin this chapter by outlining and illustrating the general mechanisms through which global environmental change may alter pairwise species interactions (Section 10.1, Fig. 10.1). We then provide an overview of the ways in which different interaction types (i.e. mutualism, competition, parasitism, and consumer–resource interactions) respond to GEC drivers (Section 10.2). In Section 10.3 we discuss how network architecture modifies the effects of GEC on species interactions, and the role of networks in determining the community-wide effects of any particular global change driver.

Behavioural Responses to a Changing World. First Edition. Edited by Ulrika Candolin and Bob B.M. Wong.
© 2012 Oxford University Press. Published 2012 by Oxford University Press.

In Section 10.4, we address how interactions between multiple global change drivers alter interspecific interactions in ways that cannot be predicted from studies of single drivers in isolation. Finally, in Section 10.5, we discuss what conclusions can be drawn from current studies and identify critical areas for future research.

10.1.1 General mechanisms of impact

Global change can alter interactions among animals both directly (e.g. through effects on behaviour, range, phenology, and physiology), or indirectly (e.g. through effects on other trophic levels or habitat quality) (Fig. 10.1). The geographic range and phenology of animal species are directly linked to climate, potentially generating temporal and spatial mismatches as climate changes (also see Chapters 5, 6, 14). Biotic invasions by definition involve a change in the range of species, and habitat fragmentation at different scales alters species distributions (Tscharntke and Brandl 2004; Tylianakis et al. 2008). Shifts in range and phenology have two potential impacts on species interactions; first, organisms may be exposed to novel species from which they were formerly spatially or temporally separated, and second, historically important interaction partners may become unavailable. In addition to spatial and temporal range shifts, GEC drivers can directly affect ontogenetic development, potentially altering the strength of key behavioural interactions at different life stages. Finally, environmental

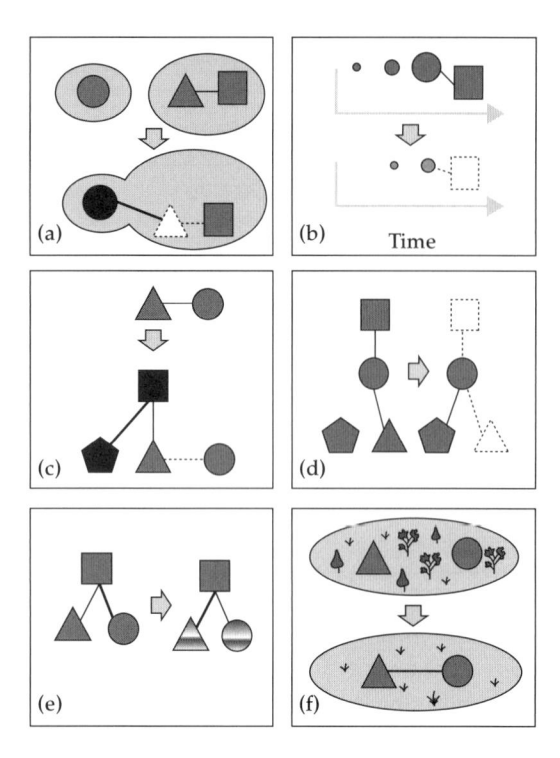

Figure 10.1 Mechanisms through which global environmental changes affect species interactions. Examples of six broad and not mutually exclusive mechanisms through which environmental changes can affect species interactions. Symbols (circles, squares, triangles, pentagons) represent individual species, and links between them indicate an interaction, with the width of the line indicating the strength of the interaction. Grey fill indicates species that are already present in the habitat, black fill indicates species that are new to that location in space or time. Dotted lines and symbols indicate lost interactions or species respectively. Gradient fill represents a change in a species' behaviour with environmental change.

(a) Geographic range shifts. Species may move into or out of the range of their interacting partners: recent warming has facilitated the reintroduction of crabs to the Antarctic Peninsula, and increasing predation is predicted to vastly reduce populations of suspension-feeding echinoderms, homogenizing them with near-shore communities at higher latitudes (Aronson et al. 2009).

(b) Altered phenology. Interactions may be altered by different phenological responses of one or more participants to environmental change: hatching dates of avian predators were found to be uncorrelated with the availability of passerine fledgling prey, whose abundance was determined by their caterpillar prey, which in turn responded to oak budburst (Both et al. 2009).

(c) Introduction of new interaction partners. Species may come into contact with new interacting partners, through invasion or altered spatial/temporal range (see (a) and (b) above): the introduction of feral pigs allowed mainland golden eagles to colonize the Californian Channel Islands. In addition to pigs, eagles ate foxes, reducing competition between foxes and skunks, releasing skunk populations to prey on other animals (Roemer et al. 2002).

(d) Loss of species. The opposite of (c) above, one or more participants may be lost from an interaction following a spatial or temporal range shift or local extinction: loss of large herbivores caused breakdown in an ant–acacia mutualism, because acacia trees no longer provided nectar and shelter, so mutualist ants were replaced by antagonist species (Palmer et al. 2008).

(e) Altered behaviour or physiology of one or more interaction partner. Species may be more or less likely to interact if their behaviour is altered by environmental change: timber harvesting in Australian forests changed the foraging behaviour of two native marsupials, increasing the strength of their interaction with avian predators (Stokes et al. 2004).

(f) Alteration of physical environment. Physical habitat characteristics may facilitate or inhibit interactions, which are then altered when the habitat is modified: reduced habitat complexity during the conversion of forest to open agriculture in coastal Ecuador allowed parasitic wasps and flies to more easily locate their bee/wasp hosts (Laliberté and Tylianakis 2010).

change can affect interactions between species through changes to the physical environment (e.g. habitat structure), which may affect species encounter rates, or by altering the benefits obtained through a specific interaction (e.g. food quality).

10.1.2 Range shifts

Climate-induced range shifts have been discussed elsewhere in this volume (e.g. Chapter 5), though their rate and extent vary markedly among species (Walther et al. 2002). This not only alters the relative abundances of interacting partners in a given area, but also restructures interactions within ecological networks. For example, increasing sea temperature has resulted in shifts in the ranges of invertebrate fauna in the northeast Atlantic. Due to successful recruitment in more frequent warm years, many of the southern barnacle species are extending their ranges northward (Hawkins et al. 2008). However, major range shifts of northern species of barnacles have not been recorded. Rather, they tend to persist in the same location, although at reduced densities. Thus, in areas where their distributions overlap, the community composition has shifted to include both northern and southern barnacle species (Hawkins et al. 2008). This restructuring of the crustacean community is expected to affect primary production of algae, and have considerable impacts on rocky intertidal community structure and subsidies to coastal ecosystems. In addition to latitudinal shifts in species ranges, GEC can also drive altitudinal shifts in the distribution of species. Moritz and colleagues (2008) quantified the effects of nearly a century (1914–2003) of climate change on the distribution of small mammals in Yosemite National Park (USA). Half of the 28 species observed showed a substantial upward shift in their elevation limits (by on average ~500 m), in response to an approximately 3°C increase in minimum temperatures. Low-elevation species expanded their ranges to include higher altitudes, whereas high elevation species contracted their ranges; this led to a change in the species assembly at mid and high elevations. Such changes in community composition can lead to altered interactions among species by providing novel interaction partners in areas where species did not previously overlap, and by altering interaction strength (e.g. predation rates and intensity of competition) where species abundances have changed (Tylianakis et al. 2008).

10.1.3 Temporal shifts

The phenology of organisms, the seasonal timing of life-cycle events, is especially responsive to one of the principal drivers of GEC: change in temperature regime. As each species will respond differently to climate change, it is predicted that the phenological shifts of individual species will cause temporal asynchrony among interaction partners. The most widely cited examples of altered phenology in response to climate change include earlier onset of spring behaviours such as nesting in birds, arrival of migrant birds and butterflies, and spawning in amphibians, which have occurred progressively earlier since the 1960s (reviewed in Walther et al. 2002). Asynchrony in the seasonal responses of adjacent trophic levels may prevent organisms or their offspring from taking advantage of periods of high resource availability (see also Chapter 6).

For example, over a 20 year period from 1985 to 2005, Both et al. (2009) examined shifts in phenology across four levels of a food web that included oak trees, caterpillars, four species of passerine birds feeding on the caterpillars, and sparrowhawks preying on the passerine chicks. While both caterpillar peak biomass and passerine hatching dates advanced significantly, they found no significant changes in the timing of oak budburst or sparrowhawk hatching. The authors found that the phenological response of the consumer to climate change was less than that of the prey at all trophic levels, and concluded that the temporal match between food demand and availability declined over time for the passerines and the sparrowhawk predators (Both et al. 2009), suggesting that higher trophic levels may be more adversely affected.

Complex behaviours such as egg-laying are governed by multiple physiological processes, each of which may respond differently to environmental change drivers. For example, peak abundances of insect prey at the breeding grounds of the pied flycatcher *Ficedula hypoleuca* have advanced with warmer spring temperatures (Both and Visser 2001). Consequently, the flycatcher has advanced its egg-laying date by shortening the time spent at the breeding grounds prior to egg-laying. However, this temporal shift has been insufficient to keep pace with advances in peak insect biomass, because egg-laying is constrained by the timing of migration from their overwintering habitat, which is unaffected by environmental changes at the breeding grounds.

Despite these examples, spring advancement does not always result in asynchrony between interacting partners. For instance, the orange tip butterfly *Anthocharis cardamines* has successfully tracked the phenological shifts of its host plants by appropriately advancing hatching dates (Sparks and Yates 1997). However, further climate change may upset even these systems, as basal producers tend to advance more readily with warming than do consumers. Furthermore, in contrast to the relatively consistent effects of spring events, changes in autumnally timed events have been found to be much more heterogeneous, with some species advancing, delaying, or not changing the timing of life-history events (Walther et al. 2002).

10.1.4 Ontogenetic changes

Behavioural interactions among species often occur at specific life-history stages. Ontogenetic development is often characterized by changes in fundamental traits such as body size, foraging and defence behaviours, which can alter the nature of interactions with other species (Yang and Rudolph 2010). For example, larval hawkmoths *Manduca sexta* are herbivores on the same plant species that they pollinate as adults. Adult moths not only feed on nectar of the flowers of *Datura wrightii*, but they also deposit eggs on the foliage. The emerging larvae subsequently forage on the leaves, changing the nature of the interaction between the two species from mutualism to antagonism (Bronstein et al. 2009). Thus, GEC-driven changes to moth behaviour or plant physiology could alter the balance between mutualism and antagonism. Predator–prey relationships can also change due to ontogenetic shifts of either predators or prey. Juvenile predators often compete with their future prey for resources, while prey can grow into a size refuge, causing a shift from predation to competition through ontogeny (Werner and Gilliam 1984; Yang and Rudolph 2010). These ontogenetic shifts in feeding behaviour can be highly sensitive to environmental change, and may serve as warning of impending large-scale changes to entire ecosystems. Therefore, examining the effects of environmental change on species interactions at a single life stage can be misleading, and the net effects of phenological shifts must be considered over the complete lifetime of organisms (Yang and Rudolph 2010).

10.1.5 Altered behaviour

Finally, behavioural shifts following environmental changes may alter the strength of interactions between species. For example, habitat degradation can reduce refuge availability, thereby increasing prey susceptibility to predators. For example, Stokes and colleagues (2004) found that timber harvesting in Australian eucalypt forests changed the foraging behaviour of two native marsupials, which preferentially forage in habitats with complex physical structure. Predation by avian and mammalian predators on these marsupials is higher in the open habitats, and animals suffer increased mortality when complex habitats with sufficient refuges are unavailable. Thus, direct behavioural responses of species to GEC may alter encounter frequencies and the strength of interactions with other species.

The above examples illustrate that the propensity of species to engage in interactions will clearly depend on the balance of costs (including difficulty of locating interaction partners) to benefits obtained. Below we outline—for several broad classes of interactions—how changes to these costs and benefits have been observed to influence the strength or frequency of interactions.

10.2 Effects of GEC on different types of behavioural interactions

10.2.1 Mutualisms

Mutualistic interactions are generally negatively affected by environmental change (Tylianakis et al.

2008), because negative effects on one partner are frequently to the detriment of the other. Environmental change can affect mutualisms in three general ways: (1) shifts from mutualism to antagonism, (2) switches to new participants, and (3) mutualism loss (Kiers et al. 2010). Global change drivers such as habitat degradation can indirectly alter mutualisms, for example by altering the costs and benefits for each participant. In the ant–*Acacia* system studied by Palmer et al. (2008), *Acacia* trees offer rewards (extra-floral nectar and thorn domatia for shelter) to ant colonies that compete to live in their branches and stems, whereas ants reciprocate by defending against insects and large mammalian herbivores. When large herbivores were lost from the habitat, the *Acacia* trees invested less in food and shelter rewards for ant mutualists, which led to increasing antagonistic behaviour and shifting competitive dominance within the ant community to a non-mutualist species. Trees occupied by non-mutualistic ants suffered increased attack by insect pests, had reduced growth and increased mortality compared with trees occupied by the mutualistic ants (Palmer et al. 2008). Thus, the elimination of large herbivores from the system caused both a switch to increasingly antagonistic behaviours, as well as a shift in competitive dominance that led to a partial loss of the mutualism.

Pollination is a mutualism of particular concern to human populations, as much of the world's food production is dependent on this ecosystem service (Klein et al. 2007). Environmental change can directly affect pollinator behaviour through numerous pathways, including changes to floral displays and rewards, competition with introduced pollinator species, land-use intensification and removal of natural habitat, and climate-induced phenological shifts (Tylianakis et al. 2008). In some cases, exotic species may act as a 'magnet' by attracting pollinators to patches of flowers that contain both native and non-native blossoms (Carvalheiro et al. 2008). However, the seed set of native plants may be reduced when pollinators frequently switch between flower species, and this may reduce the amount of native forage available. For example, pollinators that probe intervening flowers of the invasive purple loosestrife *Lythrum salicaria* can, in the process, lose monkeyflower *Mimulus ringens* pollen, thereby reducing monkeyflower seed set (Flanagan et al. 2010). Therefore, GEC-driven changes to pollinator behaviour may affect the reproductive success of the entire plant community, with cascading effects on the higher trophic levels that depend on basal plant resources.

While the vast majority of studies describe mutualisms that are negatively affected by environmental change, there are some examples of mutualisms that are robust to perturbation or even benefit from changes. For example, seed dispersal of *Prunus africana* in Kenya was found to be higher in fragmented and disturbed sites, even though species richness of frugivorous birds and monkeys that normally disperse seeds was lower (Farwig et al. 2006). Similarly, in Germany, avian seed dispersers were able to increase their flight distance to maintain their mutualism with cherry trees, despite having lower species richness and abundance in areas with higher land-use intensity (Breitbach et al. 2010). It remains uncertain, however, the extent to which animals can alter their behaviour to maintain mutualistic interactions, and whether these behavioural changes will involve only plastic changes, long-term genetic changes to populations, or both.

10.2.2 Competition

Differences in the responses of individual taxa to environmental change may lead to shifts in the competitive balance among species (Tylianakis et al. 2008). For example, colour and body size affect thermal constraints on insects. Pereboom and Biesmeijer (2003) demonstrated that warmer climates favour larger, lighter-coloured bees over smaller, darker bees. Therefore, as temperature regimes shift, bees already adapted to warmer temperatures may gain a competitive advantage. However, to some extent, bees are able to compensate behaviourally for adverse thermal conditions, with dark-coloured bees exhibiting clear preference for sucrose-rich nectar in shaded patches (Biesmeijer et al. 1999).

Global change may also alter competitive interactions among consumers by modifying the nutritional content of a common resource. For example, barnacle geese *Branta leucopsis* prefer to forage on nitrogen-enriched plots where plants have higher nutritional value, and avoid plots with high biomass (Stahl et al. 2006). In contrast, their competitor, the brown hare Lepus europaeus, preferentially forages in areas with both high biomass and high plant quality. However, when both species are present, the hares select areas with high biomass and fewer goose competitors over areas with the highest plant quality (Stahl et al. 2006). Thus, indirect competition, through forage depletion, plays a significant role in the forage choice of the hares. Anthropogenic nitrogen deposition is expected to increase the nutritional content of forage plant species, undoubtedly shifting the competitive balance among many primary consumers (Tylianakis et al. 2008). Other environmental change drivers such as increased atmospheric CO_2 may have similar effects on basal plant resources. However, the response of competitive interactions to change will depend not only on the relative local effects of each global change driver and the response of individual species to food quality, but will also be mediated by interactions with other organisms, such as parasites and predators (Tylianakis et al. 2008).

10.2.3 Parasitism/pathogens

Parasites and pathogens regulate the abundance of their host populations, influence the composition and structure of animal communities, and impact ecosystem stability and functioning (Hudson et al. 2006; Mouritsen and Poulin 2010). Climate change, land-use change, and biotic invasions are all frequently found to increase the parasite load of animals, their exposure to disease, and the availability of vectors (Tylianakis et al. 2008), potentially affecting wildlife, livestock, and even human health.

Changes in temperature can affect host resistance and recovery, as well as pathogen virulence and transmission rates (reviewed in Harvell et al. 2002; Tylianakis et al. 2008). Climate change has been found to facilitate the transmission of nematode parasites in many animal taxa, including mammals (Kutz et al. 2005) and birds (Cattadori et al. 2005). For example, climate warming and increased weather anomalies have been implicated in recent outbreaks of the mosquito-borne filarioid nematode *Setaria tundra*, a parasite of reindeer and moose in Fennoscandia, causing substantial ungulate mortality (Laaksonen et al. 2010). The synchrony between peak periods of nematode infection, activity of the mosquito vector, shedding of adult reindeer fur, and low immunity of new calves promotes the transmission of the parasite from adults to calves. During warm periods, adult reindeer tend to congregate in dense herds in mosquito-rich wetlands, which further contributes to parasite transmission. As such, warming temperatures increase not only the transmission of the parasite, but also the longevity and activity of the intermediate arthropod host and the synchrony between parasites and key life stages of the vertebrate host (Laaksonen et al. 2010). Changes to thermal regimes that alter host–parasite synchrony (by modifying the physiology or behaviour of hosts or vectors) can dramatically alter epidemiological trajectories of parasite populations, with significant consequences for host populations.

Species introductions, a frequent consequence of climate change and anthropogenic disturbance, can influence disease and parasite dynamics in native populations by increasing vector populations, acting as a source population, or by introducing new pathogens to native species. Throughout evolutionary time, parasitic host switching has frequently occurred in conjunction with climate change, biotic range shifts, and altered species interactions (Hoberg and Brooks 2008). Such host switching has been demonstrated for the parasites of many organisms, including muskoxen *Ovibos moschatus* (Kutz et al. 2004), carnivores (Zarlenga et al. 2006), and humans (e.g. Mu et al. 2005). Brooks et al. (2006) reported that since the introduction of American bullfrogs *Rana catesbeiana* to Costa Rica, a lung fluke normally found in bull frogs is now also found in leopard frogs *Rana pipiens*. This has caused substantial leopard frog population declines. Bullfrogs have since gone extinct in Costa Rica but the parasite has established itself in the local host. Thus, the legacy of interaction changes can persist even after an invasive vector has disappeared.

10.2.4 Consumer–resource interactions (predation and herbivory)

Feeding behaviours of herbivores can be altered directly as a consequence of environmental change, indirectly through changes in the behaviour of other species (e.g. predators and competitors), or through changes to plant quality, quantity, or composition (Tylianakis et al. 2008). In general, herbivores increase their feeding rate to compensate for reduced food quality under high CO_2 conditions. For example, Massad and Dyer (2010) found that increasing CO_2, light availability and nutrients all consistently increase herbivory, particularly by generalists. Nitrogen enrichment affects herbivore consumption via the nutritional quality of plants (Cornelissen and Stiling 2006), generally making plants more attractive to herbivores. This can enhance herbivore population sizes, performance, and consumption rates, though these effects can be highly variable (Tylianakis et al. 2008).

Climate warming can affect herbivore behaviour through shifts in host plant preference or range (Tylianakis et al. 2008), or cause spatial and temporal mismatches between plants and herbivores (Visser and Both 2005, but see Sparks and Yates 1997). In highly seasonal environments, herbivore reproduction is timed to coincide with peak resource availability to maximize offspring survival. For example, in the Arctic, development of trophic asynchrony is particularly likely for vertebrate herbivores, because the timing of their reproduction may be cued by changes in day length, whereas plant-growth may be cued by temperature. As mean temperatures have risen in Greenland, the migration of caribou *Rangifer tarandus* has not advanced with plant phenologies at their calving grounds. Between 2002 and 2006, for example, the onset of the plant growing season at the calving ground advanced by 14.8 days, in stark contrast to the meagre 1.3 day advance in the onset of calving (Post and Forchhammer 2008). Offspring mortality has consequently risen, while offspring production has dropped fourfold (Post and Forchhammer 2008). This example emphasizes that that the effects of environmental change on behaviour will depend, to a large extent, on the plasticity of responses and the nature of specific behavioural cues (e.g. photoperiod vs. temperature).

Predation plays a significant role in determining the structure and dynamics of ecological communities, and the nature, frequency, and strength of predator interactions with adjacent trophic levels is frequently altered by environmental change. Drivers of GEC such as climate change, competition from invasive species, and habitat modification can disproportionately affect higher trophic levels (Tscharntke and Brandl 2004), although effects are frequently variable across drivers and taxa (Tylianakis et al. 2008). While loss of predators from ecosystems can potentially benefit herbivore populations, this effect appears to be highly system-specific and dependent on the degree of specialization of the predator (Rand and Tscharntke 2007). For example, some predators are able to move extensively between modified and natural habitats, and even benefit from resources present in modified landscapes (Rand et al. 2006).

Evidence suggests that climate change can influence predator–prey dynamics by altering seasonal patterns of prey abundance, behaviour, and foraging efficiency of both predator and prey. For example, in Norwegian forests, changes to winter precipitation regimes affect the ability of roe deer *Capreolus capreolus* to escape from their lynx predator *Lynx lynx*. Nilsen et al. (2009) demonstrated that even small increases in the snow pack can reduce the escape speed of roe deer, thereby favouring lynx. They predict that, in regions where climate change results in more snow, lynx predation on deer populations will increase due to both the reduced efficiency of the deer escape behaviour and the resultant increase in lynx populations (Nilsen et al. 2009). In contrast to this example, reduced snow cover can indirectly alter predator–prey interactions by changing vegetation structure, thereby affecting prey vulnerability to attack. Over a 20 year study in the high-elevation forests of Arizona (USA), Martin (2007) found that declining snowfall allowed elk *Cervus canadensis* to remain in the study area over the winter. Browsing by elk subsequently reduced the amount of deciduous vegetation available to nesting birds. These vegetation changes, in turn, resulted in increased rates of nest predation which, in some species, was a significant determinant of population size in the subsequent year (Martin 2007).

The introduction of novel predators to ecosystems (through invasion or range shifts) can produce a wide variety of context-dependent ecological outcomes. In addition to their direct effects on prey, introduced predators can themselves be prey for higher trophic levels. They may also displace, compete with, or alter the behaviour of native predators, introduce new pathogens, or induce changes to antipredator behaviour in prey (Snyder and Evans 2006). Introduced prey species can allow predators to colonize new areas. For example, in the California (USA) Channel Islands, the introduction of feral pigs *Sus scrofa* provided an abundant source of food, which allowed mainland golden eagles *Aquila chrysaetos* to colonize the islands (Roemer et al. 2002). However, once on the islands, the eagles also preyed on the endemic island fox *Urocyon littoralis* causing a rapid and dramatic decline in native fox populations. Populations of the competitively inferior spotted island skunk *Spilogale gracilis amphiala* were subsequently released from competition with foxes. Thus, the introduction of pigs indirectly caused eagle predation to replace competition between skunks and foxes as the dominant interaction shaping the island's highest trophic levels (Roemer et al. 2002).

Range expansions of predators also comprise a considerable threat to marine ecosystems. As sea temperatures rise, temperate and sub-polar marine species are shifting their ranges pole-ward, reaching areas such as near-shore Antarctica, which has been inaccessible to these species since the Eocene (Aronson et al. 2009). For example, the reintroduction of shell-breaking crabs to the Antarctic Peninsula, following warming of the Southern Ocean, is predicted to dramatically increase predation and devastate populations of endemic suspension-feeding echinoderms. This will, in turn, homogenize polar communities with near-shore communities at higher latitudes (Aronson et al. 2009).

Although many changes to predator–prey interactions are likely to involve plastic responses, predator behaviour could potentially alter the evolutionary responses of prey to environmental change. To test this, Harmon et al. (2009) showed that behavioural differences between two predatory coccinellid beetles affected the population response of their aphid prey *Acyrthosiphon pisum* to extreme climate events (heat shocks). The attack rates of the two predators differed in relation to aphid abundance, demonstrating that interaction strengths are affected through changes in species densities. Harmon and colleagues found that heat-tolerant strains of aphids increased in frequency during the experiment (indicating environmental selection), but even though predator–prey interactions altered population growth following heat shocks, a model based on field data showed that predator–prey interactions did not affect the evolution of heat-shock tolerance (Harmon et al. 2009).

In addition to altering predator–prey interactions between individual species pairs, the entire structure of food webs can be altered by GEC. This has implications for community stability and for resisting further change (see Section 10.3 below). Recent research provides evidence that predator–prey food webs can be altered by land-use intensification (Tylianakis et al. 2007), habitat fragmentation (Cagnolo et al. 2009; but see Kaartinen and Roslin 2011), and infiltration by exotic species (Henneman and Memmott 2001). The effects of many other GEC drivers remain to be tested. However, the frequent alteration of pairwise interactions could potentially operate together to produce significant effects on network structure. It also remains to be tested how these changes might affect the ecosystem functions carried out within a food web.

10.3 Consequences of network architecture for the effects of GEC on species interactions

The above sections have summarized a number of examples of global change altering pairwise interactions between species, and even altering the structure of the networks in which these interactions are embedded. Yet, the architecture of interaction networks can, itself, confer emergent properties (e.g. stability; Dunne et al. 2002; de Ruiter et al. 2005; Bascompte et al. 2006), that determine its susceptibility to perturbation. These emergent properties cannot be predicted by studies of single species or interactions in isolation, yet they may moderate or mediate the effects of global environmental change on ecosystems.

Mutualistic networks differ significantly in their structure from antagonistic networks (Thébault and Fontaine 2010), the former having a nested structure (specialists tend to interact with generalists) with asymmetric links (interactions) between species (Bascompte et al. 2003, 2006). These attributes make mutualistic networks (such as those between plants and pollinators or seed dispersers) more dynamically stable (Thebault and Fontaine 2010), and resistant to the loss of species and their interactions (Memmott et al. 2004; Bascompte et al. 2006). This is because specialist species are the most likely to go extinct, while the species that depend on them are generalists that can continue interacting with many other species, thereby maintaining vital functions such as pollination (Bascompte et al. 2003). In contrast, antagonistic (e.g. predator–prey) networks tend to have a more compartmentalized structure, with 'cliques' of species that interact frequently with each other, but infrequently with species in other compartments (Thebault and Fontaine 2010). This compartmentalization or 'modularity' can reduce the persistence and resilience of the network (Thebault and Fontaine 2010), and cause rapid fragmentation of the network if key species (that hold species together within a compartment, or compartments together within a network) go extinct (Olesen et al. 2007). Nevertheless, depending on the type of disturbance (i.e. whether it affects certain species or the network as a whole), compartmentalization of networks may increase stability (McNaughton 1978; Krause et al. 2003), as perturbations to one subsection of the network cannot cascade easily across compartments.

The stabilizing effect of a nested interaction structure arises through a degree of redundancy in the functional importance of specialist species, with the bulk of the interactions taking place among the core of generalists (Bascompte et al. 2003). Similarly, weak interactions within a network can serve as a buffer, stabilizing the effects of strong interactions, which cause extreme oscillations in the population dynamics of predators and their prey (McCann et al. 1998). These weak interactions may become stronger at different times in response to, for example, an unusually abundant resource, thereby providing a buffer against population fluctuations of individual resource species within the network (McCann 1998). Thus, the structure of interaction networks should be viewed as spatially and temporally variable (de Ruiter 2005; Laliberté and Tylianakis 2010), with the presence or strength of individual interactions potentially being highly changeable.

These stabilizing properties of interaction networks are particularly important in the context of global change, with different drivers affecting the network in different ways. For example, Marcelo Aizen and colleagues (2008) found that plant–pollinator–network structure was altered by invasive species, with invaders being exceptionally generalist in terms of the species with which they would interact. Invaders also engaged in the most asymmetric interactions, and because interaction strengths and asymmetry can be important determinants of network stability (McCann 1998; Bascompte et al. 2006), invaders may actually generate network structures that are more stable and, hence, more difficult to return to the uninvaded state (Tylianakis 2008).

Unlike invaders, which actually participate in interactions, drivers such as land-use change alter the playing field on which interactions take place. For example, simplification of habitat structure during land-use intensification has been shown to facilitate the location of prey by predators, such that predators find a greater proportion of the prey that they are capable of utilizing (Laliberté and Tylianakis 2010; Fig. 10.1F). A corollary of this was that nearly all potential interactions were realized at all times, and the network became homogenized in space and time (Laliberté and Tylianakis 2010). Network homogenization could reduce the buffering potential provided by a dynamically variable structure. In this case, it was driven in part by changes to predator and prey mean body size (e.g. large predators disperse further, which allows them to locate more of their preferred prey), which can be selectively altered by land-use change (Larsen et al. 2005). In fact, the removal of species with certain traits from the network may even be deliberate. Bascompte and colleagues (2005) showed that combinations of interaction strengths in a marine food web reduce the likelihood of trophic cascades after the overfishing of top predators. However, predators involved in strongly interacting food chains were selectively removed by fishing, potentially generating larger community-wide effects than random predator removal.

Finally, global change drivers such as CO_2 enrichment or nitrogen deposition are likely to affect communities primarily through indirect effects via altered plant growth and physiology (Tylianakis et al. 2008). These cascading indirect effects of changing basal resource quality or quantity have the potential to dramatically alter community structure. The ability of the food web as a whole to withstand such change may depend on the extent to which effects are dispersed across the network, and the ability of consumers to respond rapidly to changing resource availability. Top predators may be particularly important in regard to the latter, as they serve to couple energy channels that arise from distinct basal resources (McCann et al. 2005; Rooney et al. 2006). If one group of resource species (e.g. nitrophilic plants) benefits from a change, such as nitrogen deposition, top predators will need to shift their focus to prey species (i.e. herbivores or their direct predators) that have increased in their abundance in response to the abundant plant resource.

Thus, the structure of interaction networks may not only be affected by global change, but it may be an important determinant of the community-wide effects of any particular global change driver. In particular, effects of extinction of species with key roles in the network (e.g. top predators, highly-connected species) may be disproportionately large, and yet these key species may be particularly sensitive to certain drivers. Even more concerning is the finding that phylogenetic relationships among species in mutualistic networks can be a strong predictor of both their generality and the taxa with which they interact, suggesting that extinctions of certain taxa may cause coextinction cascades of related species, and non-random removal of functional groups from the system (Rezende et al. 2007). Understanding the network context of altered interactions under global change will undoubtedly be a major challenge for the future.

10.4 Interactive effects of multiple drivers on species interactions

The vast majority of studies have examined the effects of single GEC drivers in isolation. However, it is becoming increasingly apparent that many drivers act synergistically or antagonistically on ecological processes, producing effects that differ from those predicted based on results of single-factor experiments. In fact, evidence of higher-order interactions between multiple drivers of environmental change is emerging so frequently that these effects may be as important as those of the individual drivers in isolation (Didham et al. 2007). Research on the effects of single GEC drivers has produced highly-variable results (Tylianakis et al. 2008), highlighting the context-dependency of individual driver effects, which may, in part, be explained by interactions among multiple drivers. While studies of the effects of multiple drivers of GEC remain relatively rare, they are essential if we are to make accurate predictions of the net effects of multiple drivers of GEC, and avoid over- or underestimating the effects of GEC on ecosystems and their services.

Drivers of GEC (e.g. temperature, nitrogen deposition, and CO_2 enrichment) have frequently been shown to have interactive effects on plant physiology, reproduction, and community composition (Ollinger et al. 2002; Shaw et al. 2002; Zavaleta et al. 2003). Yet, despite this, multiple-driver studies on animals and their interactions remain rare (Tylianakis et al. 2008). However, interactive effects of GEC on vegetation will affect animal behaviour indirectly by altering food-resource quality and quantity, as well as habitat structure. For example, elevated CO_2 and temperature independently accelerate flowering in bird's-foot trefoil *Lotus corniculatus*, but their synergistic interaction accelerated the flowering date by a dramatic 16 days. Such substantial, multiplicative effects on plant phenology may outstrip the ability of consumers to keep pace with their food resource (Carter et al. 1997).

In addition to indirect effects mediated via plant nutritional quality, multiple GEC drivers will have direct interactive effects on animals. In particular, the interaction between habitat modification and introduced species occurs so frequently that the effects of the two drivers can be difficult to isolate (Didham et al. 2007). For example, elimination of invasive fire ants *Solenopsis invicta* alone does not restore native ant community structure. Instead, conservation of native species also requires mitigation of the habitat disturbance that simultaneously drives invasion by fire ants and native species decline (King and Tschinkel 2006). Real-world ecosystems are facing many simultaneous drivers of environmental change, yet little is known about how interactive drivers will affect animal interactions, and this greatly hinders prediction of future ecosystem responses.

10.5 Conclusions

The effects of GEC drivers (both in isolation and in combination) on species interactions are complex and substantial. Virtually every species interaction studied thus far has demonstrated a response to one or more drivers of GEC. The ubiquity and magnitude of these effects serves as a warning of future changes to ecosystems that will occur due to the dramatic environmental changes rapidly occurring over much of the planet. In order to facilitate policy decisions, ecologists must be able to make accurate predictions of how GEC will affect ecosystems and their services. Whilst system complexities and higher-order effects of multiple drivers make prediction difficult, general conclusions about how environmental changes will affect specific components of ecosystems can begin to be drawn. However, the responses of many species appear to be context-dependent. The challenge for ecologists lies in determining (1) how biotic and abiotic context influence the magnitude and direction of effects of GEC on biotic interactions, and (2) how modification of pairwise interactions translates into altered ecological networks. Most scenarios of future change do not incorporate interaction networks or their relationships with ecosystem services, resilience, and human well-being. Therefore, there is an urgent need not only to continue to focus on the impacts of global environmental change on the individual species that comprise ecological networks, but also to more intensively study the linkages between them. We must explicitly acknowledge that biotic interactions result in interdependent, non-linear and sometimes abrupt responses to environmental change.

Acknowledgements

Funding was provided by a Rutherford Discovery Fellowship from The Royal Society of New Zealand (to Jason Tylianakis) and The Natural Sciences and Engineering Research Council of Canada (to Shelly Hoover).

References

Aizen M.A., Morales C.L., and Morales J.M. (2008). Invasive mutualists erode native pollination webs. *PLOS Biology*, 6, e31.

Aronson, R.B., Moody R.M., Ivany L.C. et al. (2009). Climate change and trophic response of the Antarctic bottom fauna. *PLOS ONE*, 4, e4385.

Bascompte, J.C., Jordano, P., Melian, C.J., and Olesen J.M. (2003). The nested assembly of plant-animal mutualistic networks. *Proceedings of the National Academy of Sciences of the USA*, 100, 9383–7.

Bascompte, J.C., Jordano, P., and Olesen, J.M. (2006). Asymmetric coevolutionary networks facilitate biodiversity maintenance. *Science*, 312, 431–3.

Bascompte, J.C., Melian, C.J., and Sala, E. (2005). Interaction strength combinations and the overfishing of a marine food web. *Proceedings of the National Academy of Sciences of the USA*, 102, 5443–7.

Biesmeijer, J.C., Richter, J.A.P., Smeets, M.A.J.P., and Sommeijer, M.J. (1999). Niche differentiation in nectar-collecting stingless bees: the influence of morphology, floral choice and interference competition. *Ecological Entomology*, 24, 380–8.

Both, C., Bouwhuis, S., Lessells, C.M., and Visser, M.E. (2009). Climate change and unequal phenological changes across four trophic levels: constraints or adaptations? *Journal of Animal Ecology*, 78, 73–83.

Both, C. and Visser, M.E. (2001). Adjustment to climate change is constrained by arrival date in a long-distance migrant bird. *Nature*, 411, 296–8.

Breitbach, N., Laube, I., Steffan-Dewinter, I., and Böhning-Gaese, K. (2010). Bird diversity and seed dispersal along a human land-use gradient: high seed removal in structurally simple farmland. *Oecologia*, 162, 965–76.

Bronstein, J.L., Huxman, T., Horvath, B., Farabee, M., and Davidowitz, G. (2009). Reproductive biology of *Datura wrightii*: the benefits of a herbivorous pollinator. *Annals of Botany*, 103, 1435–43.

Brooker, R.W. (2006). Plant-plant interactions and environmental change. *New Phytologist*, 171, 271–84.

Brooks, D.R., McLennan D.A., León-Règagnon, and Hoberg E.P. (2006). Phylogeny, ecological fitting and lung flukes: helping solve the problem of emerging infectious diseases. *Revista Mexicana de Biodiversidad*, 77, 225–33.

Cagnolo, L. Valladares, G., Salvo, A., Cabido, M., and Zak, M. (2009). Habitat fragmentation and species loss across three interacting trophic levels: Effects of life-history and food-web traits. *Conservation Biology*, 23, 1167–75.

Carter, E.B., Theodorou, M.K., and Morris, P. (1997). Responses of *Lotus corniculatus* to environmental change. *New Phytologist*, 136, 245–53.

Carvalheiro L. G., Barbosa E. R. M., and Memmott J. (2008). Pollinator networks alien species and the conservation of rare plants: *Trinia glauca* as a case study. *Journal of Applied Ecology*, 45, 1419–27.

Cattadori, I.M., Haydon, D.T., and Hudson, P.J. (2005). Parasites and climate synchronize red grouse populations. *Nature*, 433, 737–41.

Cornelissen T. and Stiling P. (2006). Responses of different herbivore guilds to nutrient addition and natural enemy exclusion. *Ecoscience*, 13, 66–74.

de Ruiter, P.C., Wolters, V., Moore, J.C., and Winemiller, K.O. (2005). Food web ecology: Playing Jenga and beyond. *Science*, 309, 68–71.

Didham, R.K., Tylianakis, J., Gemmell, N.J., Rand, T.A., and Ewers, R.M. (2007). Interactive effects of habitat modification and species invasion on native species decline. *Trends in Ecology and Evolution*, 22, 489–96.

Dunne, J.A., Williams, R.J., and Martinez, N.D. (2002). Network structure and biodiversity loss in food webs: robustness increases with connectance. *Ecology Letters* 5, 558–5567.

Farwig, N., Böhning-Gaese, K., and Bleher, B. (2006). Enhanced seed dispersal of *Prunus africana* in fragmented and disturbed forests? *Oecologia*, 147, 238–52.

Flanagan, R.J., Mitchell, R.J., and Karron, J.D. (2010). Increased relative abundance of an invasive competitor for pollination, *Lythrum salicaria*, reduces seed number in Mimulus ringens. *Oecologia*, 164, 445–54.

Harmon, J.P., Moran, N.A., and Ives, A.R. (2009). Species response to environmental change: Impacts of food web Interactions and evolution. *Science* 323, 1347–50.

Harvell, C.D., Mitchell, C.E., Ward J.R. et al. (2002). Climate warming and disease risks for terrestrial and marine biota. *Science*, 296, 2158–62.

Hawkins, S.J., Moore P.J., Burrows M.T., et al. (2008). Complex interactions in a rapidly changing world: responses of rocky shore communities to recent climate change. *Climate Research*, 37, 123–33.

Henneman, M.L. and Memmott, J. (2001). Infiltration of a Hawaiian Community by Introduced Biological Control Agents. *Science*, 293, 1314–16.

Hoberg, E.P. and Brooks, D.R. (2008). A macroevolutionary mosaic: episodic host-switching, geographical colonization and diversification in complex host-parasite systems. *Journal of Biogeography*, 35, 1533–50.

Hudson, P.J., Dobson, A.P., and Lafferty, K.D. (2006). Is a healthy ecosystem one that is rich in parasites? *Trends in Ecology and Evolution*, 21, 381–5.

Kaartinen, R. and Roslin, T. (2011). Shrinking by numbers: landscape context affects the species composition but not the quantitative structure of local food webs. *Journal of Animal Ecology*, 80, 622–31.

Kiers, T. E., Palmer, T.M., Ives, A.R., Bruno, J.F., and Bronstein J.L. (2010). Mutualisms in a changing world: an evolutionary perspective. *Ecology Letters*, 13, 1459–74.

King, J.R. and Tschinkel, W.R. (2006). Experimental evidence that the introduced fire ant, *Solenopsis invicta*, does not competitively suppress co-occurring ants in a disturbed habitat. *Journal of Animal Ecology*, 75, 1370–8.

Klein, A-M., Vaissiere, B.E., Cane, J.H. et al. (2007). Importance of pollinators in changing landscapes for world crops. *Proceedings of the Royal Society of London B, Biological Sciences*, 274, 303–13.

Krause, A.E., Frank, A.E., Mason, D.M., Ulanowicz, R.E., and Taylor W.W. (2003). Compartments revealed in food-web structure. *Nature*, 426, 282–5.

Kutz, S.J., Hoberg, E.P., Nagy, J., Polley, L., and Elkin, B. (2004) 'Emerging' parasitic infections in Arctic ungulates. *Integrative and Comparative Biology*, 44, 109–18.

Kutz, S.J., Hoberg, E.P., Polley, L., and Jenkins, E.J. (2005). Global warming is changing the dynamics of Arctic host-parasite systems. *Proceedings of the Royal Society of London B, Biological Sciences*, 272, 2571–6.

Laaksonen, S., Pusenius, J., Kumpula, J. et al. (2010). Climate change promotes the emergence of serious disease outbreaks of filarioid nematodes. *EcoHealth*, 7, 7–13.

Laliberté, E. and Tylianakis, J.M. (2010). Deforestation homogenizes tropical parasitoid–host networks. *Ecology*, 91, 1740–7.

Larsen, T. H., Williams, N. M., and Kremen, C. (2005). Extinction order and altered community structure rapidly disrupt ecosystem functioning. *Ecology Letters*, 8, 538–47.

Martin, T.E. (2007). Climate correlates of 20 years of trophic changes in a high-elevation riparian system. *Ecology*, 88, 367–80.

Massad, T.J. and Dyer, L.A. (2010). A meta-analysis of the effects of global environmental change on plant-herbivore interactions. *Arthropod-Plant Interactions*, 4, 181–8.

McCann, K., Hastings, A., and Huxel, G.R. (1998). Weak trophic interactions and the balance of nature. *Nature*, 395, 794–8.

McCann, K.S., Rasmussen, J.B., and Umbanhowar, J. (2005). The dynamics of spatially coupled food webs. *Ecology Letters*, 8, 513–23.

McNaughton, S.J. (1978). Stability and diversity of ecological communities. *Nature*, 274, 251–3.

Memmott, J., Waser, N.M., and Price, M.V. (2004). Tolerance of pollination networks to species extinctions. *Proceedings of the Royal Society of London B, Biological Sciences*, 271, 2605–11.

Millennium Ecosystem Assessment. (2005). *Ecosystems and Human Well-being: Current State and Trends*. Washington, Island Press.

Moritz, C., Patton, J.L., Conroy, C.J., Parra, J.L., White, G.C., and Beissinger, S.R. (2008). Impact of a century of climate change on small-mammal communities in Yosemite National Park, USA. *Science*, 322, 261–4.

Mouritsen, K.N. and Poulin, R. (2010). Parasitism as a determinant of community structure on intertidal flats. *Marine Biology*, 157, 201–13.

Mu, J., Joy, D.A., Duan, J. et al. (2005). Host switch leads to emergence of *Plasmodium vivax* malaria in humans. *Molecular Biology and Evolution*, 22, 1686–93.

Nilsen, E.B., Linnell, J.D.C., Odden, J., and Andersen, R. (2009). Climate, season, and social status modulate the functional response of an efficient stalking predator: the Eurasian lynx. *Journal of Animal Ecology*, 78, 741–51.

Olesen, J.M., Bascompte, J.C., Dupont, Y.L., and Jordano, P. (2007). The modularity of pollination networks. *Proceedings of the National Academy of Sciences of the USA*, 104, 19891–6.

Ollinger, S.V., Aber, J.D., Reich, P.B., and Freuder, R.J. (2002). Interactive effects of nitrogen deposition, tropospheric ozone, elevated CO2 and land use history on the carbon dynamics of northern hardwood forests. *Global Change Biology*, 8, 545–62.

Palmer, T.M., Stanton, M.L., Young, T.P., Goheen, J.R., Pringle, R.M., and Karban, R. (2008). Breakdown of an ant-plant mutualism follows the loss of large herbivores from an African savanna. *Science*, 319, 192–5.

Pereboom, J.J.M. and Biesmeijer, J.C. (2003). Thermal constraints for stingless bee foragers: the importance of body size and coloration. *Oecologia*, 137, 42–50.

Post, E. and Forchhammer, M.C. (2008). Climate change reduces reproductive success of an Arctic herbivore through trophic mismatch. *Philosophical Transactions of the Royal Society B:, Biological Sciences*, 363, 2367–73.

Rand, T.A. and Tscharntke, T. (2007). Contrasting effects of natural habitat loss on generalist and specialist aphid natural enemies. *Oikos*, 116, 1353–62.

Rand, T.A., Tylianakis, J.M., and Tscharntke, T. (2006). Spillover edge effects: the dispersal of agriculturally subsidized insect natural enemies into adjacent natural habitats. *Ecology Letters*, 9, 603–14.

Rezende, E.L., Lavabre, J.E., Guimaraes, P.R., Jordano, P., and Bascompte, J. (2007). Non-random coextinctions in phylogenetically structured mutualistic networks. *Nature*, 448, 925–8.

Roemer, G.W., Donlan, C.J., and Courchamp, F. (2002). Golden eagles, feral pigs, and insular carnivores: How exotic species turn native predators into prey. *Proceedings of the National Academy of Sciences of the USA*, 99, 791–6.

Rooney, N., McCann, K., Gellner, G., and Moore, J.C. (2006). Structural asymmetry and the stability of diverse food webs. *Nature*, 442, 265–9.

Sala, O.E., Chapin, F.S. III, Armesto, J.J., et al. (2000). Global biodiversity scenarios for the year 2100. *Science*, 287, 1770–4.

Shaw, M.R., Zavaleta, M.R., Chiariello, N.R. et al. (2002). Grassland responses to global environmental changes suppressed by elevated CO2. *Science*, 298, 1987–90.

Snyder, W.E. and Evans, E.W. (2006). Ecological effects of invasive arthropod generalist predators. *Annual Review of Ecology, Evolution, and Systematics*, 37, 95–122.

Sparks, T.H. and Yates, T.J. (1997). The effect of spring temperature on the appearance dates of British butterflies 1883–1993. *Ecography*, 20, 368–74.

Stahl J., Van Der Graaf A.J., Drent R.H., and Bakker J.P. (2006). Subtle interplay of competition and facilitation among small herbivores in coastal grasslands. *Functional Ecology*, 20, 908–15.

Stokes, V.L., Pech, R.P., Banks, P.B., and Arthr, A.D. (2004) Foraging behaviour and habitat use by *Antechinus flavipes* and *Sminthopsis murina* (Marsupialia: Dasyuridae) in response to predation risk in eucalypt woodland. *Biological Conservation*. 117, 331–42.

Suttle, K.B., Thomsen, M.A., and Power, M.E. (2007). Species interactions reverse grassland responses to changing climate. *Science*, 315, 640–2.

Thébault, E. and Fontaine, C. (2010). Stability of ecological communities and the architecture of mutualistic and trophic networks. *Science*, 329, 853–6.

Tscharntke, T. and Brandl, R. (2004). Plant-insect interactions in fragmented landscapes. *Annual Review of Entomology*, 49, 405–30.

Tylianakis, J.M., (2008). Understanding the Web of Life: the Birds, the Bees and Sex with Aliens. *PLOS Biology*, 6, e47. doi: 10.1371/journal. pbio.0060047.

Tylianakis, J.M., Didham, R.K., Bascompte, J., and Wardle, D.A. (2008). Global change and species interactions in terrestrial ecosystems. *Ecology Letters*, 11, 1351–63.

Tylianakis, J.M., Tscharntke, T., and Lewis, O.T. (2007). Habitat modification alters the structure of tropical host-parasitoid food webs. *Nature*, 445, 202–5.

Visser, M.E. and Both, C. (2005). Shifts in phenology due to global climate change: the need for a yardstick. *Proceedings of the Royal Society of London B, Biological Sciences*, 272, 2561–9.

Walther, G.-W. (2010). Community and ecosystem responses to recent climate change. *Philosophical Transactions of the Royal Society B:, Biological Sciences*, 365, 2019–24.

Walther, G.-W., Post, E., Convey, P. et al. (2002). Ecological responses to recent climate change. *Nature*, 416, 389–95.

Werner, E.E. and Gilliam, J.F. (1984). The ontogenetic niche and species interactions in size-structured populations. *Annual Review of Ecology and Systematics*, 15, 393–425.

Yang, L.H. and Rudolf, V.H.W. (2010). Phenology, ontogeny and the effects of climate change on the timing of species interactions. *Ecology Letters*, 13, 1–10.

Zarlenga, D.S., Rosenthal, B.M., La Rosa, G., Pozio, E., and Hoberg E.P. (2006). Post-Miocene expansion, colonization, and host switching drove speciation among extant nematodes of the archaic genus *Trichinella*. *Proceedings of the National Academy of Sciences of the USA*, 103, 7354–9.

Zavaleta E.S., Shaw, M.R., Chiariello, N.R., Mooney, H.A., and Field, C.B. (2003). Plants reverse warming effect on ecosystem water balance. *Proceedings of the National Academy of Sciences of the USA*, 100, 9892–3.

PART III
Implications

CHAPTER 11

Behavioural plasticity and environmental change

Josh Van Buskirk

⮊ Overview

Most organisms exhibit phenotypic plasticity as an evolved response to environmental variation; hence there is widespread hope that adaptive plasticity might lessen the detrimental impacts of environmental change on individuals and populations. Here, I discuss the special role that plasticity in behaviour can play under rapid environmental change. This role arises because behavioural modes of plasticity are common and permit relatively rapid and reversible responses to changing conditions. Key issues are the degree to which behavioural plasticity improves individual fitness, and the impact that plasticity has on population persistence. Behavioural plasticity is quite often beneficial for individuals, and in some cases accounts for most of the observed phenotypic response to environmental change. However, there are several reasons for expecting that maladaptive behavioural plasticity may be especially important in the context of anthropogenic impacts, and the many reports of ecological traps suggest that maladaptive responses to environmental changes caused by humans are widespread. The population-level consequences of plasticity are not well studied, but limited evidence suggests that behavioural responses have enabled persistence or reduced population declines. I conclude by highlighting several topics on which further research is needed.

11.1 Introduction

The Earth is presently committed to dramatic anthropogenic environmental change. Large increases in greenhouse gas emissions and warming of land and oceans are projected to occur even if construction of new carbon-producing infrastructure were to halt immediately (Davis et al. 2010). These changes will have dramatic impacts on biological diversity (Pereira et al. 2010). Indeed, quantitatively important effects on the phenology of populations, geographic distributions, and the structure of local communities have already been observed in numerous taxa (Parmesan 2006).

Some populations may fail to adjust as environmental conditions move outside the range of tolerance of individuals, and therefore will face extinction at least locally (Schwartz et al. 2006). But persistence in the face of rapid change is likely to be common. Organisms exhibit three kinds of response to environmental change that potentially facilitate persistence: dispersal to follow moving environmental conditions, evolutionary (genetic) adaptation, and phenotypic plasticity (Davis et al. 2005, Hoffmann and Willi 2008; Chevin et al. 2010). Of course, the three modes are not entirely independent: plasticity will evolve as the environment changes if selection indirectly targets reaction norms or if trait means are correlated with plasticity, and shifts in geographic distributions will modify selection for evolutionary adaptation.

Plasticity has attracted special attention in the context of environmental change because dispersal and adaptation are often limited. Dispersal can be obstructed by habitat discontinuities, especially in

Behavioural Responses to a Changing World. First Edition. Edited by Ulrika Candolin and Bob B.M. Wong.
© 2012 Oxford University Press. Published 2012 by Oxford University Press.

landscapes with high human population density (Thomas 2011). Evolutionary adaptation to environmental change is an ongoing process in all populations, but may at times be unable to keep pace with high rates of environmental change (Gomulkiewicz and Holt 1995; Chevin and Lande 2010). This is especially true in organisms with a long generation time or low reproductive rate (Lynch and Lande 1993; Reznick and Ghalambor 2001). In a few cases, evolutionary adaptation is slow because genetic variation is nearly absent, or is limiting specifically in the direction undergoing selection (Blows and Hoffmann 2005; Willi and Hoffmann 2009). Thus, phenotypic plasticity may enable adjustment to changing environments even when dispersal is impossible and genetic variation is insufficient to accommodate the rate of change.

11.1.1 The special role of behavioural plasticity

In this chapter, I evaluate the special role of plasticity in behaviour under rapid transition in the environment, including the kinds of changing conditions expected under anthropogenic climate change. This role, as articulated by West-Eberhard (1989), arises in part because behavioural responses to changing conditions are so often adaptive. Behaviour has evolved particularly high levels of adaptive plasticity because it is less susceptible to two sources of selection that frequently oppose the evolution of plasticity. One is a fitness cost arising from the time lag of expression. Models show that the period of time between a change in the environment and the expression of an adaptive inducible character creates a cost that opposes the evolution of plasticity (Moran 1992; Padilla and Adolph 1996). The longer this lag, the longer the period of time during which the organism has not yet produced an appropriate phenotype, and the greater the probability that the environment will change once again and render the expressed phenotype inappropriate. This cost is usually small for behaviour because responses are relatively rapid, sometimes nearly instantaneous (West-Eberhard 1989). For example, prey can retreat immediately to refuge when they detect a predator. Physiological or morphological modes of predator-induced plasticity may be equally effective, but they take somewhat longer to develop and are therefore somewhat more costly (Padilla and Adolph 1996). A side-effect of the lability of behaviour is that a relatively wide range of environmental features is available to act as cues that actuate behavioural plasticity. Many cues can indicate the current state of the environment, whereas only a more limited set of cues reliably predicts conditions in the future.

There is also a second source of selection against plasticity that is of reduced importance for behaviour. This arises from the reversibility of character expression. According to theory, plasticity is more likely to evolve when reversibility is possible, especially under fine-grained temporal variation (Gabriel 1999). Characters that, once deployed, cannot be reversed are susceptible to being trapped in an inappropriate state if the environment changes once again. Behavioural traits tend to be highly labile in fluctuating environments (West-Eberhard 1989). In the example of prey hiding within a refuge, once the predator has moved away the prey can immediately move back out of the refuge and recommence foraging. For these reasons, plasticity in behaviour is taxonomically widespread, affects all sorts of activities, and is induced by an especially broad range of kinds of environmental variability and cues.

One reason that behavioural plasticity could be important under rapid environmental change is that conditions can change in a saltational rather than gradual fashion. Over the next century, temperatures will certainly increase in many parts of the Earth, and specific regions will become wetter or drier. Long term genetic responses to these changes will often occur. But conditions will also exhibit intermittent reversals, sudden switches, and extreme events (Easterling et al. 2000; Min et al. 2011). Behavioural plasticity confers an advantage under these conditions because it is relatively rapid, always induced within a single generation and sometimes within a matter of seconds or minutes. At the same time, plasticity can reduce the evolutionary response to selection under variable environments by uncoupling the phenotype from the genotype (West-Eberhard 2003, p. 178). The efficacy of selection is reduced by any mechanism that decreases the correlation between selected phenotypes and breeding values (Falconer and MacKay 1996, p. 243). However, less efficient selection and a weaker response may prove beneficial under some kinds of environmental change: high levels of genetic variation turn out to be detrimental for mean population fitness under some forms of fluctuating selection. When the environment is highly variable, and especially when variation is stochastic, an evolutionary response to shifting conditions reduces population mean fitness because the current phenotypic optimum does not predict the future optimum (Kawecki 2000). Thus, a plastic response can be preferable to an evolved response when the environment changes unpredictably and non-directionally. If plasticity is adaptive, then the organism is ensured of an immediate improvement in fitness without being committed to expression of any specific phenotype under an uncertain future environment.

Another reason for optimism about the contribution of behavioural plasticity is that it is usually already present in the population. Indeed, phenotypic plasticity in general is very common in most organisms, involving all sorts of behavioural, morphological, and physiological characters (Travis 1994). Presumably, this is because organisms have been exposed to heterogeneous environments over long periods of time, and plasticity has evolved as an adaptive response to variation (Bradshaw 1965; Stearns 1989). Moreover, variation in environmental conditions at the scale of days and months is often much greater than the long-term trends, so we may naïvely expect that plastic responses evolved under fluctuations in conditions will be adequate to deal with directional change in the environment. Thus, there is widespread hope that plasticity already present within populations will allow organisms to cope with impending environmental change (Hendry et al. 2008; Chevin et al. 2010).

11.1.2 Potential fitness effects of behavioural plasticity

A central question here is: can we expect behavioural plasticity to alleviate negative impacts of environmental change on populations? The key issue is whether plasticity improves individual fitness enough to maintain positive population growth. In other words, is plasticity adaptive and is the magnitude of plasticity sufficient to accommodate changing conditions (Fig. 11.1)? Ghalambor et al. (2007) pointed out that plasticity need not be perfect to be beneficial. They defined non-adaptive plasticity as that which makes the organism less fit in the new environment than it would have been had it expressed no plasticity at all. But, between non-adaptive plasticity and the optimal response, there is plenty of scope for responses that are beneficial but not optimal (shaded region in Fig. 11.1). Incomplete, beneficial plasticity can produce a quantitative improvement in fitness and thereby greatly reduce the chance of extinction under rapid environmental change (Chevin and Lande 2010). This leads to three categories of behavioural response to novel environments: optimal, beneficial, and maladaptive. The question is: how often are behavioural responses to changing environments at least beneficial?

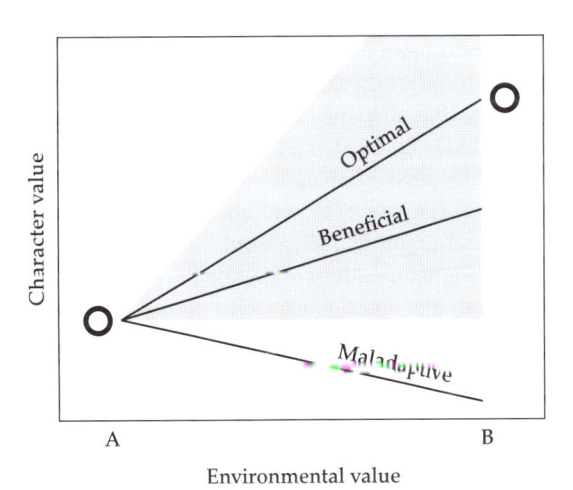

Figure 11.1 Optimal, beneficial, and maladaptive plasticity in response to environmental change. A and B are two environments, and lines illustrate three kinds of reaction norm. The vertical axis represents a behavioural character, and character values that confer maximal fitness are indicated by open circles. Optimal plasticity, in the strict sense, occurs when the individual or genotype exhibits the optimal behaviour in both environments. Beneficial plasticity occurs when the reaction norm confers higher fitness in environment B than a non-plastic genotype would confer; many possible beneficial reaction norms can occur within the shaded region. Maladaptive plasticity causes lower fitness in environment B than if there were no plasticity.

We cannot simply assume that plasticity induced by environmental change will improve individual fitness. Not all forms of plasticity are beneficial for the individual, and some good examples of maladaptive plasticity are induced by variation in temperature and precipitation, the kinds of environmental variation associated with climate change. For instance, plant responses to drought are usually beneficial in the sense that they reduce water loss, but they also cause reduced investment in growth, reproduction, or defence (Holzer et al. 1988; Desprez-Loustau et al. 2006). The response can be especially costly in the presence of insect herbivores. Another example is poikilothermic animals that exhibit higher development rates and smaller body sizes when subjected to warm conditions (Angilletta 2009, p. 158). This response could also be costly if fitness is related to body size, or if rapid early growth is negatively correlated with later performance or structural quality (e.g. Munch and Conover 2003). In both cases, phenotypic plasticity may have evolved as an adaptive response that is effective under the usual inducing environments, heat or drought. But both are involved in trade-offs that become costly under specific circumstances.

Plasticity induced by temperature could be especially likely to be detrimental because it is often caused by unavoidable properties of physiological systems. The case of development rate cited above is a good example: more rapid development with increasing temperature in poikilotherms is to some extent an inevitable consequence of the Boltzmann factor describing the temperature dependence of biochemical reaction kinetics (Gillooly et al. 2001). But a higher development rate in these circumstances is not necessarily adaptive. In fact, the widespread evolution of reduced temperature-specific development rates within populations inhabiting warm climates, known as counter-gradient variation (Levins 1969; Conover et al. 2009), indicates that ever-faster development under warm conditions is not optimal. Depending on the relative magnitudes of plasticity and evolved population differences, temperature-induced plasticity in development could be either beneficial or maladaptive (Fig. 11.2). Examples of strongly maladaptive responses to temperature, similar to that in Fig. 11.2C, have been detected in *Drosophila* (James et al. 1995; Azevedo et al. 1998). Weakly maladaptive responses, as in Fig. 11.2B, are probably common. It is not known whether these scenarios apply to behavioural traits as well. I am aware of only two studies that could have detected counter-gradient variation in behaviour, and in these cases it was not important (Lindgren and Laurila 2005; Laurila et al. 2008).

Another reason to question whether behavioural plasticity will be beneficial under environmental change is that the environments and species interactions experienced by many organisms in the future could be quite different from those currently encountered (Williams and Jackson 2007). Even if plasticity is known to be adaptive under current conditions, this is no guarantee that responses to completely novel conditions in the future will also be beneficial. In general, predictions about the adaptive basis of plasticity cannot be extrapolated beyond the range of environments in which plasticity evolved (de Jong 2005; Ghalambor et al. 2007). Wilczek et al. (2010) point out that better information about the mechanisms underlying responses to changing environments might be helpful for anticipating reactions to future environments, including climate change. Indeed, behaviour may be fundamentally different from life history traits such as growth and development because the underlying physiology is less temperature-sensitive. Nevertheless, these examples instruct us to at least consider that plasticity, especially when induced by temperature, may not be beneficial.

There are known neurobehavioural mechanisms that could account for direct impacts of environmental change, and climate change in particular, on behaviour, and these need not be favourable for affected individuals. For example, brain development and adult behaviour are altered by temperature during early development in insects and some vertebrates (Crews 2003). In honeybees *Apis mellifera* developmental temperature affects the size and number of microglomeruli within the mushroom bodies in the central nervous system (Groh et al. 2004). This area of the brain is involved in the memory and learning that underlies complex behavioural tasks, and temperature-induced plasticity is

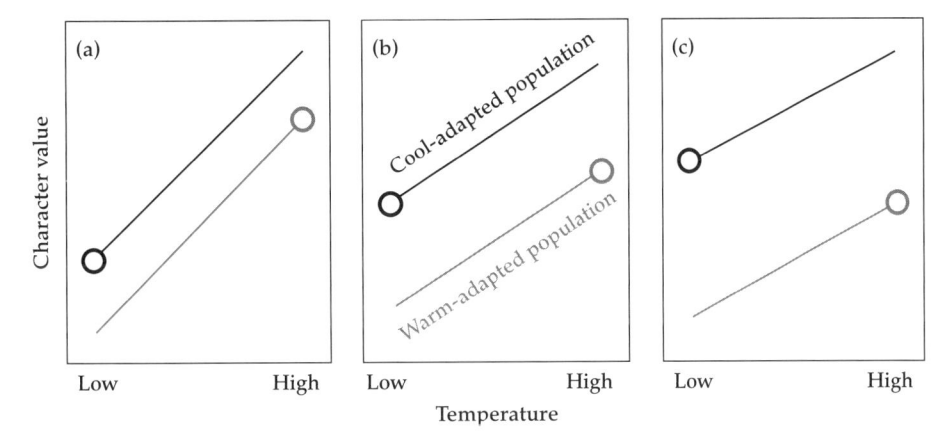

Figure 11.2 Maladaptive plasticity can be caused by counter-gradient variation with respect to temperature. Lines illustrate reaction norms measured in a common garden experiment for two populations originating from warm and cold climates. Open circles represent character values expressed by the two populations when reared under temperatures mimicking their source localities. In panels (a) and (b), genetic divergence between populations partly compensates for environmental effects on the phenotype; in panel (c), genetic divergence overcompensates for temperature-induced plasticity (Conover et al. 2009). Plasticity is inferred to be maladaptive in panels (b) and (c) because organisms originating from both populations experience lower fitness in the foreign environment than they would had they exhibited no plasticity. The vertical axis could represent any character, but known examples of overcompensatory counter-gradient variation involve development rate (e.g. Azevedo et al. 1998).

connected directly to behaviour later in life (Tautz et al. 2003). These findings describe developmental mechanisms that could plausibly connect environmental changes expected under climate change with behaviour. The responses to typical variation in temperature may or may not be adaptive, and the impacts of more extreme conditions on neuronal development and behaviour are not known (see also Chapter 4).

These arguments are all rather theoretical, and they may conflict with our intuition and experience that animals typically respond appropriately to changing conditions. What does the empirical evidence suggest? Below I review studies illustrating that behavioural plasticity associated with environmental change is rarely optimal and often not even beneficial.

11.2 Assessing the fitness consequences of behavioural plasticity

11.2.1 Optimal plasticity

The condition of optimality is difficult to demonstrate for any trait, and optimal phenotypic plasticity is perhaps more difficult to establish than for static traits. The best evidence would come from consistent stabilizing selection in more than one environment favouring the maintenance of the behaviour at its observed values in each. This may not occur very often. However, Charmantier et al.'s (2008) study of the timing of avian breeding appears to be an example of a plastic response to variable weather conditions that maintains the population mean phenotype near the optimum regardless of the temperature. The population of great tits *Parus major* at Wytham Woods near Oxford, UK, has advanced its mean egg-laying date by 12 days since the early 1960s. This amounts to about 0.074 days per °C in summed maximum daily temperatures after 1 March. Individual females have responded to year-to-year variation in temperature by laying eggs 0.071 days earlier per °C. Thus, plasticity by individual birds in the population has been entirely sufficient to explain the observed trend in breeding phenology. But this alone does not demonstrate that plasticity is optimal. Annual estimates of selection acting on reproductive timing have not changed through time, which indicates that the plastic response has, so far, kept pace with the rate of environmental change. This situation is apparently not general, because other long-term estimates of selection on behaviour sometimes show temporal trends, which indicates that the optimum is increasingly different from the behaviour expressed by most individuals (e.g. Visser et al. 1998) (see also Chapter 6).

11.2.2 Beneficial plasticity

Beneficial plasticity improves individual fitness, but nevertheless falls short of the optimal response most of the time. It is relatively easy to test whether a reaction norm improves individual fitness, but, as in the case of optimal plasticity discussed above, more difficult to determine whether it is optimal or only beneficial. Indirect evidence for beneficial plasticity comes from the many quantitative genetic experiments that observe variation among genotypes within populations in their behavioural responses to a standard environmental manipulation (e.g. de Meester 1996; Bell and Sih 2007). These examples relate to predator-induced behavioural plasticity, and the average response is known to enhance individual fitness in the presence of predators. But the existence of so much standing variation can be interpreted as indicating that the optimum is not attained by many individuals. Indeed, most explanations for the maintenance of genetic variation in plasticity imply that many individuals fail to express optimal responses in any specific situation (de Jong and Gavrilets 2000; Sutter and Kawecki 2009).

The best examples of beneficial behavioural plasticity, this time in the context of climate change, come from long-term studies of birds (see also Chapter 6). This is because birds are sufficiently long-lived that repeated observations of the behaviour of individuals are possible, thus providing direct estimates of plasticity. Moreover, the fitness consequences of these behaviours can be estimated. In a well-known study, Both and Visser (2001) discovered that a population of pied flycatchers *Ficedula hypoleuca* in the Netherlands initiated breeding earlier as local spring temperatures increased. Individual birds achieved this by shortening the time between the date of their arrival from the wintering area in Africa and the date of egg-laying. This response presumably involved a strong behavioural component, although physiological adjustments must also have been involved. The majority of the phenological shift was due to plasticity because individual birds adjusted their breeding date from one year to the next in conjunction with annual fluctuations in temperature. In spite of plasticity, selection favouring early breeding also increased in strength over the same time period. In 1980, the directional selection differential targeting breeding date was only slightly negative (about −0.05 SD units), but by 2000 the differential had shifted to roughly −0.4. This suggests that the phenological response of the birds was falling behind the changing optimal breeding timing determined by the environment. Thus, pied flycatchers responded in the right direction, and the response was primarily due to plasticity, but the reaction was not sufficient to keep pace with climate change. Many cases of adaptive responses to climate change probably fall within this category.

There are other examples in which a change in behaviour associated with environmental change can be ascribed primarily to phenotypic plasticity. Recent shifts in the reproductive behaviour of amphibians are almost certainly due to facultative responses of individuals to warmer temperatures. European common frogs *Rana temporaria* move toward ponds to breed in early spring, and these movements have begun occurring earlier in recent years (Beebee 1995; Tryjanowski et al. 2003; Lappalainen et al. 2008; Scott et al. 2008). The question is: how much of this change is due to individual-level plasticity? Data presented by Phillimore et al. (2010) provide a basis for answering this question. Using observations compiled by the UK Phenology Network, Phillimore et al. estimated the magnitude of change in breeding date within populations that was associated with year-to-year variation in local temperature. This change, about three days advancement per 1 °C change in mean temperature during late winter, can be interpreted as entirely due to plasticity. I applied the estimate of Phillimore et al. to published reports of long-term trends in the breeding date of *R. temporaria* populations elsewhere in Europe, and discovered good agreement between the observed trends and the change expected if only plasticity were occurring (Table 11.1). The percentage of observed change explained by plasticity was high across the three studies in Table 11.1. Although we can assume that earlier breeding during warm years is beneficial for frogs, we do not know whether the response has been sufficient to keep up with the rate of environmental change.

In these examples, most of the behavioural response observed over many decades can be ascribed to plasticity. This may not be the usual situation. Plasticity accounts for only a fraction of the long-term change in spring arrival timing in numerous species of birds sampled at a ringing site in eastern North America (Van Buskirk et al. 2012). The date of arrival of 32 species of locally breeding birds has advanced by an average of 0.091 ± 0.048 (95% CI) days · year^{-1} over a period of 46 years (Van Buskirk et al. 2009). We estimated phenotypic plasticity in individual arrival date by recording how the first capture of marked birds changed with the temperature recorded during the 30 days just prior to arrival, using over 2500 individuals recaptured in multiple years. These plasticity estimates were then used to calculate the fraction of the observed phenological change that could be due to plasticity, given the regional temperature trend over the period of study. This method was analogous to that employed in Table 11.1. The data suggest that only about 23% ± 18% (95% CI) of observed change in arrival was due to plasticity. An even smaller value was obtained when temperature data came from a larger area encompassing the spring migratory pathway through southeastern North America. The majority of phenological change observed in these species was presumably due to a genetic response to selection acting on arrival time. Unlike Both and Visser (2001), we cannot test this hypothesis because data on the reproductive success of individuals are not available and therefore direct estimates of selection on arrival date cannot be calculated. Nevertheless, it is clear that behavioural plasticity occurs in the beneficial direction in this collection of species, but is considerably less than what would be favoured by the rate of environmental change.

What are the population-level consequences of beneficial plasticity? Incomplete, but beneficial, behavioural responses could be important for preventing population declines associated with environmental change. A possible example is Møller et al.'s (2008) analysis of migration phenology and population trends in European birds. Species that have shown the greatest advancement of spring migration timing in recent decades have also experienced the smallest reductions in population size during the same period. Although the causal association is uncertain, Møller et al.'s interpretation was that a beneficial response to warmer conditions allows individuals to breed at the correct time and this enables the more plastic species to maintain higher reproductive output than less plastic species. Of course, the ringing study of Van Buskirk et al.

Table 11.1 Long term changes in the date of first breeding in three *Rana temporaria* frog populations are within the range predicted purely under phenotypic plasticity. The observed temperature trend, recorded at nearby weather stations, was the slope of the regression of temperature during winter or early spring against year. The observed breeding trend came from regressions of date on which frogs were first heard or eggs were seen against year. The predicted trend is the change expected if frogs react to the observed temperature trend by the amount that Phillimore et al. (2010) estimated to be due to behavioural plasticity (mean: −2.98 d/°C; 95% highest posterior density (HPD): −4.84 to −1.33).

Study site	Years	Observed trends Temp. (°C/yr)	Observed trends Breeding (d/yr)	Predicted trend (d/yr) Mean	Predicted trend (d/yr) 95% HPD	% of trend due to plasticity	Source
Great Britain (50.9°N, 0.21°W)	1979–1994	0.11	−0.321	−0.328	−0.532 to −0.146	102	Beebee 1995
Poland (52.06 °N, 16.83 °E)	1978–2002	0.12	−0.32	−0.358	−0.581 to −0.160	112	Tryjanowski et al. 2003
Finland (64.22 °N, 24.88 °E)	1952–2005	0.031	−0.141	−0.0934	−0.152 to −0.042	67	Lappalainen et al. 2008

(2012) described above shows that we cannot be certain that shifts in migration phenology are primarily due to plasticity, rather than evolutionary change. Nor do we know whether phenotypic responses are keeping pace or falling short of the shifting phenotypic optimum (Both and Visser 2001). But, subject to those remaining unknowns, the study by Møller et al. (2008) may illustrate population-level benefits of adaptive behavioural plasticity.

Sinervo et al. (2010) reported a sobering example of behavioural plasticity that is insufficient to prevent extinction in the face of climate change. Lizards of many species select microhabitats to regulate their body temperature and provide opportunities for foraging and reproduction. There is no conflict between thermoregulation and other activities when weather conditions are cool: the upper surface of a rock is warmer and simultaneously has more prey and mating opportunities. However, as the environmental temperature becomes too warm, lizards must retreat to shaded microhabitats where prey are less abundant. The action of retreating to the cool underside of a rock during hot weather is an example of adaptive behavioural plasticity: it is obviously beneficial in the short term, and even necessary to avoid overheating. Yet retreat is costly in the long run, because lizards hiding under rocks cannot forage, defend territories, or reproduce. Sinervo et al. discovered that lizard populations tend to be declining and disappearing in localities where temperatures have increased such that activity is restricted for at least four hours per day during the reproductive months. In this case, the behavioural response to increasing temperature is necessary for the individual, and beneficial in the sense that it allows individuals to do better than they would otherwise, but it conflicts with other activities necessary for reproduction and, ultimately, population persistence.

11.2.3 Maladaptive plasticity

Maladaptive phenotypic plasticity is widely considered to be rare, but this may stem from a reporting bias favouring studies that find beneficial plasticity. In this context, selection estimates from experiments that were conducted for other purposes are especially useful. Uli Steiner and I recently reviewed empirical evidence for costs of plasticity (Van Buskirk and Steiner 2009), and our data suggest that maladaptive reaction norms may be quite common. The original experiments, covering 16 plant species and 7 animals, were designed to measure selection on genotypes that varied in their capacity to express plasticity. Most of the papers also reported selection acting on trait values within treatments, although these data are not important for testing plasticity costs (DeWitt et al. 1998). In all, we extracted 522 selection estimates from 25 studies, pertaining to different kinds of traits, fitness measures, and environmental treatments. Only 12% of the estimates provided strong evidence for adaptive plasticity, and fully 70% of tests found no indication that plasticity was favoured by divergent selection between environments (Fig. 11.3). The inescapable conclusion is that a great deal of phenotypic plasticity induced by standard experimental manipulations does not improve individual performance, and may often decrease fitness.

Maladaptive behavioural plasticity occurs in several types. One is caused by strong associations among diverse behavioural traits, termed 'behavioural syndromes' (reviewed in Sih et al. 2004). Correlated behavioural types are widespread, reflecting the different personality types of individual animals, or tendencies to exhibit general kinds of behaviours. Behavioural syndromes are known to carry over multiple contexts and to affect suites of seemingly unrelated activities. This implies that behaviours are not free to evolve independently; consequently, specific behaviours when viewed in isolation can be maladaptive in specific situations. This may well apply to some of the traits in Fig. 11.3. But it is not particularly worrisome under the typical range of environmental conditions, because behavioural syndromes are assumed to have evolved as combinations of behaviours that maximize lifetime fitness in those normal conditions (Sih et al. 2004).

Concern arises, however, when animals are exposed to environments that are novel or atypical. Even activities that are not involved in behavioural syndromes can show maladaptive plasticity when

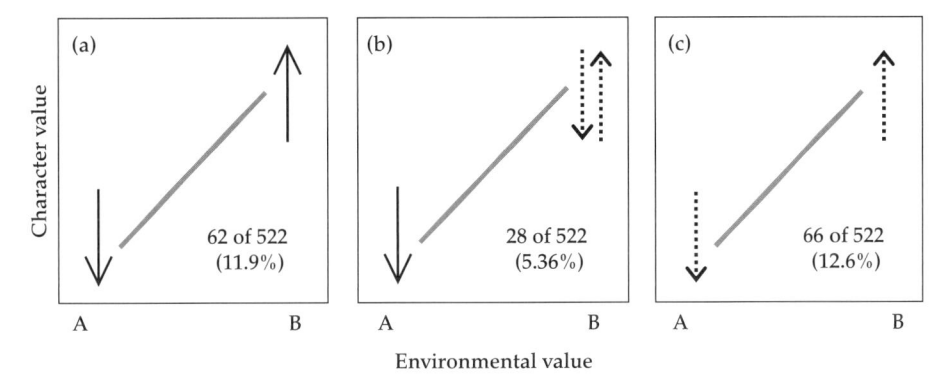

Figure 11.3 Evidence for adaptive phenotypic plasticity was infrequent in a meta-analysis of 25 studies compiled by Van Buskirk and Steiner (2009). Reporting bias should be minimal in this survey, because the authors were primarily interested in measuring costs of plasticity rather than direct selection on trait values. Grey bars represent reaction norms, and the arrows indicate the direction of natural selection (solid are significant at $\alpha = 0.05$; broken are not significant). The manipulated environments in the original studies included light level, predation, herbivory, and the level of competition, among others. The most common traits were morphology, phenology, and physiology. Panel (a): strong evidence that plasticity was adaptive occurred in those cases for which significant divergent selection was measured across environments. Panel (b): weaker evidence for adaptive plasticity was found if significant selection occurred in the same direction as plasticity in one environment, with no significant selection in the other environment. Panel (c): evidence was judged to be very weak if selection coefficients were of opposite sign but never significant. Numbers report the frequency of occurrence of the three conditions; at least one condition depicted here was fulfilled in only 29.9% of tests.

animals occur in extreme environments. When freshwater systems become eutrophic due to excessive nutrient input, visibility in the water decreases with increasing algal growth and primary productivity. This can have unexpected impacts on the mating behaviour of fish that use visual cues during courtship (Candolin 2009). For example, turbid water conditions prevent dominant male sticklebacks *Gasterosteus aculeatus* from controlling the visual signals produced by subordinate males, and therefore females frequently make poor mate choice decisions when visibility is low (Wong et al. 2007). Males, too, react inappropriately to eutrophication: courtship effort increases with water turbidity, but males receive no benefits from their extra effort because females are no more likely to mate with them (Candolin et al. 2007). This response to turbidity would qualify as maladaptive under the definition in Fig. 11.1 because males experience lower fitness than they would have experienced had they not reacted at all. Such maladaptive behavioural responses to novel environments are predicted by reaction norm models of plasticity (de Jong 2005), and we can only assume that these situations will be increasingly common under atypical environments of the future.

A widespread example of maladaptive behavioural plasticity is the so-called 'ecological trap' (also see Chapter 5). This term refers to maladaptive habitat choice causing lower fitness than would occur in an alternative available habitat (Dwernychuk and Boag 1972). As the term implies, the animal is seen to be 'trapped' into an inappropriate behavioural response by a mismatch between habitat quality and the stimuli that it uses to judge quality. This can happen if environmental change reduces the quality of the habitat while leaving the stimulus unchanged, or if environmental change increases the level of the stimulus while leaving the habitat unchanged (Robertson and Hutto 2006). The former situation has been observed when birds suffer poor nesting success after settling in non-native vegetation that appears to be suitable (Lloyd and Martin 2005). A well-known case of the latter type of trap is that of aquatic insects ovipositing on artificial surfaces, such as oil, asphalt, or glass, that reflect polarized light and therefore mimic the surface of water (Kriska et al. 1998; Horvath et al. 2007). Several ecological traps involving mate choice, such as buprestid beetles that copulate with beer bottles resembling female beetles (Gwynne and Rentz 1983), would be downright humorous if they were not so destructive for the animals. In all these examples, the trap can be seen as resulting from behavioural plasticity in habitat or resource use, resulting in lower individual fitness than would have accrued had the animal chosen a less-preferred nearby resource.

The occurrence of ecological traps illustrates that some changes in the environment effected by humans have already induced maladaptive behavioural responses in animals. But what about the kinds of environmental change specifically expected to be associated with climate change, such as locally higher temperature and more variable precipitation? Perhaps reassuringly, most ecological traps are not known to be associated with temperature or moisture availability (Schlaepfer et al. 2002; Robertson and Hutto 2006). The reason is obvious. Organisms that have evolved behavioural responses to temperature typically use temperature itself as a direct cue reflecting the current state of the environment. Therefore, in the case of temperature and perhaps moisture, it is less likely that the actual condition of the environment can become decoupled from the organism's assessment of the environment. Temperature and moisture should therefore not be involved in ecological traps, except in situations where they themselves function as indicators of some other feature of the habitat.

11.3 Outlook

The preceding survey illustrates that plasticity in behaviour is widespread and very often beneficial for individual animals. But this broad generalization will not be sufficient for anticipating and mitigating impacts of humans on vulnerable populations. A major goal of future research must be to understand and predict where plasticity lies on the continuum from beneficial through maladaptive. We have examples of all types of behavioural response, including plasticity that is apparently optimal. But our mechanistic understanding of environments, specific behaviours, and whole-organism plastic responses is inadequate for anticipating future impacts. Charmantier et al.'s (2008) study of great tits in the UK indicates that plasticity can be so strong as to hold the population on a moving optimum for many generations. We do not know how long this can continue, but the behavioural responses of individual birds may eventually reach a limit. Other studies of similar systems suggest that behavioural plasticity is in fact limited (Both and Visser 2001; Van Buskirk et al. 2012). The main issue is whether plasticity is sufficient to maintain fitness and forestall population declines (Chevin et al. 2010).

In order to better understand and predict the fitness consequences of plasticity, we need research into the kinds of environmental changes that will be most important in coming decades. The expected global temperature and rainfall patterns are relatively clear (Solomon et al. 2007). But what do the IPCC projections imply for environmental change at spatial and temporal scales that are relevant for populations? Fluctuations within and among years will be important especially for plastic responses of individuals, but their frequency and magnitude cannot easily be inferred from results of climate models (Min et al. 2011). Temporal variation on the scale of years to decades will be relevant for the persistence of populations. Regional-scale spatial variation is important for defining the envelope within which plasticity and genetic changes are relevant, rather than wholesale movement of individuals and resulting shifts in distributions. Clearly, much additional work is needed to develop a clear idea of environmental changes on the scales that are relevant for most organisms.

One issue that appears repeatedly in discussions of plasticity and environmental change is the impact of novel or rare environments. Theory states clearly that unpredictable and potentially maladaptive phenotypes can be induced by conditions that are evolutionarily unfamiliar, and that it can be difficult for populations to evolve adaptive plasticity in response to unusual or extreme conditions (Lynch and Gabriel 1987; Moran 1992; de Jong 2005; Snell-Rood et al. 2010). Environments of the future are expected to be more extreme than at present, and unlike existing environments at least within the usual spatial range of dispersal (Easterling et al. 2000; Williams and Jackson 2007). This implies that plasticity could become progressively less beneficial with time, as conditions become increasingly unlike current conditions. Clearly, we need additional research on theoretical and empirical topics in the evolution of plasticity in low-frequency environments.

Counteracting the unknown impacts of novel environments is the certainty that beneficial plasticity in behaviour will continue to evolve in the future; selection favouring adaptive plasticity is inevitable as environments change. There are examples of this already (e.g. Nussey et al. 2005). As for any other trait, though, phenotypic plasticity is likely to respond to selection only in certain dimensions (Walsh and Blows 2009). Constraints on the evolution of plasticity are, in general, poorly understood (Via and Lande 1985; DeWitt et al. 1998; Van Buskirk and Steiner 2009). Genetic correlations between plasticity and mean trait values, costs of plasticity, or trade-offs among different modes of plasticity could seriously affect the potential for adaptation of heightened behavioural plasticity under environmental change.

There has been much discussion of the limits to plasticity and their implications under conditions of long-term directional environmental change (Visser 2008; Snell-Rood et al. 2010; Reed et al. 2011). DeWitt et al. (1998) and others thereafter have listed and described the kinds of limits that may occur. Plasticity limits are suspected to be important in nature, but in fact we know very little about limits to plasticity in any organism. Indeed, limits may be relatively unimportant for plasticity in phenology, which has thus far been especially important under climate change (Parmesan 2006). Birds may be able to migrate and breed earlier, and plants may initiate flowering earlier, indefinitely as long as the cues that trigger migration and reproduction continue to advance. The only truly meaningful limits will be those that arise from inescapable functional trade-offs established by the environment. Sinervo et al.'s (2010) study of lizards, discussed earlier, may be an example of this. In principle, there is of course no limit to the behavioural response of a lizard to increasing temperature; it could spend up to 24 hours per day enjoying cool temperatures on the underside of rocks. But the animals face a trade-off because time spent on thermoregulation cannot be used for foraging or mating. This creates a functional limit to behavioural plasticity beyond which fitness is compromised and populations inevitably decline. Further work is needed on the nature and general importance of these kinds of limits to plasticity in general, and particularly in the context of climate change.

Better answers to all these questions will certainly improve our ability to anticipate the contributions of behavioural plasticity to population persistence under environmental change. But the risk underlying academic discussions of this sort is that we are simply arranging the deck chairs on the Titanic. The relative importance of evolutionary change, dispersal, and plasticity for populations could become irrelevant if environmental change is as strong and abrupt as some have projected for the future (Parry et al. 2007). Indeed, this is beginning to seem likely in the absence of restraints on human behaviour that would control resource consumption, habitat destruction, overharvesting, and invasive species. Thus, while evolutionary ecologists need to increase research efforts to understand topics relevant to mitigating environmental change impacts, we may want to devote some time on the side to modifying the behaviour of our fellow humans (Pace et al. 2010).

Acknowledgements

Thanks to Yvonne Willi, the editors, and two reviewers for comments on the manuscript and to the Swiss National Science Foundation for funding.

References

Angilletta, M.J. (2009). *Thermal Adaptation: A Theoretical and Empirical Synthesis*. New York, Oxford University Press.

Azevedo, R.B.R., James, A.C., McCabe, J., and Partridge, L. (1998). Latitudinal variation of wing: thorax size ratio and wing-aspect ratio in *Drosophila melanogaster*. *Evolution*, 52, 1353–62.

Beebee, T.J.C. (1995). Amphibian breeding and climate. *Nature*, 374, 219–20.

Bell, A.M. and Sih, A. (2007). Exposure to predation generates personality in threespined sticklebacks (*Gasterosteus aculeatus*). *Ecology Letters*, 10, 828–34.

Blows, M.W. and Hoffmann, A.A. (2005). A reassessment of genetic limits to evolutionary change. *Ecology*, 86, 1371–84.

Both, C. and Visser, M.E. (2001). Adjustment to climate change is constrained by arrival date in a long-distance migrant bird. *Nature*, 411, 296–8.

Bradshaw, A.D. (1965). Evolutionary significance of phenotypic plasticity in plants. *Advances in Genetics*, 13, 115–55.

Candolin, U. (2009). Population responses to anthropogenic disturbance: lessons from three-spined sticklebacks *Gasterosteus aculeatus* in eutrophic habitats. *Journal of Fish Biology*, 75, 2108–21.

Candolin, U., Salesto, T., and Evers, M. (2007). Changed environmental conditions weaken sexual selection in sticklebacks. *Journal of Evolutionary Biology*, 20, 233–9.

Charmantier, A., McCleery, R.H., Cole, L.R., Perrins, C., Kruuk, L.E.B., and Sheldon, B.C. (2008). Adaptive phenotypic plasticity in response to climate change in a wild bird population. *Science*, 320, 800–3.

Chevin, L.-M. and Lande, R. (2010). When do adaptive plasticity and genetic evolution prevent extinction of a density-regulated population? *Evolution*, 64, 1143–50.

Chevin, L.-M., Lande, R., and Mace, G.M. (2010). Adaptation, plasticity, and extinction in a changing environment: towards a predictive theory. *PLOS Biology*, 8, e1000357.

Conover, D.O., Duffy, T.A., and Hice, L.A. (2009). The covariance between genetic and environmental influences across ecological gradients: reassessing the evolutionary significance of countergradient and cogradient variation. *Annals of the New York Academy of Sciences*, 1168, 100–29.

Crews, D. (2003). The development of phenotypic plasticity: where biology and psychology meet. *Developmental Psychobiology*, 43, 1–10.

Davis, M.B., Shaw, R.G., and Etterson, J.R. (2005). Evolutionary responses to changing climate. *Ecology*, 86, 1704–14.

Davis, S.J., Caldeira, K., and Matthews, H.D. (2010). Future CO_2 emissions and climate change from existing energy infrastructure. *Science*, 329, 1330–3.

de Jong, G. (2005). Evolution of phenotypic plasticity: patterns of plasticity and the emergence of ecotypes. *New Phytologist*, 166, 101–17.

de Jong, G. and Gavrilets, S. (2000). Maintenance of genetic variation in phenotypic plasticity: the role of environmental variation. *Genetical Research*, 76, 295–304.

de Meester, L. (1996). Evolutionary potential and local genetic differentiation in a phenotypically plastic trait of a cyclical parthenogen, *Daphnia magna*. *Evolution*, 50, 1293–8.

Desprez-Loustau, M.L., Marcais, B., Nageleisen, L.M., Piou, D., and Vannini, A. (2006). Interactive effects of drought and pathogens in forest trees. *Annals of Forest Science*, 63, 597–612.

DeWitt, T.J., Sih, A., and Wilson, D.S. (1998). Costs and limits of phenotypic plasticity. *Trends in Ecology & Evolution*, 13, 77–81.

Dwernychuk, L.W. and Boag, D.A. (1972). Ducks nesting in association with gulls: an ecological trap? *Canadian Journal of Zoology*, 50, 559–63.

Easterling, D.R., Meehl, G.A., Parmesan, C., Changnon, S.A., Karl, R.R., and Mearns, L.O. (2000). Climate extremes: observations, modeling, and impacts. *Science*, 289, 2068–74.

Falconer, D.S. and Mackay, T.F.C. (1996). *Introduction to quantitative genetics, 4th edition*. Longman, London.

Gabriel, W. (1999). Evolution of reversible plastic responses: inducible defenses and environmental tolerance. In Tollrian, R. and Harvell, D. (eds) *The Ecology and Evolution of Inducible Defenses*, pp. 286–305. Princeton, Princeton University Press.

Ghalambor, C.K., McKay, J.K., Carroll, S.P., and Reznick, D.N. (2007). Adaptive versus non-adaptive phenotypic plasticity and the potential for contemporary adaptation in new environments. *Functional Ecology*, 21, 394–407.

Gillooly, J.F., Brown, J.H., West, G.B., Savage, V.M., and Charnov, E.L. (2001). Effects of size and temperature on metabolic rate. *Science*, 293, 2248–51.

Gomulkiewicz, R. and Holt, R.D. (1995). When does natural selection prevent extinction? *Evolution*, 49, 201–7.

Groh, C., Tautz, J., and Rössler, W. (2004). Synaptic organization in the adult honey bee brain is influenced by brood-temperature control during pupal development. *Proceedings of the National Academy of Sciences of the USA*, 101, 4268–73.

Gwynne, D.T. and Rentz, D.C.F. (1983). Beetles on the bottle: male buprestids mistake stubbies for females (Coleoptera). *Journal of the Australian Entomological Society*, 22, 79–80.

Hendry, A.P., Farrugia, T.J., and Kinnison, M.T. (2008). Human influences on rates of phenotypic change in wild animal populations. *Molecular Ecology*, 17, 20–9.

Hoffmann, A.A. and Willi, Y. (2008). Detecting genetic responses to environmental change. *Nature Reviews Genetics*, 9, 421–32.

Holtzer, T.O., Archer, T.L., and Norman, J.M. (1988). Host plant suitability in relation to water stress. In Heinrichs, E.A. (ed.) *Plant Stress-Insect Interactions*, pp. 111–37. New York, Wiley.

Horvath, G., Malik, P., Kriska, G., and Wildermuth, H. (2007). Ecological traps for dragonflies in a cemetery: the attraction of *Sympetrum* species (Odonata,

Libellulidae) by horizontally polarizing black gravestones. *Freshwater Biology*, 52, 1700–9.

James, A. C., Azevedo, B.R., and Partridge, L. (1995). Cellular basis and developmental timing in a size cline of *Drosophila melanogaster*. *Genetics*, 140, 659–66.

Kawecki, T.J. (2000). The evolution of genetic canalization under fluctuating selection. *Evolution*, 54, 1–12.

Kriska, G., Horvath, G., and Andrikovics, S. (1998). Why do mayflies lay their eggs en masse on dry asphalt roads? Water imitating polarized light reflected from asphalt attracts Ephemeroptera. *Journal of Experimental Biology*, 201, 2273–86.

Lappalainen, H.K., Linkosalo, T., and Venalainen, A. (2008). Long-term trends in spring phenology in a boreal forest in central Finland. *Boreal Environment Research*, 13, 303–18.

Laurila, A., Lindgren, B., and Laugen, A.T. (2008). Antipredator defenses along a latitudinal gradient in *Rana temporaria*. *Ecology*, 89, 1399–413.

Levins, R. (1969). Thermal acclimation and heat resistance in *Drosophila species*. *American Naturalist*, 103, 483–99.

Lindgren, B. and Laurila, A. (2005). Proximate causes of adaptive growth rates: growth efficiency variation among latitudinal populations of *Rana temporaria*. *Journal of Evolutionary Biology*, 18, 820–8.

Lloyd, J.D. and Martin, T.E. (2005). Reproductive success of chestnut-collared longspurs in native and exotic grassland. *Condor*, 107, 363–74.

Lynch, M. and Gabriel, W. (1987). Environmental tolerance. *American Naturalist*, 129, 283–303.

Lynch, M. and Lande, R. (1993). Evolution and extinction in response to environmental change. In Kareiva, P., Kingsolver, J., and Huey, R. (eds) *Biotic Interactions and Global Change*, pp 234–50. Sunderland, MA, Sinauer Associates.

Min, S.-K., Zhang, X., Zwiers, F.W., and Hegerl, G.C. (2011). Human contribution to more-intense precipitation extremes. *Nature*, 470, 378–81.

Møller, A.P., Rubolini, D., and Lehikoinen, E. (2008) Populations of migratory bird species that did not show a phenological response to climate change are declining. *Proceedings of the National Academy of Sciences of the USA*, 105, 16195–200.

Moran, N.A. (1992). The evolutionary maintenance of alternative phenotypes. *American Naturalist*, 139, 971–82.

Munch, S.B. and Conover, D.O. (2003). Rapid growth results in increased susceptibility to predation in *Menidia menidia*. *Evolution*, 57, 2119–27.

Nussey, D.H., Postma, E., Gienapp, P., and Visser, M.E. (2005). Selection on heritable phenotypic plasticity in a wild bird population. *Science*, 310, 304–6.

Pace, M.L., Hampton, S.E., Limburg, K.E., et al. 2010. Communicating with the public: opportunities and rewards for individual ecologists. *Frontiers in Ecology and the Environment*, 8, 292–8.

Padilla, D.K. and Adolph, S.C. (1996). Plastic inducible morphologies are not always adaptive: the importance of time delays in a stochastic environment. *Evolutionary Ecology*, 10, 105–17.

Parmesan, C. (2006). Ecological and evolutionary responses to recent climate change. *Annual Review of Ecology, Evolution, and Systematics*, 37, 637–69.

Parry, M.L., Canziani, O.F., Palutikof, J.P., van der Linden, P.J., and Hanson, C.E., (eds) (2007). *Climate Change 2007: Impacts, Adaptation, and Vulnerability*. Contribution of Working Group II to the Fourth Assessment Report of the Intergovernmental Panel on Climate Change. Cambridge University Press.

Pereira, H.M., Leadley, P.W., Proenca, V. et al. (2010). Scenarios for global biodiversity in the 21st Century. *Science*, 330, 1496–501.

Phillimore, A.B., Hadfield, J.D., Jones, O.R., and Smithers, R.J. (2010). Differences in spawning date between populations of common frog reveal local adaptation. *Proceedings of the National Academy of Sciences of the USA*, 107, 8292–7.

Reed, T.E., Schindler, D.E., and Waples, R.S. (2011). Interacting effects of phenotypic plasticity and evolution on population persistence in a changing climate. *Conservation Biology*, 25, 56–63.

Reznick, D.N. and Ghalambor, C.K. (2001). The population ecology of contemporary adaptations: what empirical studies about the conditions that promote adaptive evolution. *Genetica*, 112, 183–98.

Robertson, B.A. and Hutto, R.L. (2006). A framework for understanding ecological traps and an evaluation of existing evidence. *Ecology*, 87, 1075–85.

Schlaepfer, M.A., Runge, M.C., and Sherman, P.W. (2002). Ecological and evolutionary traps. *Trends in Ecology & Evolution*, 17, 474–80.

Schwartz, M.W., Iverson, L.R., Prasad, A.M., Matthews, S.N., and O'Connor, R.J. (2006). Predicting extinctions as a result of climate change. *Ecology*, 87, 1611–15.

Scott, W.A., Pithart, D., and Adamson, J.K. (2008). Long-term United Kingdom trends in the breeding phenology of the common frog, *Rana temporaria*. *Journal of Herpetology*, 42, 89–96.

Sih, A., Bell, A.M., Johnson, J.C., and Ziemba, R.E. (2004). Behavioral syndromes: an integrative overview. *Quarterly Review of Biology*, 79, 241–77.

Sinervo, B., Mendez-de-la-Cruz, F., Miles, D.B., et al. (2010). Erosion of lizard diversity by climate change and altered thermal niches. *Science*, 328, 894–9.

Snell-Rood, E.C., Van Dyken, J.D., Cruickshank, T., Wade, M.J., and Moczek, M.P. (2010). Toward a population genetic framework of developmental evolution: the costs, limits, and consequences of phenotypic plasticity. *BioEssays*, 32, 71–81.

Solomon, S., Qin, D., Manning, M., et al., (eds) (2007). *Climate Change 2007: The Physical Science Basis*. Contribution of Working Group I to the Fourth Assessment Report of the Intergovernmental Panel on Climate Change. Cambridge University Press.

Stearns, S.C. (1989). The evolutionary significance of phenotypic plasticity. *BioScience*, 39, 436–45.

Sutter, M. and Kawecki, T.J. (2009). Influence of learning on range expansion and adaptation to novel habitats. *Journal of Evolutionary Biology*, 22, 2201–14.

Tautz, J., Maier, S., Groh, C., Rössler, W., and Brockmann, A. (2003). Behavioral performance in adult honey bees is influenced by the temperature experienced during their pupal development. *Proceedings of the National Academy of Sciences of the USA*, 100, 7343–7.

Thomas, C.D. (2011). Translocation of species, climate change, and the end of trying to recreate past ecological communities. *Trends in Ecology & Evolution*, 26, 216–21.

Travis, J. (1994). Evaluating the adaptive role of morphological plasticity. In Wainwright P.C. and Reilly, S.M. (eds) *Ecological Morphology*, pp. 99–122. University of Chicago Press.

Tryjanowski, P., Rybacki, M., and Sparks, T. (2003). Changes in the first spawning dates of common frogs and common toads in western Poland in 1978–2002. *Annales Zoologici Fennici*, 40, 459–64.

Van Buskirk, J., Mulvihill R. S., and Leberman, R. C. (2009). Variable shifts in spring and autumn migration phenology in North American songbirds associated with climate change. *Global Change Biology*, 15, 760–71.

Van Buskirk, J., Mulvihill, R. S., and Leberman, R. C. (2012). Phenotypic plasticity cannot explain climate-induced changes in the timing of bird migration. In review.

Van Buskirk, J. and Steiner, U. K. (2009). The fitness costs of developmental canalization and plasticity. *Journal of Evolutionary Biology*, 22, 852–60.

Via, S., and Lande, R. (1985). Genotype-environment interaction and the evolution of phenotypic plasticity. *Evolution*, 39, 505–22.

Visser, M. E. (2008). Keeping up with a warming world: assessing the rate of adaptation to climate change. *Proceedings of the Royal Society of London B, Biological Sciences*, 275, 649–59.

Visser, M. E., van Noordwijk, A. J., Tinbergen, J. M., and Lessells, C. M. (1998). Warmer springs lead to mistimed reproduction in great tits (*Parus major*). *Proceedings of the Royal Society of London B, Biological Sciences*, 265, 1867–70.

Walsh, B., and Blows, M. W. (2009). Abundant genetic variation plus strong selection = multivariate genetic constraints, a geometric view of adaptation. *Annual Review of Ecology, Evolution, and Systematics* 40, 41–59.

West-Eberhard, M.J. (1989). Phenotypic plasticity and the origins of diversity. *Annual Review of Ecology and Systematics*, 20, 249–78.

West-Eberhard, M.J. (2003). *Developmental Plasticity and Evolution*. New York, Oxford University Press.

Wilczek, A.M., Burghardt, L.T., Cobb, A.R., Cooper, M.D., Welch, S.M., and Schmitt, J. (2010). Genetic and physiological bases for phenological responses to current and predicted climates. *Philosophical Transactions of the Royal Society B: Biological Sciences*, 365, 3129–47.

Willi, Y. and Hoffmann, A.A. (2009). Demographic factors and genetic variation influence population persistence under environmental change. *Journal of Evolutionary Biology*, 22, 124–33.

Williams, J.W. and Jackson, S.T. (2007). Novel climates, no-analog communities, and ecological surprises. *Frontiers in Ecology and the Environment*, 5, 475–82.

Wong, B.B.M., Candolin, U., and Lindstrom, K. (2007). Environmental deterioration compromises socially-enforced signals of male quality in three-spined sticklebacks. *American Naturalist*, 170, 184–9.

CHAPTER 12

Population consequences of individual variation in behaviour

Fanie Pelletier and Dany Garant

⊃ Overview

The potential links between behaviour and population dynamics have repeatedly been emphasized in the literature. This chapter discusses the interplay between behavioural phenotypes and population processes by illustrating how human-induced environmental changes can promote these feedbacks. We have divided our review into five sections. First, we explain why one would expect a feedback between behaviour and population dynamics. In the second section, we suggest hypotheses on which behaviour are more likely to affect population growth. Third, we present a short review of research exploring the more classic links of the population–behaviour feedback, showing that population dynamics can affect individual behaviour. In the fourth section, we highlight empirical studies that support the more recent suggestion of a reverse link between population and behaviour. Finally, we provide examples of study systems where researchers have documented the complete feedback loop between behaviour and population dynamics.

'Civilization, particularly in its recent history, has been a major source of geologic change on the earth, equivalent in the magnitude of its effect to the natural geologic forces of rain and frost. The constructions arising from the efforts of man differ from other features on the face of the earth in their relative lack of stability. Without the activity of man most of these changes would never appear at all, and without man's care and maintenance most of them would disappear.'

(Slobodkin 1961)

12.1 Introduction

Why are some species thriving in highly modified human landscapes while others go extinct when their habitat is affected by human activities? Why are some populations of a species coping with human presence in some areas but not in others? For example, brown bears *Ursus arctos* are in decline in parts of North America (Servheen et al. 1999), while their numbers are increasing throughout Scandinavia (Zedrosser et al. 2001). Obviously, there can be many reasons for such geographical differences, but human behaviour is likely to play a key role, as it represents the main cause of mortality for bears (Bischof et al. 2008). It has been suggested that differences in behavioural response to humans could partly explain the differences observed in population trends among regions for a given species and among species within a region. Differences in aggressive behaviour toward humans between North American and European bears could affect the ability of these populations to cope with human presence (Swenson 1999). Indeed, as North American bears are typically more aggressive toward humans than bears are in Scandinavia, the

Behavioural Responses to a Changing World. First Edition. Edited by Ulrika Candolin and Bob B.M. Wong.
© 2012 Oxford University Press. Published 2012 by Oxford University Press.

latter will usually be perceived as 'less dangerous', which in turn could affect human attitudes toward their conservation. This difference in behavioural response might partly explain why protection efforts in Scandinavia have been much more effective than in North America. Examples of similar species showing contrasting population trends in response to human-driven change also exist. In eastern North America, the rapid decline in golden-winged warblers *Vermivora chrysoptera* seems to be driven by the loss of early successional scrub habitat due to human-induced changes, such as habitat transformation through reforestation and possibly climate change (Gill 1997). In contrast, the blue-winged warbler *Vermivora cyanoptera*, a closely-related species, is expanding its range northward (Gill 1997). Combined with breeding habitat losses, hybridization in sympatric zones between these two species is now threatening the persistence of the golden-winged warbler (Vallender et al. 2007). How could two similar bird species show such contrasted patterns in their response to habitat changes that ultimately lead to opposite population trends? Is it possible that behavioural phenotypes of blue-winged warblers are more plastic, allowing them to use more diversified types of habitats?

The aim of population ecology is to understand why populations of animals fluctuate in size over time. Although population dynamics has been a topic of interest for demographers and ecologists for at least the last two centuries, we still have very little empirical understanding of how genetically based traits, such as behaviour, affects population growth. Researchers have argued for a long time over the possibility that population composition (e.g. age, sex, or size) could affect its growth rate. Chitty (1967) was one of the first to suggest that behaviour can affect population dynamics beyond the effect of density, weather, disease, and predation. Based on the observation that the survival value of certain behaviours changes with population density in small mammal populations, he suggested that 'all species have a form of behaviour that can prevent unlimited increase in density' (Chitty 1967). In other words, Chitty was suggesting that demographic changes occurring in small mammal cycles are mediated by natural selection operating on the genetic composition of the population. He called this process 'the natural selection of self-regulatory behaviour'. The rationale for his hypothesis was (in his own words):

'Behaviour is taken to include the numerous manifestations of hostility discussed by Tinbergen (1957), and the first assumption that all species have some form of dispersion mechanism, or method of spacing themselves out by avoiding, threatening, or otherwise influencing certain other members of the same species. The pattern of dispersion is assumed to depend partly on the properties of the individuals (genotype, age, experience, hormone balance, etc.), and partly on the properties of their environment (weather; amount, kind and distribution of food and cover; number and kind of animals present, etc.). The second assumption is that this behaviour persists only because it has survival value for the individual and is constantly being selected for.' (p. 51)

Although his hypothesis was later rejected based on the argument that most behavioural traits mentioned by Chitty (1967) were not heritable (Boonstra and Boag 1987), his idea of a link between genetically based phenotype (including behaviour) and population growth is still valid. Recently, studies focusing on eco-evolutionary dynamics have shown that, under some circumstances, changes in phenotype can affect population processes (see reviews in Saccheri and Hanski 2006; Kokko and Lopez-Sepulcre 2007; Pelletier et al. 2009). Such studies have laid the groundwork for further investigation of the feedback between individual variation and population dynamics. Interestingly, part of this renewed interest on the interplay between ecology and evolution has been stimulated by differences in individual and population responses to anthropogenic activities (Stockwell et al. 2003). These studies led to the realization that evolution can occur on contemporary time scales (within a few generations, Hendry and Kinnison 1999; see also Chapter 16) and that humans can outpace any other predators as selective agents of wild species (Darimont et al. 2009). As several wild populations face new environmental conditions imposed by humans, researchers need to know whether induced changes in behaviour, phenotype or life history can affect population dynamics and, ultimately, impair population persistence on the short, medium, and long-term scales (Stockwell et al. 2003). Yet, even if we could cite numerous studies suggesting that differences in behavioural tactics might explain why some individuals, populations, and species do better than others, the magnitude of the effect of behavioural variation on population growth remains largely undocumented.

This chapter will discuss the potential interplay between behavioural phenotypes and population processes by illustrating how human-induced environmental changes can promote these feedbacks. We do not seek to provide an extensive literature review of all the articles and books written on this topic, but instead limit our synthesis to studies that set out to quantify, at least partially, these feedbacks. To achieve this goal, we have divided the chapter into five different sections. First, we present the rationale for why one would expect a feedback between behaviour and population dynamics (Section 12.2). Second, we suggest hypotheses as to what behaviours are more likely to affect population growth (Section 12.3). Third, we briefly review research exploring the classic links between population-level processes and behaviour, and how population dynamics can affect individual behaviour (Section 12.4). Fourth, we highlight empirical studies that support the more recent suggestion of a reverse link between population and behaviour (Section 12.5). Finally, we provide examples of study systems where researchers have documented the complete feedback loop (Section 12.6). In each section, we specifically chose examples where humans have affected the interaction between these processes. Where such studies were unavailable, we provided examples of how behaviours affect population dynamics in natural contexts.

12.2 Should we expect a link between behaviour and population dynamics?

Several previous contributions have suggested that individual behaviour can affect population growth (Caro 1998; Sutherland 1996; Festa-Bianchet and Apollonio 2003). Indeed, it seems sensible to expect that more territorial animals should be less likely to colonize a new environment after translocation, that cutting of trees is more likely to affect animals using tree cavities, or that species with higher dispersal are more likely to move northward with climate change. To test these expectations, however, one requires detailed long-term data on behaviour, individual reproduction, and survival, as well as population dynamics. Moreover, even if such data are available, additional difficulties in linking these two fields come from the inherent challenge of quantifying the effects of animal behaviour on population growth. Recent advances in population biology, however, are likely to help integrate these disciplines. Indeed, it is now possible to quantify the effects of a given phenotype on population dynamics (Coulson et al. 2006; Coulson and Tuljapurkar 2008). Although these approaches have not yet been applied to behavioural phenotypes, they are potentially suitable for assessing how much variation in population growth can be explained by variation, for example, in aggressiveness.

The rationale behind the framework used to link individual traits and population dynamics is simple (see Fig. 12.1) and not necessarily novel (see for example Pimentel 1968). However, data and methods needed to quantify this interplay have only recently become available (Coulson et al. 2006; Coulson and Tuljapurkar 2008). Behaviours are traits with underlying genetic variation (Stirling et al. 2002) and individual variation in genetically-based traits, such as activity, aggressiveness, or dispersal, is often correlated with some aspects of individual fitness (survival and reproduction, see previous chapters). One could then argue that directional or stabilizing selection should favour individuals expressing fitter behaviours, resulting in a very small effect of behaviour on population growth. However, the strength of selection acting on any trait is typically low (Kingsolver et al. 2001; Hoekstra et al. 2001) and fluctuates over time in the wild (Siepielski et al. 2009). Such weak and fluctuating selection could allow different behavioural phenotypes to coexist at the population level and act positively or negatively on its growth. Thus, it is logical to expect a signature of a distributional change in behavioural traits at the population level (Kokko and Lopez-Sepulcre 2007; Saccheri and Hanski 2006). Given that population growth is the result of the combined effects of mean survival,

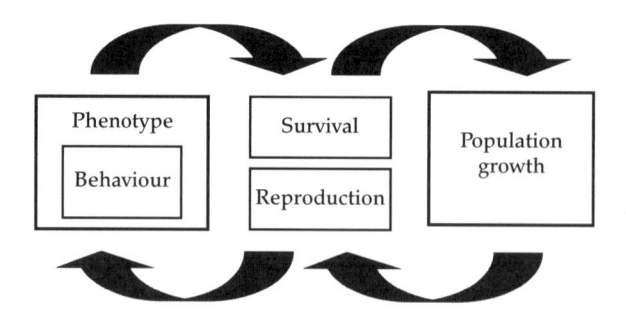

Figure 12.1 A schematic representation of the links between behaviour and population growth.

recruitment, emigration, and immigration, if behaviour (or any genetically-based trait) affects any of these components it could also potentially affect population growth (Fig. 12.1). Although all behaviours associated with individual performance are likely to affect population growth, it seems sensible to suggest that changes in behaviour affecting vital rates (such as survival and recruitment) with high elasticity (elasticity of a vital rate is defined as the proportional change in the population growth rate for a proportional change in a vital rate) (Benton and Grant, 1999) are likely to be more important for population responses to human activities. As selective pressures imposed by human exploitation can lead to changes in a trait's distribution within a decade (Hendry et al. 2008), we might expect that these changes will affect not only the distribution of phenotypes but also the dynamics of the exploited population (Stockwell et al. 2003). In this context, these feedbacks could become very important as certain behavioural phenotypes or life history traits might be selected for differently in populations facing human-mediated changes.

12.3 Whose behaviour might be more likely to affect population dynamics?

Typically, population dynamic models consider the female portion of the population. As females produce the offspring and, in several species, a single male can inseminate several females, one is generally more concerned by the number of females in a population when predicting population growth. Considering only the female portion also allows for simpler population models (Rankin and Kokko 2007). Thus, one might expect that change in female behaviour is more likely to affect population growth compared to change in male behaviour, although this remains to be tested. There are, however, two ways in which males can affect population growth. First, males still represent a component of the population density which can affect female survival and reproduction; second, males can actively affect the demographic rates of females (Mysterud et al. 2002). The brown bear populations mentioned previously can serve as an example to illustrate this point.

Brown bears are non-social carnivores with overlapping home ranges. The mating season occurs from mid May to early July. During that period, males and females remain together for a period ranging from a few hours to several days. The long lactation period of bears (often two years) decreases the availability of fertile females, but a female can become receptive within a few days if she loses her young during the breeding season. As a result, evidence of infanticide by non-resident males has been reported in that species (Swenson et al. 1997). In Scandinavia, brown bears are hunted. Although the hunter does not necessarily target males, the high rate of mortality among male bears can frequently disrupt the social structure. Swenson et al. (1997) have documented the consequences of infanticide on bear population growth by comparing cub survival in sites where the adult males were killed during the previous hunting seasons versus sites where males were not killed. They found that cub survival was much lower in areas where adult males were hunted in at least one of the two previous hunting seasons. This lower cub survival is due to non-resident males coming in and killing cubs. Their study suggested that the killing of one adult male can have a population effect equivalent to killing 0.5 to 1 adult females due to its effect on cub survival (Swenson et al., 1997), providing strong support to the contention that individual behaviour (including male behaviour) can affect population growth (see also Chapter 8).

As mentioned previously, quantifying the influence of behavioural traits on population growth is not a simple task. Indeed, to show that behaviour alone leaves a signature on population dynamics requires data indicating that—every other process being constant—an animal expressing one behaviour or a group of individuals using a specific behavioural tactic, makes a greater contribution to population growth than individuals (or groups) using a different behaviour. Yet, to provide such evidence using only observational data is difficult since, in nature, differences in animal behaviour are generally correlated with differences in other aspects of phenotype, including body size or shape, reproductive strategy, colours, and so on (see Réale et al. 2010 for further discussion). Such correlations make it difficult to dissect which component of population growth is specifically affected by differences in behaviour, independent of other aspects of phenotype. For this reason, in the remaining part of this chapter, we will not only discuss feedback between behaviour and population processes *per se*, but also broadly consider how variation at the individual level can leave a signature at the population level.

12.4 From the population to the individual level

Here we take a step back and focus on empirical studies that have first emphasized the importance of the interplay between population dynamics and individual behaviour. Perhaps the best examples of this relationship come from studies documenting 'Allee effects'. An Allee effect can broadly be defined as any component of an individual's fitness positively correlated to population density or size (Courchamp et al. 2008). A positive association between population size and individual fitness can occur, for instance, in gregarious species where individuals benefit from the presence of conspecifics for the improved detection of predators. An Allee effect could also be observed if very low population densities disrupt mating systems and lead to difficulties in finding mates (Courchamp et al. 2008). A typical example comes from the Saiga antelope *Saiga tatarica tatarica*, a polygynous species where males defend harems and can mate with several females. Monitoring of a population in the Kalmykia region of Russia between 1992–2002, Milner-Gulland et al. (2003) revealed that intense poaching of the male Saiga for their horns drastically reduced the number of males in the population. In early 2000, the population sex ratio became so skewed towards females that the number of pregnancies in females declined significantly and led to population collapse in 2001 (Milner-Gulland et al. 2003). As human activities (e.g. exploitation) can alter population size of several species, anthropogenic Allee effects are likely to become more common.

High population density can also influence behavioural tactics used by individuals. As population density increases, competition for resources will become more important and individuals might alter their behaviour accordingly. As pointed out by Kokko and Lopez-Sepulcre (2007), if density influences all individual performances in the same way, then this is analogous to the usual concept of density-dependence. If, however, individual variation for genetically-based traits, such as behaviour or life history tactics, also modulates individual performance at high density, this would consequently affect population growth. We would then have the potential for a feedback loop between individual and population processes (Kokko and Lopez-Sepulcre 2007).

Similarly, we could expect more aggressive or exploratory individuals to perform better in specific ecological situations. For example, during colonization events or at high density, individuals expressing greater aggressiveness and/or exploratory behaviours could be favoured during competition for food or brooding sites (Box 12.1). As discussed in previous chapters, such variation can result from behavioural plasticity. However, recent studies in behavioural ecology have emphasized that individuals can show consistent behavioural differences over time or across situations (often called

Box 12.1 Aggressiveness and human activities: the case of the bluebird *Sialia* sp.

Western bluebirds *Sialia mexicana* are secondary cavity nesters that typically occur in open forests. Around the 1930s, changes in forestry and agricultural practices led to a reduction in the availability of natural cavities, resulting in extirpation of Western bluebirds in many areas. Since then, installation of nest boxes throughout the

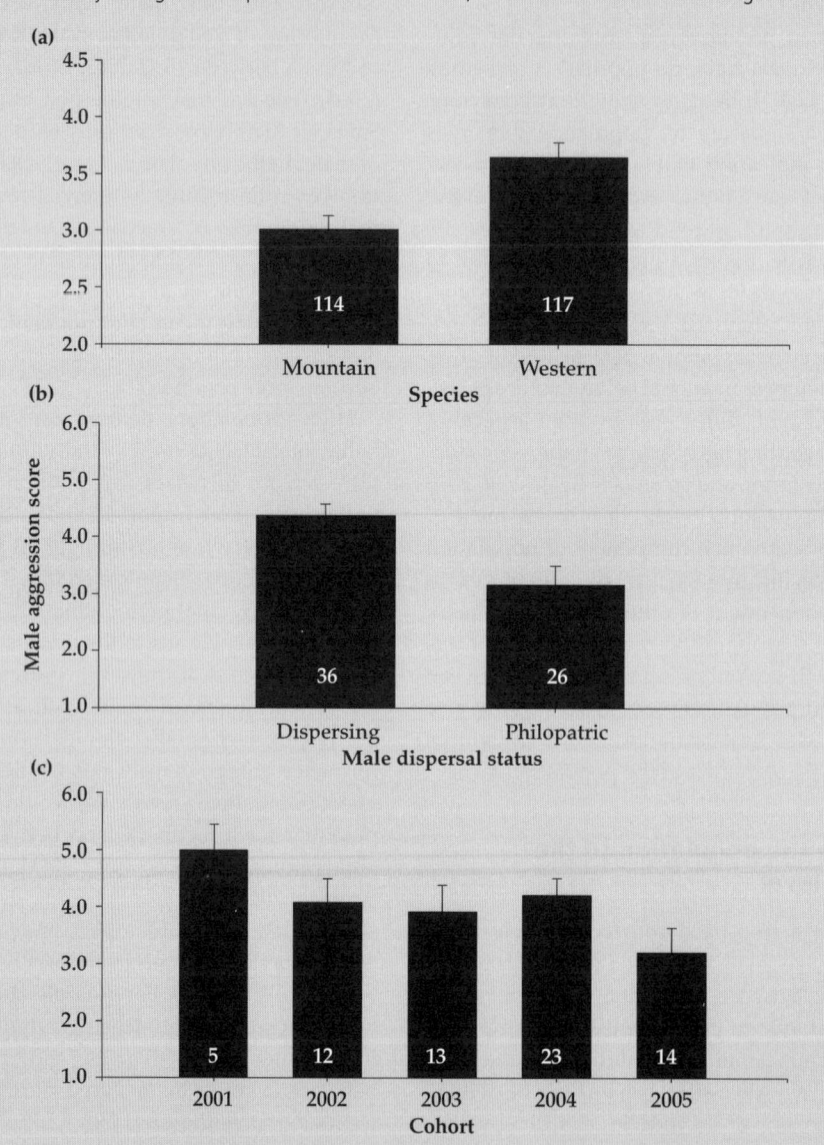

Figure 12.2 Variation in aggression in male bluebirds. (a) Male aggression score is higher in Western than in mountain bluebirds. (b) Male Western bluebirds that disperse to breed further away from their natal population are more aggressive than males that remain in their natal population. (c) Aggression score in Western bluebirds decreases across cohorts after colonization of a new habitat. Numbers represent the number of birds monitored and the bars are the means ± SE. (Modified from Duckworth and Badyaev, 2007.)

northwestern United States has resulted in the rapid recolonization of the species' historical range. At the same time, the addition of artificial nest boxes provided a unique opportunity for researchers to investigate how differences in behaviour can favour the success of migrants with different phenotypes, and how such differences in behaviour modulate interactions with bird species already breeding in the nest boxes.

There is abundant individual variation in aggressive behaviour expressed by male Western bluebirds (Duckworth 2006). These differences are consistent within individuals and across contexts, and have a heritable genetic basis (Duckworth and Kruuk 2009). Interestingly, aggressiveness is also positively correlated with dispersal, both at the phenotypic and genetic levels (Duckworth and Kruuk 2009). These differences have potential consequences for Western bluebirds' population dynamics, both through intra and interspecific interactions.

At the interspecific level, Western bluebirds have regularly been in contact with a closely-related species, the mountain bluebird *Sialia currucoides*, during their range expansion. Mountain bluebirds were less affected than Western bluebirds by changes in human activities during the 1930s, so that many of them were already present (and occupied nest boxes) in areas that Western bluebirds recolonized. Therefore, in many areas, the two species competed for the artificial breeding sites. Studies have found that the more aggressive Western bluebirds, as they expanded their range eastwards, were displacing the mountain bluebird (Duckworth and Badyaev 2007; Fig. 12.2 a).

At the intraspecific level, it has been shown that aggressive males show greater dispersal abilities and, thus, are more prone to colonize new habitats; whereas non-aggressive males disperse less, but perform best in their original habitats (Fig. 12.2 b (Duckworth 2008)). As a result, non-aggressive males perform better in older populations while aggressive males have a higher fitness in newly established populations. Aggressiveness therefore decreases through time for a given population (Fig. 12.2 c). The population composition (aggressive vs. non-aggressive phenotypes) also potentially affects the level of competition with mountain bluebirds, as the level of aggression among birds is lower in older populations, where mountain bluebirds could persist.

'personality', Réale et al. 2010). These differences can be heritable (Dingemanse et al. 2002; Bell, 2005) and selected for differently in contrasted ecological contexts (Boon et al. 2007). Thus, we can expect that human activities influencing the environment will modify the type and strength of selective pressures on, and the extent of the heritability of, these 'personality traits'.

Investigation into the indirect effect of predators on prey populations could be useful to predict what consequences stressful environments imposed by humans can have on wild species. Indeed, it has been suggested that animals perceive human disturbance similarly to predation risk (Frid and Dill 2002), with predation and non-lethal disturbance leading to similar trade-offs in behaviour (Fig. 12.3). Although it might be difficult to test whether animals perceive humans as predators, empirical tests of at least some parts of the conceptual framework suggested by Frid and Dill (2002) have been made. In bighorn sheep *Ovis canadensis*, researchers have found that in years where predation by cougar *Puma concolor* was intense, lambs were smaller in the autumn compared to years where predation was infrequent (Bourbeau-Lemieux et al. 2011). This effect of predation is probably mediated by a difference in behaviour of the sheep in years with high and low predation. This species is known to spend more time vigilant on cliffs, where the vegetation is rare, to avoid predators. In turn, this difference in body size affected lamb survival during the winter, exacerbating the effect of predation on population growth. In another study on the same species, it was shown that individual variation in behaviour affected the probability of survival through predation of adult females (Box 12.2). In these two examples, human activities were not to blame, although some researchers have suggested that the rise in white-tailed deer *Odocoileus virginianus* density due to agricultural and silvicultural activities (Côté et al. 2004) might have triggered an increase in cougar populations in parts of western North America. This may have led to depensatory predation pressure on alternative prey (Robinson et al. 2002).

Similar to the effects of predation, in populations that are frequently disturbed by ecotourism or other

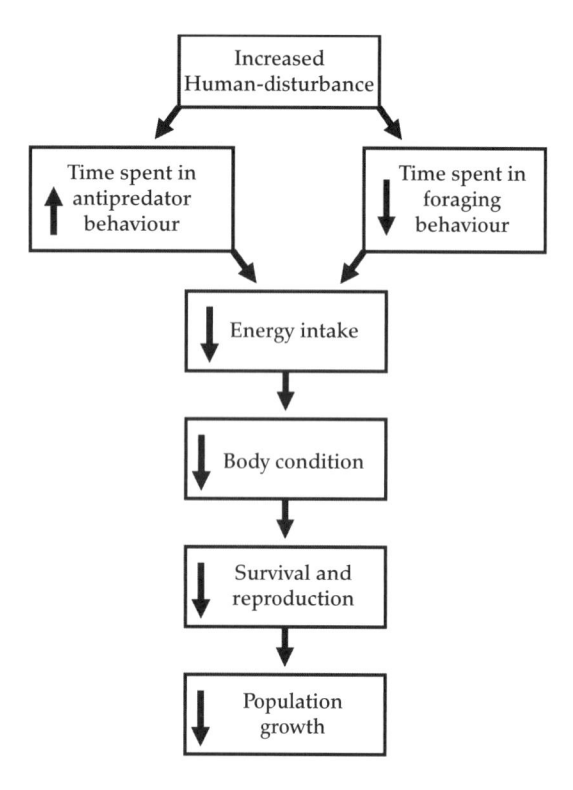

Figure 12.3 Conceptual model outlining the behavioural mechanisms by which increased rates of human disturbance could reduce population growth. Adapted from Frid and Dill (2002).

non-lethal activities, individuals might spend more time being vigilant and less time foraging, which over the long term can reduce survival or reproduction and, ultimately, population growth (Frid and Dill 2002; see also Chapter 17). Blumstein et al. (2005) have investigated interspecific variation in behavioural response to human presence (including the distance at which an individual detects human presence, hereafter referred to as alert distance) with variation in intensity of human visitation. They showed that the quantity of food consumed by a species is negatively influenced by the distance at which animals detect humans, the frequency of disturbance by humans, and the interaction between these factors. As alert distance was affected by a species' size (large species can detect humans further away), they suggest that large species are more likely to be disturbed by human activities (Blumstein et al. 2005). Although these studies support the contention that animals respond similarly to human disturbance as they do to predation risk, it is important to tease out underlying differences between these two factors. For example, species with large body size are better at detecting humans from farther away, leading to the prediction that they will experience greater disturbance from human activity. However, in the context of predation, if the same species are also better at detecting predators from a long distance, they might actually be less vulnerable to predation. More research effort is required before we can draw general conclusions about the link between behavioural responses to human activities and population growth.

12.5 From the individual to the population level

Theoretical biologists have done a great deal of work contrasting the consequences of different life history strategies and behavioural plasticity on population parameters. On the other hand, only a handful of empirical studies have assessed the importance of individual variation in behaviour, phenotype, or life history for population growth in nature. Recently, however, by applying new approaches to the analyses of long-term data sets of marked animals, it has been possible to evaluate the relative contribution of phenotype and environment to explain population dynamics. For example, Pelletier et al. (2007) calculated individual contributions to population growth (Coulson et al. 2006) from a long-term time series of marked Soay sheep *Ovis aries*. They showed that the distribution of body size in the population in each year explains, on average, 5% (and up to 18% in bad environmental conditions) of the population growth (Pelletier et al. 2007). Although this example pertains to how morphological attributes can affect population growth, the same analytical approach could easily be used to quantify the interplay between behaviour and population processes.

Despite the importance of understanding how wild species will cope with human-driven changes, most studies have been conducted in captivity and with domesticated species (Pelletier et al. 2009). Human effects on behaviour are, indeed, most

Box 12.2 Individual behaviour and survival in a risky environment

Ram Mountain is an isolated mountainous outcrop located around 30 km east of the main Canadian Rockies. Bighorn sheep using this area have been intensively monitored since 1971. For over 40 years, all sheep have been individually marked and monitored from birth to death. Sheep are captured in a corral trap baited with salt. Predation by cougar is relatively rare in this population. There were, however, short episodes of intense predation where individual cougars specialized on bighorn sheep leading to a decline in population size (Festa-Bianchet et al. 2006). From 1996 to 1999, researchers measured sheep boldness. Ewes were characterized using a boldness index based on their propensity to being trapped (Réale et al. 2000), which is assumed to reflect individual differences in willingness to accept the risk involved in licking the salt bait. Ewes that were rarely captured were considered shy and ewes frequently captured were considered bold. During years of high predation, adult ewe mortality increased dramatically to 27% a year, which is almost three times higher than the long-term average. Using the long-term data, Réale et al (2000) estimated the strength of selection on female boldness and docility in years of low and high predation. They showed that bold ewes were more likely to survive in years of high predation (Fig. 12.4). This result suggests that bold ewes contribute more to population growth rate (through their survival) in those years. Although, it is still unknown why bold ewes were less likely to be killed by cougars (one hypothesis is that bold ewes might be more willing to use alternative—or novel—habitats and thus be less predictable to the predator), this example underscores two important points. First, it is critical to note that even if the sheep population is declining during years of high predation, some behavioural traits can be preferentially selected. It also suggests that, in some cases, behavioural plasticity might not be enough to ensure population persistence. Second, as a result of individual variation in behaviour, one or few specialist predators can have strong consequences for the population dynamics of prey species. Although the sheep–cougar dynamics reported here are natural, it has been suggested that human driven-changes can be comparable to predation risk ((Frid and Dill, 2002) and see text). Similar patterns could thus emerge in the case of anthropogenic disturbances.

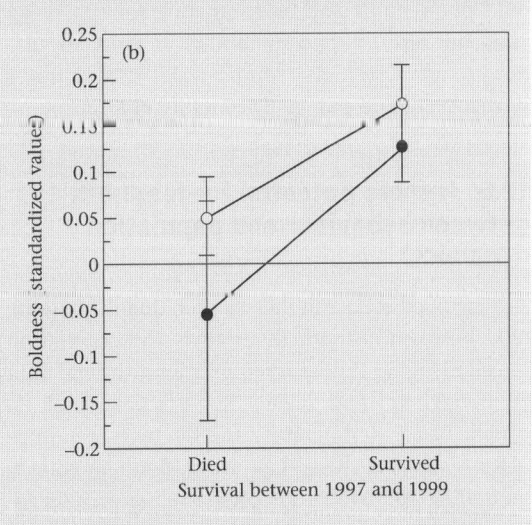

Figure 12.4 Boldness of bighorn ewes that died and survived during years of high predation (mean ± SE, dark circles <7 years old; white circles >7 years old). Adapted from Réale and Festa-Bianchet (2003), with permission from Elsevier.

noticeable in domestic species where individuals are subjected to drastically different environmental conditions and selective pressures from those usually documented under natural conditions (see also Chapter 8). Fish species used for aquaculture purposes, for example, are typically maintained for several generations in captivity where they are kept at high density and selected for rapid growth (Gross 1998). Such environmental characteristics have inadvertently also resulted in important modifications to domestic fish behaviour. Several studies conducted in salmonids have shown that domestic fish show higher levels of aggressiveness and are more prone to take risks than their wild counterparts (Biro et al. 2004; Blanchet et al. 2008). Domestic fish also usually show reduced antipredator responses and defective reproductive behaviours when placed in natural environments (Fleming et al. 2000). These differences in behaviour are ultimately threatening native populations as farmed fish are commonly escaping into the wild. Furthermore, in several instances, captive-bred fish are used to supplement natural populations. Problems may arise directly through competition with wild fish or indirectly through interbreeding, which potentially results in disruption of local adaptation, lower fitness, and eventually reduced population growth (Fleming et al. 2000; Araki et al. 2007). For example, Araki et al. (2007) showed a strong decline in lifetime reproductive success of steelhead trout *Oncorhynchus mykiss* that were captive-bred and released into the wild. This decline occurred after only a few generations of captive breeding. More recently, it has been shown that escaped Atlantic salmon *Salmo salar* not only depressed population recruitment by breeding in wild populations and by competing for spawning sites, but also affected the ability of wild populations to adapt to higher winter water temperatures linked to climate change (McGinnity et al. 2009).

12.6 Is there potential for feedback between behaviour and population dynamics?

As discussed earlier in this chapter, quantifying the links between individual variation and population is a challenging task as it requires longitudinal data on behaviour (and ideally some measure of its genetic basis), demography and fluctuation in population size. As these data are difficult to collect, modelling approaches are often used as an alternative to evaluate how selection in genetically-based behavioural strategies may feed back to the population level (see, for example, McNamara and Dall, 2011; Zheng et al. 2009). In this section, we focus on empirical examples where these feedbacks have been documented in the field.

An excellent example of feedback between selection and population growth has been found in a population of side-blotched lizards *Uta stansburiana* in California, USA (Sinervo et al. 2000). In this population, two heritable female reproductive strategies have been documented: yellow-throated females produce few large offspring (K-strategist) and orange-throated lizards produce many small offspring (r-strategist). The authors have elegantly shown that orange-throated females are favoured at low density because they produce a large number of small offspring. As environmental conditions are not restrictive due to low competition, most orange-throated offspring survive and the population thus becomes dominated by this morph. These r-strategists eventually cause an overshoot of the habitat carrying capacity within a single year, which triggers a population crash. At higher density, the yellow-throated K-strategy is favoured because the females' large offspring have higher survival rates during the crash years (Fig. 12.5). A large frequency of 'K-strategists' in the population reduces the rate of population growth. These lizards' reproductive strategies fit well with the r and K paradigm (MacArthur and Wilson 1967; Pianka 1970), yet these strategies are adopted by individuals of the same population. Selection on reproductive strategy differs according to population density (Fig. 12.5), leading to changes in their relative frequency, which causes the population density to fluctuate over a two-year period. This example also suggests that individual variation in behaviour might negatively impact population growth, although lizards adopting the K-reproductive strategy will contribute more to population growth in years of high density compared to those adopting the r strategy.

How does this example relate to human driven-changes? Human activities often tend to select for r-strategists. For example, as fisheries often target the larger fish within a population, those reproducing at a smaller size (and thus a younger age) are more likely to pass on their genes to the next

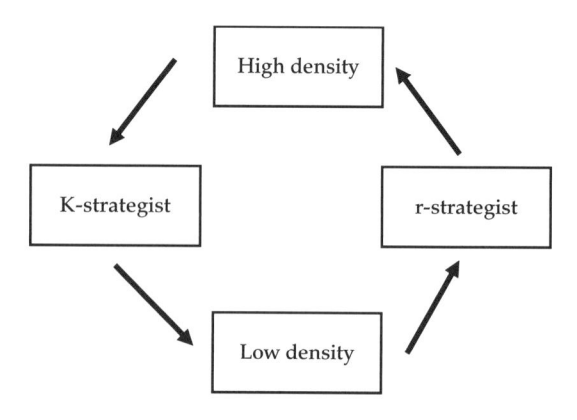

Figure 12.5 Graphical representation of strategies favoured by lizards according to population density. Based on Sinervo et al. (2000).

POPULATION CONSEQUENCES OF INDIVIDUAL VARIATION IN BEHAVIOUR 169

Box 12.3 Yellow bellied marmots: feedbacks between climate change, weaning dates, and population growth

Human-driven global climate changes are occurring at an unprecedented rate. Based on the last report published in 2010 by the Intergovernmental Panel on Climate Change, in the last 100 years, the planet's average temperature has increased by 0.74 °C. Phenological responses to climate change have been reported for a wide range of species (Walther et al. 2002; Parmesan and Yohe 2003). More recently, consequences of such changes for population dynamics have also emerged in the scientific literature. A good example comes from the longitudinal monitoring of individually marked yellow-bellied marmots *Marmota flaviventris* in Colorado, USA. A study by Ozgul et al. (2010) has revealed that, over the past 25 years, marmots have been emerging earlier from hibernation. Earlier emergence dates can have several implications for the marmot's biology, but can a change in phenology also have measurable feedbacks on population growth?

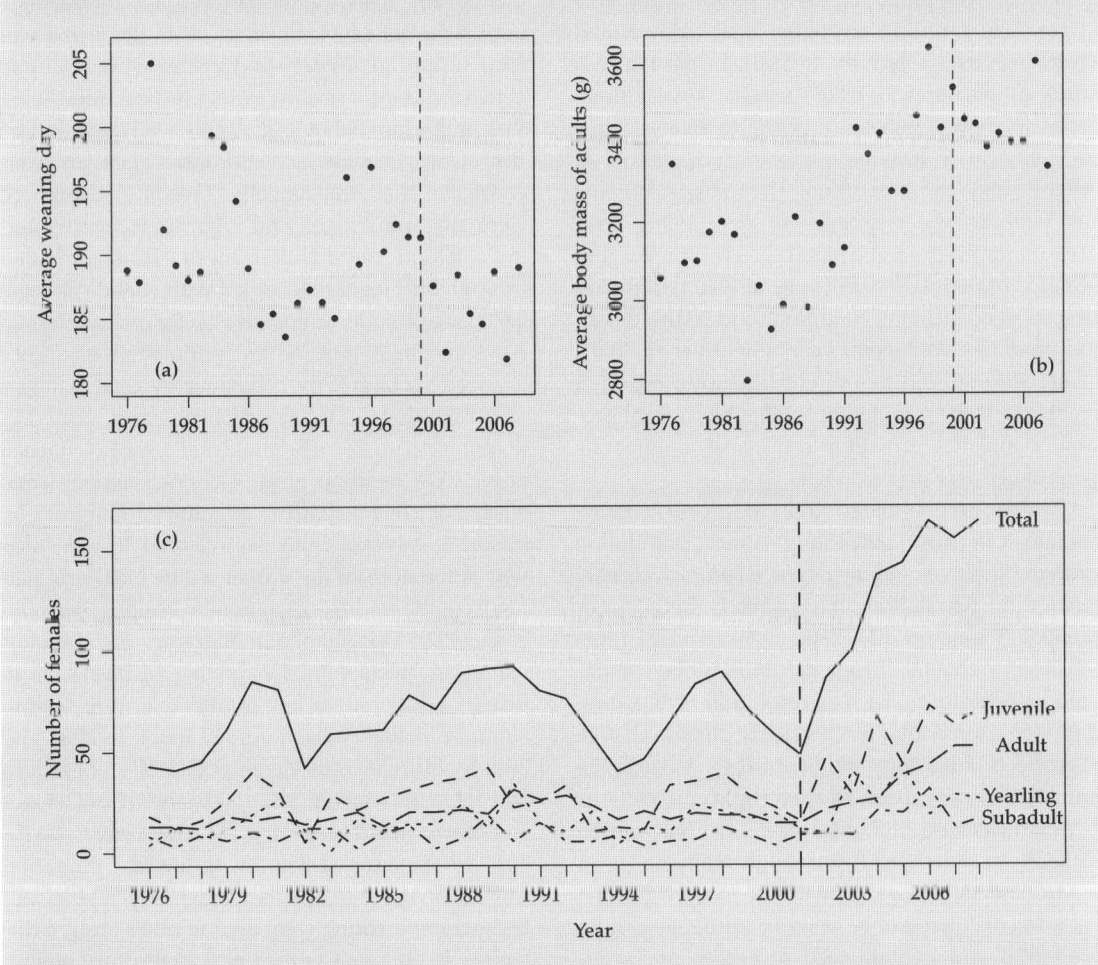

Figure 12.6 Trends in weaning dates and demography for female yellow-bellied marmots from 1976 to 2008 in Colorado, USA. (a) Average weaning date, (b) average body mass of adult females, and (c) size and composition of the marmot's population (females only). Note that the vertical dashed lines delineate different phases of population dynamics. Figure provided by A. Ozgul.

(Cont.)

> **Box 12.3 (Cont.)**
>
> Ozgul et al. (2010) were able to tackle this question. They applied this framework to the marmot's long-term data set (33 years, 1976–2008) to evaluate how environmental change has affected phenotypic traits and population dynamics over the study period. They showed that by advancing their emergence date from hibernation, marmots were able to wean their young earlier (Fig. 12.6 a). This early start allowed adults and juveniles to enjoy a longer growing season and to achieve larger body masses before hibernation (Fig. 12.6 b). Importantly, most of this phenotypic change could be attributed to a plastic response to environmental change, rather than a microevolutionary response (Ozgul et al. 2010). Consequently, the change in phenotype has led to an increase in adult survival, which has resulted in an abrupt increase in marmot population growth in recent years (Fig. 12.6 c). Such an increase in population growth could eventually feedback on individuals through density-dependent effects.

generation compared to those reproducing at a larger size (and at an older age) (Rowe and Hutchings 2003; Olsen et al. 2004). In particular, a study by Swain et al. (2007) used a 30-year time-series to show a genetic response to size-selective mortality in a heavily exploited population of Atlantic cod *Gadus morhua*. They suggested that the evolutionary responses to overfishing may account for the observed small age-specific size and absence of population recovery in this population despite good growing conditions and reduced fishing pressures over recent years (Swain et al. 2007). Thus, artificial selection might lead to unexpected feedbacks at the population level, which might promote population growth in some cases or affect its persistence over time in others.

Change in phenology due to global climate change can also potentially affect population growth. Positive and negative effects of climate change on populations have already been documented (Parmesan and Yohe 2003). Several traits affected by climate change can broadly be considered as behaviours, such as migration or breeding date (see also Chapter 6). Obviously, climate change also affects physiology and life history. As such, the animal responses observed are complex interactions between different biological processes that can be difficult to detangle. Anthropogenically induced environmental changes, however, create 'large scale experiments' that can be used to understand how ecological and evolutionary processes are intertwined (Box 12.3).

Other human activities, such as artificial food supplementation, can also be seen as large-scale experiments. For instance, it has become increasingly popular to set up bird feeders. With the increase of food supply in winter—and milder winters due to climate change—several birds are now modifying their migratory behaviour. A very well documented case of such changes is the European blackcap *Sylvia atricapilla* that traditionally breeds in southern Germany and Austria and migrates to overwintering sites in the Mediterranean. Food provided through bird feeders (coupled with climate change) has made other locations suitable for overwintering. As a result, a part of the population has since changed its migratory behaviour by establishing a new migratory route to wintering areas located in the United Kingdom (Berthold et al. 1992; Rolshausen et al. 2009; Bearhop et al. 2005). An experimental study has shown that the migratory route used by offspring hatched from captive-bred British birds was different from the route used by birds originating from the traditional migration regions (Berthold et al. 1992). This change in migratory direction has led to an increase in the number of blackcaps in Britain: while only a few individuals were seen in summer in the 1960s, a survey made in 2003 and 2004 by birdwatchers suggests that 31% of gardens had frequent visits of this species between October and March (Bearhop et al. 2005). There were also fitness differences associated with the new migration route: British blackcaps had higher clutch sizes and fledged more young compared to other birds, a difference likely due to differences in territory quality of male blackcaps, as birds migrating to Britain arrived earlier than others on their breeding ground (Bearhop et al. 2005). Thus, this change in behaviour affected individual performance, promoted the establishment of a new population, and affected population growth. Additionally, as blackcaps form seasonal breeding pairs on their breeding ground, this change in migratory orientation also led to temporal segregation and promoted assortative mating between birds from the same wintering habitat (Bearhop et al. 2005), eventually leading to rapid microevolutionary changes in morphology between birds with different migratory routes (Bearhop et al. 2005; Rolshausen et al. 2009).

12.7 Concluding remarks and future directions

Throughout this chapter, we aimed to emphasize the possible ways in which behaviour and population dynamics might interact. One of the obvious conclusions is that while several studies have shown the influence of population dynamics on behaviour, still very few have managed to convincingly assess the effect of behaviour on population dynamics and even fewer studies have thus managed to document an actual feedback. Empirical studies quantifying the effects of variation in behaviour are therefore needed if we want to understand the consequences of individual variation in behaviour at the population level. We also require more studies documenting the effect size of behaviours on population growth, beyond the effects of density, weather, disease, and predation. How much more variation in population growth can we actually explain by including behaviour in population dynamics models? No one has convincingly addressed this question under field conditions. While it is possible that behaviour is important under some circumstances, it may not affect population growth in other contexts, but so far no generality can be drawn. It is also important to emphasize that in some specific situations, individuals adopting a particular behavioural tactic could contribute more to population growth, even if populations are stable or declining. A second conclusion is that studies conducted in a human-perturbed context that have also documented individual fitness are still relatively rare. However, as human-driven changes are often large scale in magnitude, we might expect that studies conducted in these contexts are more likely to bring insights into the links between behaviour and population dynamics in the near future. Most importantly, to understand the effect of humans on animal populations, studies also need to be conducted in habitats where humans and animals persist together and at different spatial and temporal scales. Obviously, if a link between behavioural phenotypes and fitness exists, and if these behavioural phenotypes are heritable, it is almost inevitable that human-mediated changes in phenotypic distribution should affect population growth. Very short-term individual variations in behaviour, due to human disturbances that are of small magnitude, are likely to reduce energy intake, but if this disturbance is short in duration or localized in space, it should have little influence on population dynamics. It would therefore be very helpful to have a framework allowing us to predict under what circumstances we might expect changes in behaviour to affect population processes. Such frameworks should be developed in combination with studies aiming to depict the extent to which the contribution of a change in a given behavioural phenotype will affect population growth compared to other environmental drivers (climate, density) and human activities.

References

Araki, H., Cooper, B., and Blouin, M. S. (2007). Genetic effects of captive breeding cause a rapid, cumulative fitness decline in the wild. *Science*, 318, 100–3.

Bearhop, S., Fiedler, W., Furness, R. W., Votier, S. C., Waldron, S., Newton, J., Bowen, G. J., Berthold, P., and Farnsworth, K. (2005). Assortative mating as a mechanism for rapid evolution of a migratory divide. *Science*, 310, 502–4.

Bell, A. M. (2005). Behavioural differences between individuals and two populations of stickleback (Gasterosteus aculeatus). *Journal of Evolutionary Biology*, 18, 464–73.

Benton, T. G. and Grant, A. (1999). Elasticity analysis as an important tool in evolutionary and population ecology. *Trends in Ecology and Evolution*, 14, 467–71.

Berthold, P., Helbig, A. J., Mohr, G., and Querner, U. (1992). Rapid microevolution of migratory behaviour in a wild bird species. *Nature*, 360, 668–70.

Biro, P. A., Abrahams, M. V., Post, J. R., and Parkinson, E. A. (2004). Predators select against high growth rates and risk-taking behaviour in domestic trout populations. *Proceedings of the Royal Society of London B, Biological Sciences*, 271, 2233–7.

Bischof, R., Fujita, R., Zedrosser, A., Soderberg, A., and Swenson, J. E. (2008). Hunting patterns, ban on baiting, and harvest demographics of brown bears in Sweden. *Journal of Wildlife Management*, 72, 79–88.

Blanchet, S., Páez, D. J., Bernatchez, L., and Dodson, J. J. (2008). An integrated comparison of captive-bred and wild atlantic salmon (salmo salar): Implications for supportive breeding programs. *Biological Conservation*, 141, 1989–99.

Blumstein, D. T., Fernandez-Juricic, E., Zollner, P. A., and Garity, S. C. (2005). Inter-specific variation in avian responses to human disturbance. *Journal of Applied Ecology*, 42, 943–53.

Boon, A. K., Réale, D., and Boutin, S. (2007). The interaction between personality, offspring fitness and food abundance in North American red squirrels. *Ecology Letters*, 10, 1094–104.

Boonstra, R. and Boag, P. T. (1987). A test of the Chitty hypothesis: inheritance of life-history traits in meadow voles Microtus pennsylvanicus. *Evolution*, 41, 929–47

Bourbeau-Lemieux, A., Festa-Bianchet, M., Gaillard, J. M., and Pelletier, F. (2011). Predator-driven component Allee effects in a wild ungulate. *Ecology Letters*, 14, 358–63.

Caro, T. (1998). *Behavioral Ecology and Conservation Biology*, New York, Oxford University Press.

Chitty, D. (1967). The natural selection of self-regulatory behaviour in animal populations *Proceedings of the Ecological Society of Australia*, 2, 51–78.

Côté, S. D., Rooney, T. P., Tremblay, J.-P., Dussault, C., and Waller, D. M. (2004). Ecological impacts of deer overabundance. *Annual Review of Ecology, Evolution and Systematic*, 35, 113–47.

Coulson, T., Benton, T. G., Lundberg, P., Dall, S. R. X., Kendall, B. E., and Gaillard, J. M. (2006). Estimating individual contributions to population growth: evolutionary fitness in ecological time. *Proceedings of the Royal Society of London B, Biological Sciences*, 273, 547–55.

Coulson, T. and Tuljapurkar, S. (2008). The dynamics of a quantitative trait in an age-structured population living in a variable environment. *American Naturalist*, 172, 599–612.

Courchamp, F., Berec, L., and Gascoigne, J. (2008). *Allee Effects in Ecology and Conservation*. Oxford, UK, Oxford University Press.

Darimont, C. T., Carlson, S. M., Kinnison, M. T., Paquet, P. C., Reimchen, T. E., and Wilmers, C. C. (2009). Human predators outpace other agents of trait change. *Proceedings of the National Academy of Sciences of the USA*, 106, 952–4.

Dingemanse, N. J., Both, C., Drent, P. J., Van Oers, K., and Van Noordwijk, A. J. (2002). Repeatability and heritability of exploratory behaviour in great tits from the wild. *Animal Behaviour*, 64, 929–38.

Duckworth, R. A. (2006). Behavioral correlations across breeding contexts provide a mechanism for a cost of aggression. *Behavioural Ecology*, 17, 1011–19.

Duckworth, R. A. (2008). Adaptive dispersal strategies and the dynamics of a range expansion. *American Naturalist*, 172 (Suppl. 1), S4–S17.

Duckworth, R. A. and Badyaev, A. V. (2007). Coupling of dispersal and aggression facilitates the rapid range expansion of a passerine bird. *Proceedings of the National Academy of Sciences of the USA*, 104, 15017–22.

Duckworth, R. A. and Kruuk, L. E. B. (2009). Evolution of genetic integration between dispersal and colonization ability in a bird. *Evolution*, 63, 968–77.

Festa-Bianchet, M. and Apollonio, M. (2003). General introduction. In Festa-Bianchet, M. and Apollonio, M. (eds) *Animal Behavior and Wildlife Conservation*. Washington, DC, Island Press.

Festa-Bianchet, M., Coulson, T., Gaillard, J. M., Hogg, J. T., and Pelletier, F. (2006). Stochastic predation events and population persistence in bighorn sheep. *Proceedings of the Royal Society of London B, Biological Sciences*, 273, 1537–43.

Fleming, I. A., Hindar, K., Mjolnerod, I. B., Jonsson, B., Balstad, T., and Lamberg, A. (2000). Lifetime success and interactions of farm salmon invading a native population. *Proceedings of the Royal Society of London B, Biological Sciences*, 267, 1517–23.

Frid, A. and Dill, L. (2002). Human-caused disturbance stimuli as a form of predation risk. *Conservation Ecology*, 6(1), 11.

Gill, F. B. (1997). Local cytonuclear extinction of the Golden-winged Warbler. *Evolution*, 51, 519–25.

Gross, M. R. (1998). One species with two biologies: Atlantic salmon (salmo salar) in the wild and in aquaculture. *Canadian Journal of Fisheries and Aquatic Sciences*, 44, 131–44.

Hendry, A. P., Farrugia, T. J., and Kinnison, M. T. (2008). Human influences on rates of phenotypic change in wild animal populations. *Molecular Ecology*, 17, 20–9.

Hendry, A. P. and Kinnison, M. T. (1999). The pace of modern life: measuring rates of contemporary microevolution. *Evolution*, 53, 1637–53.

Hoekstra, H. E., Hoekstra, J. M., Berrigan, D., Vignieri, S. N., Hoang, A., Hill, C. E., Beerli, P., and Kingsolver, J. G. (2001). Strength and tempo of directional selection in the wild. *Proceedings of the National Academy of Sciences of the USA*, 98, 9157–60.

Kingsolver, J. G., Hoekstra, H. E., Hoekstra, J. M., Berrigan, D., Vignieri, S. N., Hill, C. E., Hoang, A., Gibert, P., and Beerli, P. (2001). The strength of phenotypic selection in natural populations. *American Naturalist*, 157, 245–61.

Kokko, H. and Lopez-Sepulcre, A. (2007). The ecogenetic link between demography and evolution: can we bridge the gap between theory and data? *Ecology Letters*, 10, 773–82.

Macarthur, R. H. and Wilson, E. O. (1967). *Theory of Island Biogeography*, Princeton, Princeton University Press.

Mcginnity, P., Jennings, E., Deeyto, E., Allott, N., Samuelsson, P., Rogan, G., Whelan, K., and Cross, T. (2009). Impact of naturally spawning captive-bred atlantic salmon on wild populations: Depressed recruitment and increased risk of climate-mediated extinction. *Proceedings of the Royal Society of London B, Biological Sciences*, 276, 3601–10.

Mcnamara, J. M. and Dall, S. R. X. (2011). The evolution of unconditional strategies via the 'multiplier' effect. *Ecology Letters*, 14, 237–43.

Milner-Gulland, E. J., Bukreeva, O. M., Coulson, T. N., Lushchekina, A. A., Kholodova, A. B., and Grachev, I. A. (2003). Reproductive collapse in saiga antelope harems. *Nature*, 422, 135.

Mysterud, A., Coulson, T., and Stenseth, N. C. (2002). The role of males in the dynamics of ungulate populations. *Journal of Animal Ecology*, 71, 907–15.

Olsen, E. M., Heino, M., Lilly, G. R., Morgan, M. J., Brattey, J., Ernande, B., and Dieckmann, U. (2004). Maturation trends indicative of rapid evolution preceded the collapse of northern cod. *Nature*, 428, 932–5.

Ozgul, A., Childs, D. Z., Oli, M. K., Armitage, K. B., Blumstein, D. T., Olson, L. E., Tuljapurkar, S., and Coulson, T. (2010). Coupled dynamics of body mass and population growth in response to environmental change. *Nature*, 466, 482–5.

Parmesan, C. and Yohe, G. (2003). A globally coherent fingerprint of climate change impacts across natural systems. *Nature*, 421, 37–42.

Pelletier, F., Clutton-Brock, T., Pemberton, J., Tuljapurkar, S., and Coulson, T. (2007). The evolutionary demography of ecological change: linking trait variation and population growth. *Science*, 315, 1571–4.

Pelletier, F., Garant, D., and Hendry, A. P. (2009). Eco-evolutionary dynamics: an introduction. *Philosophical Transactions of the Royal Society B: Biological Sciences*, 364, 1483–90.

Pianka, E. R. (1970). On r and K selection. *American Naturalist*, 104, 592–7.

Pimentel, D. (1968). Population regulation and genetic feedback. *Science*, 159, 1432–7.

Rankin, D. J. and Kokko, H. (2007). Do males matter? The role of males in population dynamics. *Oikos*, 116, 335–48.

Réale, D. and Festa-Bianchet, M. (2003). Predator-induced natural selection on temperament in bighorn ewes. *Animal Behaviour*, 65, 463–70.

Réale, D., Gallant, B. Y., Leblanc, M., and Festa-Bianchet, M. (2000). Consistency of temperament in bighorn ewes and correlates with behaviour and life history. *Animal Behaviour*, 60, 589–97.

Réale, D., Garant, D., Humphries, M. M., Bergeron, P., Careau, V., and Montiglio, P.-O. (2010). Personality and the emergence of the pace-of-life syndrome concept at the population level. *Philosophical Transactions of the Royal Society B: Biological Sciences*, 365, 4051–63.

Robinson, H. S., Wielgus, R. B., and Gwilliam, J. C. (2002). Cougar predation and population growth of sympatric mule deer and white-tailed deer. *Canadian Journal of Zoology*, 80, 556–68.

Rolshausen, G., Segelbacher, G., Hobson, K. A., and Schaefer, H. M. (2009). Contemporary evolution of reproductive isolation and phenotypic diuvergence in sympatry along migratory divide. *Current Biology*, 19, 2097–101.

Rowe, S. and Hutchings, J. A. (2003). Mating systems and the conservation of commercially exploited marine fish. *Trends in Ecology and Evolution*, 18, 567–72.

Saccheri, I. and Hanski, I. (2006). Natural selection and population dynamics. *Trends in Ecology and Evolution*, 21, 341–7.

Servheen, C., Herrero, S., and Peyton, B. (1999). *Bears. Status Survey and Conservation Action Plan*, Cambridge, UK, IUCN/SSC Bear and Polar Bear Specialist Groups.

Siepielski, A. M., Dibattista, J. D., and Carlson, S. M. (2009). The temporal dynamics of phenotypic selection in the wild. *Ecology Letters*, 12, 1261–76.

Sinervo, B., Svensson, E., and Comendant, T. (2000). Density cycles and an offspring quantity and quality game driven by natural selection. *Nature*, 406, 985–8.

Slobodkin, L. B. (1961). *Growth and Regulation of Animal Populations*, New York, NY, Holt, Rinehart and Winston.

Stirling, D. G., Réale, D., and Roff, D. A. (2002). Selection, structure and the heritability of behaviour. *Journal of Evolutionary Biology*, 15, 277–89.

Stockwell, C. A., Hendry, A. P., and Kinnison, M. T. (2003). Contemporary evolution meets conservation biology. *Trends in Ecology and Evolution*, 18, 94–101.

Sutherland, W. J. (1996). *From Individual Behaviour to Population Ecology*. Oxford, Oxford University Press.

Swain, D. P., Sinclair, A. F., and Hanson, J. M. (2007). Evolutionary response to size-selective mortality in an exploited fish population. *Proceedings of the Royal Society of London B, Biological Sciences*, 274, 1015–22.

Swenson, J. E. (1999). Does hunting affect the behavior of brown bears in Eurasia? *Ursus*, 11, 157–62.

Swenson, J. E., Sandegren, F., Söderberg, A., Bjärvall, A., Franzén, R., and Wabakken, P. (1997). Infanticide caused by hunting male bears. *Nature*, 386, 450–1.

Vallender, R., Robertson, R. J., Friesen, V. L., and Lovette, I. J. (2007). Complex hybridization dynamics between golden-winged and blue-winged warblers (*Vermivora chrysoptera* and *Vermivora pinus*) revealed by AFLP, microsatellite, intron and mtDNA markers. *Molecular Ecology*, 16, 2017–29.

Walther, G.-R., Post, E., Convey, P., Menzel, A., Parmesan, C., Beebee, T. J. C., Fromentin, J.-M., Hoegh-Guldberg, O., and Bairlein, F. (2002). Ecological responses to recent climate change. *Nature*, 416, 389–95.

Zedrosser, A., Dahle, B., Swenson, J. E., and Gerstl, N. (2001). Status and management of the brown bear in Europe. *Ursus*, 12, 9–20.

Zheng, C., Ovaskainen, O., and Hanski, I. (2009). Modelling single nucleotide effects in the *Pgi* gene on dispersal in the Glanville fritillary: coupling between ecological and evolutionary dynamics. *Philosophical Transactions of the Royal Society B: Biological Sciences*, 364, 1519–32.

CHAPTER 13

Ecosystem consequences of behavioural plasticity and contemporary evolution

Eric P. Palkovacs and Christopher M. Dalton

⊃ Overview

Animal behaviour is a critical, but underappreciated, link between human activity and ecosystem processes. In this chapter, we review the impacts of behavioural trait changes on ecosystems. Such effects are often the result of changes in traits related to consumption and nutrient cycling. Phenotypic plasticity and contemporary evolution are two mechanisms of rapid behavioural trait change that shape ecosystem processes. However, the effects of plasticity and evolution have typically been considered in isolation. We propose a framework to integrate the ecosystem effects of plasticity and evolution using a reaction norm approach. This method can be used to parse the contributions of plasticity, evolution, and the evolution of plasticity to ecosystem change and may be applied to predict the effects of human-driven trait change on ecosystems.

13.1 Introduction

Animal behaviour is a central theme in evolutionary ecology, population ecology, and community ecology. In contrast, the role of behaviour in shaping ecosystems has received much less attention. Some well known examples of behaviour moulding ecosystem processes do exist, such as the impacts of spawning Pacific salmon *Oncorhynchus* spp. and foraging brown bears *Ursus arctos* on ecosystem structure and function in the freshwaters and forests of the American Pacific Northwest. In these coastal habitats, migrating salmon enter freshwaters from the ocean, importing large quantities of marine derived nutrients into otherwise nutrient poor streams. These nutrients are then transported far into the forest by foraging bears, with large effects on both aquatic and terrestrial ecosystems (Schindler et al. 2003). Recent evidence supports the notion that behaviour is a widespread and potent driver of ecosystem processes (Schmitz et al. 2008). Despite such evidence, animal behaviour has yet to occupy a central place in ecosystem ecology. A greater integration of animal behaviour and ecosystem ecology can enable evolutionary theory to be applied to predict future ecosystem change (López-Sepulcre 2011). Predicting such changes is an urgent goal because ecosystems, and the services they provide, are being rapidly altered by human activity.

Two well-characterized mechanisms by which animal behaviour may shape ecosystem processes are consumption and nutrient cycling. The ecosystem effects of consumption, or the threat of being consumed, have been explored through the related concepts of keystone predation (Paine 1969; Power et al. 1996) and trophic cascades (Hairston et al. 1960; Terborgh et al. 2010). The ecosystem effects of nutrient cycling have been explored with respect to nutrient translocation, rates of recycling, and the chemical composition of excreted wastes (Vanni 2002). With the notable exception of behaviourally-mediated trophic cascade research (Schmitz et al. 1997), few ecosystem studies have historically included a detailed consideration of short-term behavioural trait change. Because human-driven environmental change is happening rapidly, it is precisely these short-term trait changes that will mediate the role of behaviour in shaping ecosystem responses to human activity.

Humans can change the behaviours of animals by influencing two mechanisms of short-term trait change—phenotypic plasticity and contemporary evolution. Because plasticity can emerge within a single generation, ecologists have long recognized the potential for plasticity to mould ecological interactions (Miner et al. 2005). The nascent study of eco-evolutionary dynamics has added evolution to the mix, establishing that contemporary evolution (sometimes called rapid evolution) can also impact ecological processes (Post and Palkovacs 2009; Schoener 2011). Despite studies showing that either plasticity or contemporary evolution can influence ecosystem processes independently, there presently exists a very limited understanding of the joint effects of plasticity *and* evolution on ecosystems.

Here we review two well-established mechanisms by which behaviours can shape ecosystems—consumption and nutrient cycling. We then describe how phenotypic plasticity and contemporary evolution can shape behavioural traits that can then impact ecosystem processes. Finally, we propose a framework to integrate the ecosystem effects of phenotypic plasticity and evolutionary trait changes using a reaction norm approach, which explicitly recognizes that plasticity and evolution operate together to shape the ecosystem consequences of behavioural trait change. This framework complements the approach of Ellner et al. (2011), which similarly decomposes sources of ecological change into their evolutionary and plastic (or other non-heritable) components.

13.2 Behavioural effects on ecosystems

For the purpose of this review, we broadly define ecosystems as bounded sets of interacting organisms and their associated abiotic environment. Ecosystem processes thus entail direct and indirect linkages between biotic and abiotic components, and an ecosystem effect occurs when a change in one part of the system drives changes throughout the system via these linkages. Ecosystem change of this sort can be detected by measuring aggregate traits of the system such as primary productivity and decomposition. Dramatic shifts in ecosystem processes often co-occur with shifts in community structure (Hooper et al. 2005). Therefore, we use the term ecosystem effect to denote both measured changes to ecosystem processes and changes in community structure, as the two are commonly inter-related.

13.2.1 Consumption

Acquisition of food resources represents one of the strongest interactions between animals and their environment. Many cases of demonstrated ecosystem effects by animals result from consumption or the threat of being consumed. Consumption by predators can alter communities directly, by reducing the abundance of a preferred prey, and indirectly, by altering how prey interact with other community members through changes in their behaviour (Schmitz et al. 2004; also see Chapter 10). These changed communities can then impact ecosystem processes. Such multi-trophic interactions and their ecosystem effects have spawned fundamental ecological concepts, including keystone species and trophic cascades.

The term keystone species was coined by Paine (1969) to describe the strong effects that a predator can have on community composition if it selectively preys upon a dominant competitor. The original example of this phenomenon is the purple sea star *Pisaster ochracues*, that fundamentally changes prey communities by selectively consuming a competitive-dominant species of mussel *Mytilus californianus*, thereby creating space for other sessile invertebrates. Communities with *Pisaster* are more diverse than those without, which became monocultures of the competitive dominant mussel (Paine 1966). Research in other systems found that regulation of consumers by predators directly affects intertidal primary production (Silliman and Bertness 2002). Subsequent adoption of the term 'keystone species' in other systems led to a broadened definition that made keystone species nearly inseparable from 'ecosystem effects', where a keystone species is any species whose effect on the ecosystem is large relative to its abundance in that system (Power et al. 1996). Under this broad definition, many other means by which predators shape ecosystems become subject to the term keystone species, including species that initiate trophic cascades.

The classic density-mediated trophic cascade results when the addition of a predator reduces the abundance of herbivores, which increases the abundance of primary producers (Hairston et al. 1960). Density-mediated trophic cascades can cause changes to diverse ecosystem processes, such as primary productivity, decomposition, and nutrient cycling (Pace et al. 1999). However, trophic cascades can also result from predator-induced change in prey behaviour, a phenomenon termed a behaviourally-mediated trophic cascade (Schmitz et al. 1997). Such effects occur when the threat of predation induces predator-avoidance behaviour in prey (a non-consumptive effect), thereby reducing allocation of prey effort to feeding. Reduced feeding activity can induce trophic cascades, even if the actual abundance of consumers does not change (Schmitz et al. 2004).

Such a pattern has been demonstrated in many systems, with one compelling example being the interaction between wolves *Canis lupus*, elk *Cervus canadensis*, and woody vegetation in the Rocky Mountains of North America. In this system, wolves have been reintroduced to areas where they had previously been extirpated. Reintroduction of wolves altered the foraging behaviour of elk, inducing them to forage less often and away from preferred aspen stands (Fortin et al. 2005). This change in elk behaviour resulted in both redistribution of woody vegetation across space and an overall increase in woody vegetation abundance (Beyer et al. 2007). Other examples of behaviourally-mediated trophic cascades abound in the literature, in ecosystems including lakes, streams, marine intertidal, open ocean, terrestrial grassland, and forest (Schmitz et al. 2004).

In addition to changes in prey behaviour, changes in predator behaviour can also impact trophic cascades. A high profile example of such an effect has been suggested in the Pacific Northwest of North America, where a shift in prey choice by killer whales *Orcinus orca* may have interrupted the trophic cascade between sea otters *Enhydra lutris*, sea urchins *Strongylocentrotus spp.*, and kelp forests. Historically, sea otters were apex predators in coastal areas and reduced the abundance of the dominant herbivore, sea urchins; however, a reduction in alternate prey sources may have led killer whales to switch prey and consume sea otters. Numerical reduction in sea otter abundance reduced predation on sea urchins and increased herbivory of kelp forests, resulting in ecosystem-altering changes at each trophic level (Estes et al. 1998).

Finally, the interacting effects of predator hunting behaviour and prey avoidance behaviour may have ecosystem consequences. Theory suggests that prey are especially likely to evolve behavioural avoidance mechanisms for sit-and-wait predators, whose sedentary and cryptic habits enable prey to use scent and chemical cues as reliable indicators of the presence of a predation threat (Preisser et al. 2007). Prey may not benefit from the same responses to active predators, whose mobility may enable them to move faster than their cues. If predator avoidance behaviour occurs more strongly in response to sit-and-wait predators than actively-hunting predators, such behavioural shifts can have ecosystem effects, for example on the strength of trophic cascades (Preisser et al. 2007). A theoretical study indicates that foraging mode of predators alters coexistence patterns among primary producers (Calcagno et al. 2011), and empirical studies demonstrate the strong role that predator foraging mode plays in structuring food webs and nutrient cycling through the behavioural and physiological responses of prey (Schmitz 2008; Lazzaro et al. 2009; Carey and Wahl 2010). In addition, prey escape mode can impact predator–prey interactions, yet no studies have explicitly examined how the escape mode of prey affects trophic cascades (Wirsing et al. 2010).

13.2.2 Nutrient cycling

The availability of nutrients can place important controls on ecosystems. If a nutrient critical for biological processes is scarce, animals can strongly influence the ecosystem by either increasing or decreasing the availability of that nutrient, which can affect ecosystem processes such as primary production and decomposition (Vitousek and Howarth 1991; Greenwood et al. 2007). One mechanism whereby animals can affect their chemical environment is through the release of waste. In this sense, animals act as chemical converters, ingesting compounds from the environment, modifying those compounds through physiological processes, and releasing excesses as waste that can impact nutrient availability in the ecosystem (McIntyre et al. 2008). For example, a study comparing nutrient excretion by two species of aquatic herbivore found that one species released higher levels of phosphorus waste than the other, increasing the availability of phosphorus, thereby increasing primary productivity (Knoll et al. 2009).

The amount and chemical composition of released wastes are controlled by numerous behavioural mechanisms including diet selection and activity level. Animals can alter the chemical composition of their waste by altering the chemical composition of their intake. Omnivorous consumers, in particular, can dramatically alter their chemical intake by switching food sources (often from autotrophic to heterotrophic sources). Such diet switches almost always result in changes in the amount and chemical form of nutrients released into the environment (Vanni 2002; Saba et al. 2009). The volume and chemical composition of waste products are also impacted by activity levels. Increases in metabolic activity, whether due to predation, increased competition, or altered foraging behaviour, may have the effect of increasing release of nitrogenous waste at the same level of phosphorus waste (Uliano et al. 2010). Since primary production in many terrestrial and aquatic systems is tightly controlled by the availability of nitrogen and phosphorus, changing the relative availability of these compounds through release of wastes can have profound ecosystem effects (Glaholt and Vanni 2005; McIntyre et al. 2008).

While much remains to be learned about how animal behaviour alters ecosystems via changed nutrient cycling within a single system, animals have repeatedly been shown to have important ecosystem effects by altering the spatial distribution of nutrients across the landscape. These effects are especially pronounced when animals move nutrients from ecosystems that are replete in nutrients to those that are nutrient starved, providing a nutrient subsidy to the recipient ecosystems (Flecker et al. 2010). For example, salmon are profoundly important to streams draining into the North Pacific. The critical role of salmon stems from their spawning migration, in which they carry nutrients from rich marine systems and deposit them in nutrient-poor streams as spawning excess and senesced somatic tissue. The nutrients relocated by salmon support a drastically altered stream community (Tiegs et al. 2009) and affect neighbouring terrestrial systems, as studies have found salmon-derived nutrients in trees and even songbirds (Christie et al. 2008).

While mass seasonal migrations like those of salmon are undoubtedly important for ecosystems, less dramatic movements within and between habitat patches can also influence critical ecosystem processes. Daily and small-scale movement patterns are behaviours that are often highly flexible and can have pronounced ecosystem effects when animals move nutrients both across and within ecosystem boundaries. Such effects can occur when animals ingest resources in one habitat and transport nutrients as waste products to another. For example, snow geese *Chen caerulescens* migrate daily from feeding areas in agricultural fields to roosting areas in relatively predator-free wetlands. By taking in nutrients from one ecosystem, agricultural fields, and depositing them as fecal matter in another, wetlands, snow geese provide massive nutrient inputs to roosting ponds equal to 40% and 75% of the total nitrogen and phosphorus loading into those wetlands (Post et al. 1998). Such cases of cross boundary subsidies are common in bird species, and the ecosystem consequences are variable and widespread (Schmitz et al. 2010). In another example, gizzard shad *Dorosoma cepedianum* consume detritus from the benthic zone of lakes. Following benthic feeding bouts, gizzard shad move into the pelagic zone, where they excrete the nitrogen and phosphorus consumed from the benthos, enhancing phytoplankton production (Vanni et al. 2005). Coral reef fishes play a similar role in marine ecosystems. Several species of grunts *Haemulon flavolineatum* and *Haemulon plumieri* make nightly foraging excursions from coral reefs into surrounding seagrass beds. The daily return of fish to the safety of the reef results in the deposition of nutrients, providing a significant subsidy which may enhance the growth rate of corals (Meyer and Schultz 1985).

Predators can also impact the spatial distribution of nutrients by depositing prey carcasses across the landscape. For example, brown bears foraging on salmon carcasses often drag meals away from stream edges and into nearby riparian habitats, and their physical movement of the carcasses increases nitrogen load and carbon flux in those terrestrial soils (Holtgrieve et al. 2009). Similarly, wolves aggregate moose *Alces alces* carcasses on Isle Royal National Park, which provide a highly concentrated nutrient supply to growing plants (Bump et al. 2009). The presence of carcasses can alter competitive dynamics and shape the community structure of plants that establish in the new canopy.

13.3 Rapid behavioural trait change

Behaviour can play a critical role in determining how ecosystems function. Identifying key behavioural traits is important for understanding how anthropogenic environmental change may alter ecosystems. For traits to be important drivers of short-term ecosystem change, however, they must also be subject to rapid change under altered environmental conditions. Here we focus on two causes of rapid behavioural trait change—phenotypic plasticity and contemporary evolution—and their impacts on ecosystems.

13.3.1 Behavioural plasticity

Short-term behavioural trait changes can be the product of phenotypic plasticity in response to environmental cues. Such behavioural shifts can have strong ecosystem effects. As discussed above, much existing research documents how predation affects consumer behaviour and how these behavioural changes can impact ecosystem function through behaviourally-mediated trophic cascades. Under the risk imposed by predators, prey spend less time foraging and do so closer to cover (Schmitz et al. 1997). These behavioural shifts reduce the number of prey consumed and alter the spatial distribution of prey foraging activity (Schmitz 1998). Because the prey are often herbivores, behavioural changes may directly increase primary productivity (Halaj and Wise 2001) and alter the spatial distribution and community composition of primary producers (Calcagno et al. 2011).

On a broader spatial scale, predation risk can drive changes in daily movement patterns, with consequences for nutrient translocation. Daily movement patterns by many bird species between foraging habitat and night roosts can move huge amounts of nutrients about the landscape (Post et al. 1998). These movements may be a response to increased predation risk (and reduced vigilance) under the low light conditions of night, and relaxation of selection pressure by predators could reduce the incidence of these behaviours.

The risk of predation modifies prey behaviour not just through risk avoidance, but also through changes in metabolic rate and diet choice, with implications for nutrient cycling. Despite being less active, risk-exposed prey are expected to have elevated metabolic rates due to risk-induced stress, which can correspond to behavioural change (Houston 2010). Increased metabolic rate heightens energy requirements and changes foraging behaviour, with risk-exposed prey seeking out more energy rich foods (Hawlena and Schmitz 2010b). This change in diet preference could itself drive changes in ecosystems by modifying predation pressure or herbivory on various prey items. Changed foraging also affects nutrient cycling, with increased energy demands reducing the relative amount of carbon recycled, with potential consequences for decomposition, nutrient cycling, and primary productivity (Hawlena and Schmitz 2010a).

In addition to predation, resource availability may also alter behaviour in ways that can change ecosystems, although the ecosystem implications of such shifts have been less thoroughly explored. Diet selectivity of a consumer is a function of the available prey items in the environment (Iwasa et al. 1981), and shifts in relative resource abundance can drive shifts in consumer behaviours, with important ecosystem effects. The diet switch of killer whales from large marine mammals to sea otters (and resulting trophic cascade) may have been driven by the decline of large marine mammals in the coastal region of Alaska (Springer et al. 2008). Other examples of diet choice impacting trophic cascades abound in systems ranging from pelagic marine systems near Antarctica (Ainley et al. 2006), vegetation in riparian areas (Henschel et al. 2001), zooplankton communities in lakes (Yako et al. 1996), and algal communities in estuaries (Geddes and Trexler 2003). More generally, predator diet breadth and selection have been shown in modelling and laboratory studies to place strong control on food web structure (Jiang and Morin 2005).

In addition to the independent effects of predation and resource availability, these factors may interact to shape the ecosystem effects of behavioural plasticity. For example, animals are less likely to alter their feeding behaviour in response to predation in low resource environments, where they must weigh the risk of starvation from reduced foraging against increased risk of predation while foraging (Sih 1980). Thus, low resource situations can drive increased risk-taking by prey, resulting in dampened behaviourally-mediated cascades, but perhaps heightened density-mediated trophic cascades as prey populations decline either from reduced resource acquisition or increased predation (Schmitz et al. 1997). Changes in physical habitat structure can also modify the balance organisms make between feeding and predator avoidance (Trussell et al. 2006). In one case, seed predation by mice *Peromyscus leucopus* under predation threat increased only when an invasive shrub *Lonicera maackii* created cover from visual predators (Mattos and Orrock 2010).

Predation and resource availability are classic ecological drivers. Thus, their importance with respect to the ecosystem effects of behavioural plasticity may be unsurprising. However, other drivers of behavioural plasticity have been shown to impact ecosystems in less obvious ways. For example, parasites on a species of New Zealand cockle *Austrovenus stutchburyi* cause the cockles to alter their behaviour and lay exposed on the surface rather than buried in the benthos. This change in behaviour has ecosystem consequences, shifting competitive dynamics between anemones and limpets on benthic substrates and causing whole-scale community changes in these habitats (Thomas et al. 1998). Crayfish *Orconectes limosus* change their movement patterns in response to the availability of refugia, and this shift changes sediment accretion in small streams with cascading effects on algal growth and biofilm cover (Statzner et al. 2000). Harvester ants *Pogonomyrmex occidental* create more extensive nests in the presence of prairie dog *Cynomys ludovicianus* burrows, disturbing more soil and altering vegetation patterns across the landscape (Alba-Lynn and Detling 2008). These are just a few examples of how behavioural plasticity can cause changes to ecosystems, and we suspect that the list of potential examples is limited primarily by the amount of effort expended to explore such effects.

13.3.2 Contemporary evolution

Ecologists have long studied the importance of phenotypic plasticity, but only recently has it been recognized that evolution can happen at the pace of ecological change and thereby impact ecological dynamics (Post and Palkovacs 2009; Schoener 2011). Perhaps not surprisingly, some of the same types of trait changes that impact ecosystems via phenotypic plasticity have also been found to shape ecosystems via contemporary evolution. Indeed, plastic changes can be the phenotypic precursors to adaptive genetic changes (Ghalambor et al. 2007).

Prey can rapidly evolve defenses against predators, which may have important impacts on community and ecosystem processes. These effects have been shown in a variety of laboratory systems (Matthews et al. 2011). However, such effects have rarely been considered in the wild. One study system where such effects have been considered in the wild is the Trinidadian guppy *Poecilia reticulata*. Above barrier waterfalls, guppies occur in the absence of fish predators (low-predation populations). Below waterfalls, guppies coexist with predators (high-predation populations). A series of introduction experiments have transplanted guppies from high predation sites to low predation sites (Reznick et al. 2008). Such experiments have shown that guppy behaviour can evolve rapidly (over a period of several years) when guppies are released from predation. Importantly for ecosystems, predator release leads to a decrease in guppy escape ability (O'Steen et al. 2002) and an increase in guppy consumption rates, as measured under laboratory conditions of

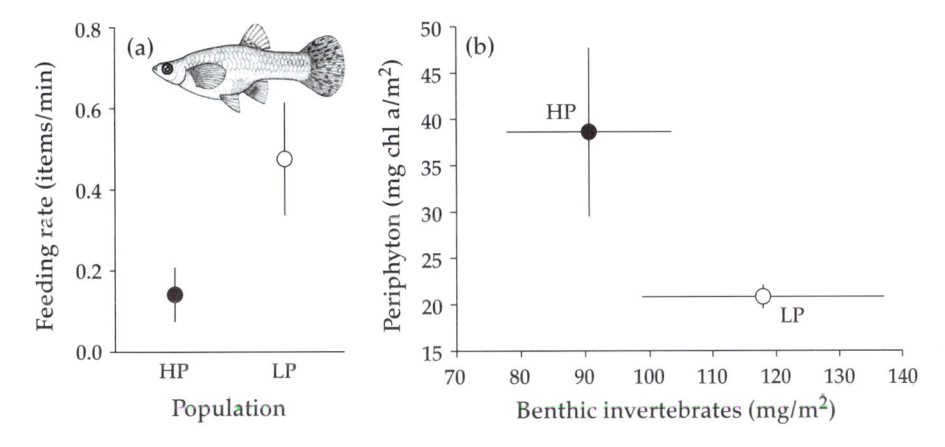

Figure 13.1 The evolution of guppy feeding traits impacts Trinidadian stream ecosystems. Panel (a) shows the feeding rates (mean ± 95% CI) expressed by guppies locally adapted to high-predation (HP) and low-predation (LP) environments. The increase in feeding rate observed in LP is likely to be an adaptation to increased guppy density, decreased resource availability, and a release from selection related to predator-escape performance (data adapted from Palkovacs et al. 2011). Panel (b) shows the ecosystem responses to HP and LP guppies in terms of periphyton and benthic invertebrate biomass (means ± SE), as measured in stream mesocosms. The decrease in periphyton biomass observed in LP results from increased consumption rates and increased ingestion of periphyton relative to invertebrates, coupled with reduced nutrient excretion rates (data adapted from Palkovacs et al. 2009).

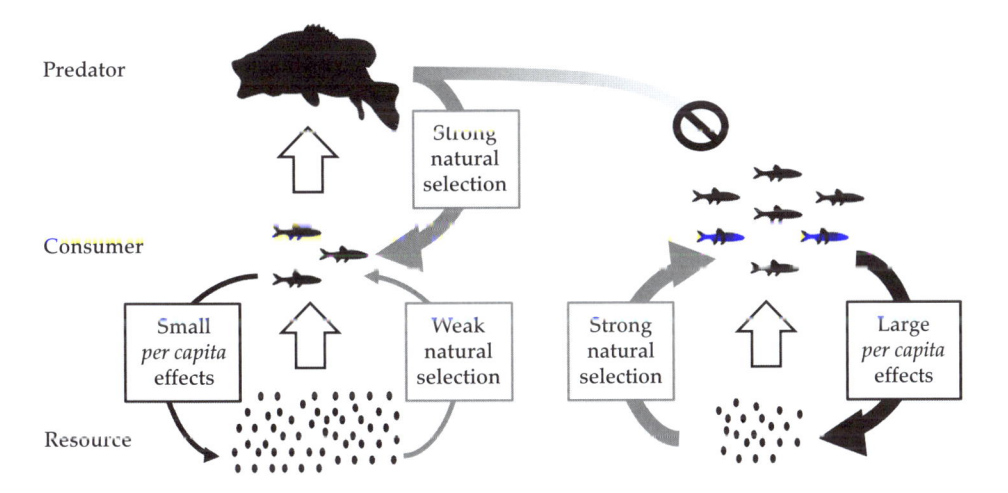

Figure 13.2 Predator removal may amplify the strength of trophic cascades via the contemporary evolution of prey. In the classic density-mediated trophic cascade, the removal of predators increases prey density, which decreases resources. However, predator removal may also shift natural selection acting on prey. When predators are present and resources are abundant (at left), natural selection acting on prey is expected to be driven primarily by predation and favour the evolution of prey traits that enhance predator detection and escape ability. When predators are absent and resources are limited (at right), natural selection acting on prey is expected to be driven primarily by competition and favour the evolution of prey traits that enhance resource acquisition. This shift in selection may cause the per capita effects of prey on resources to be greater in the absence of predators, thereby amplifying the density mediated trophic cascade.

equal prey availability (Palkovacs et al. 2011). These changes may be driven by increased guppy density (and heightened competition) at low-predation sites, coupled with a trade-off between escape ability and competitive ability (Palkovacs et al. 2011). In addition, predator release and heightened competition appear to reduce the selectivity of feeding behaviour at low-predation sites, leading to a higher consumption of abundant (but relatively nutrient poor) algal resources (Zandonà et al. 2011). At the ecosystem level, the evolution of increased guppy consumption rates (Fig. 13.1a), coupled with increased feeding on algae, combine to reduce algal standing stocks in low-predation habitats (Fig. 13.1b; Palkovacs et al. 2009; Bassar et al. 2010).

The importance of these findings in guppies is the broader idea that the evolution of prey traits may impact the strength of trophic cascades (Fig. 13.2). When predators are present, prey density is low and natural selection acting on prey may favour escape ability. Because of trade-offs between escape ability and foraging ability, predator presence may cause the per capita effects of prey on resources to be relatively low. When predators are removed, prey density increases and natural selection acting on prey may favour competitive ability and increased foraging efficiency. This shift may increase the *per capita* effects of prey on resources. Therefore, prey evolution may act to amplify the strength of trophic cascades (Palkovacs et al. 2011). More research is needed to investigate this idea theoretically and to test its generality across study systems.

In addition to prey behaviour, the evolution of predator behaviour can also change ecosystems. As discussed previously for plasticity effects, this change can come in the form of prey selectivity. For example, alewife *Alosa pseudoharengus* populations in New England lakes exhibit two life history forms, an anadromous (sea-run) form and a landlocked (freshwater resident) form. Genetic evidence suggests that landlocked populations have independently evolved in lakes, perhaps as a result of human dam construction (Palkovacs et al. 2008). Anadromous alewives spawn in freshwater, where juveniles rear for a summer before migrating to sea. In contrast, landlocked alewives spend their entire life cycle in freshwater. Alewives act as a keystone predator in New England lakes (Brooks and Dodson 1965); however, differences in prey selectivity cause

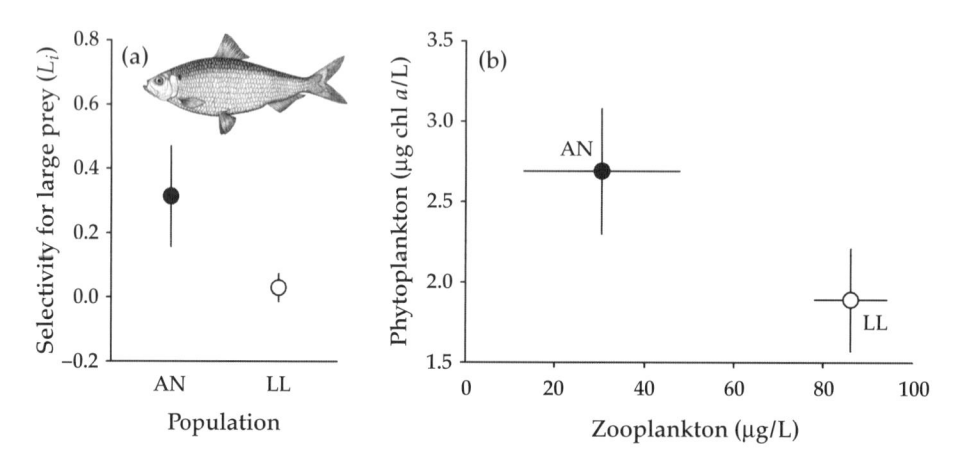

Figure 13.3 The evolution of alewife prey selectivity impacts New England lake ecosystems. Panel (a) shows feeding selectivity (mean ± 95% CI) for large zooplankton prey (>1.0 mm body length) measured using Strauss' (1979) selectivity index, L_i. This index of prey selectivity takes on values from -1 to +1, with positive values indicating preference, values near 0 indicating random feeding, and negative values indicating avoidance. Anadromous alewives (AN) show a preference for large-bodied prey items, whereas landlocked alewives (LL) show random feeding with respect to large-bodied prey (data adapted from Palkovacs and Post 2008). Panel (b) shows the ecosystem responses to AN and LL alewives in terms of phytoplankton and zooplankton biomass (means ± SE), as measured in lake mesocosms. The increase in zooplankton biomass observed in LL results from decreased prey size-selectivity, which increases the abundance of large-bodied cladocerans in the environment. This increase in large-bodied cladocerans increases grazing pressure on phytoplankton, thereby decreasing phytoplankton biomass by weakening the trophic cascade (data adapted from Palkovacs and Post 2009).

the ecosystem effects of landlocked alewives to differ from those of their anadromous ancestors. Anadromous alewives are strongly size-selective, seeking out and consuming the largest prey items available in the lake (Fig. 13.3a). In contrast, landlocked alewives are neutrally size-selective, consuming prey items in proportion to their abundance (Palkovacs and Post 2008). This difference in prey selectivity causes major changes in zooplankton communities, with anadromous alewives eliminating large-bodied grazing cladocerans and causing a stronger trophic cascade than landlocked alewives (Fig. 13.3b, Post et al. 2008; Palkovacs and Post 2009).

In addition to differences in prey selectivity, anadromous and landlocked alewives also differ—even more obviously—in migratory behaviour. This difference changes seasonal zooplankton dynamics and alters nutrient levels. The emigration of anadromous juveniles each season provides a temporal prey refuge for zooplankton in anadromous lakes, which enables some large-bodied species to persist despite strong size-selective predation during the summer months (Post et al. 2008). Additionally, spawning adult anadromous alewives import marine nutrients into freshwater ecosystems, just as do spawning Pacific salmon (Walters et al. 2009). Thus, the evolution of landlocked alewife populations has initiated a suite of behavioural changes that have had large effects on lake ecosystems, namely weaker, but temporally persistent, predation pressure on large-bodied zooplankton with cascading effects on phytoplankton, and elimination of nutrient subsidies from marine sources (Post et al. 2008). It is likely that similar effects may also be present in other coastal ecosystems where evolutionary transitions from anadromy to freshwater residency in fishes are not uncommon, especially where hydrologic connectivity has been altered (Hendry et al. 2004).

13.4 Reaction norms and ecosystem effects

The above examples of contemporary evolution impacting ecosystems involve comparisons of evolutionarily divergent populations. However, these studies do not specifically parse the evolutionary component of ecosystem change from the effects of phenotypic plasticity (but see Ellner et al. 2011). Similarly, experimental evaluations of plasticity effects have typically examined individuals from a single population, adapted to a single set of environmental conditions. In each case, the picture revealed by such studies is incomplete. Obtaining a more complete picture requires the simultaneous consideration of evolution and plasticity. It also requires a framework for linking phenotypic variation to ecosystem responses. Since phenotypic plasticity itself is a function of evolutionary history, we propose a framework for considering the effects of evolution, plasticity, and the evolution of the plastic response based on reaction norms.

Reaction norms depict the phenotypes produced by a given genotype under differing environmental conditions. While populations are typically comprised of many genotypes, and thus contain many reaction norms, it is convenient here to think about population mean reaction norms. Reaction norms depict evolutionary effects as phenotypic differences between populations tested in the same environment and plasticity effects as phenotypic differences within populations tested in different environments. Populations with a high degree of phenotypic plasticity have steeply sloping reaction norms. Evolutionary changes that alter the slope of the reaction norm indicate the evolution of phenotypic plasticity. Once the reaction norms have been formalized (based either on theoretical expectations or empirical observations), the relative contributions of evolution, plasticity, and the evolution of plasticity to total trait change can be determined. Then, the function relating phenotype values and the ecosystem responses can be applied to determine the relative contributions of evolution, plasticity, and the evolution of plasticity to total ecosystem change.

As an example, imagine a case where a predator is introduced to a habitat where the herbivore population has not experienced predation in its recent evolutionary past (e.g. wolf reintroduction into Yellowstone Park, trout stocking into mountain lakes and streams). Prior to the introduction, the herbivore population is predator 'naïve' (i.e. locally adapted to the absence of predators) and exists in a

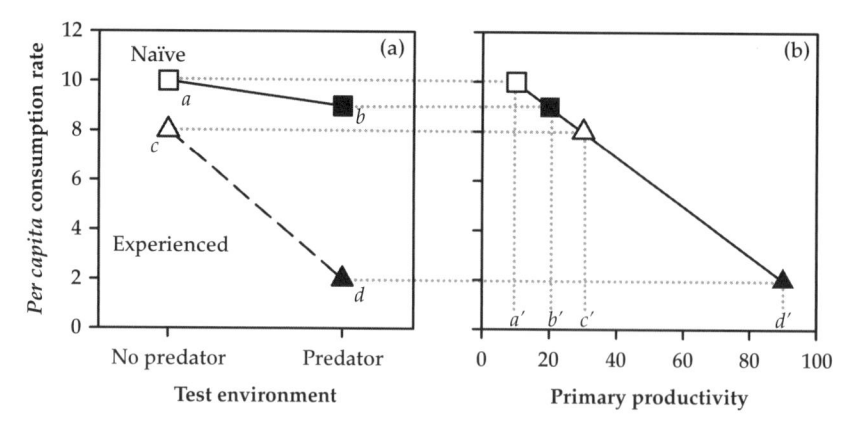

Figure 13.4 Reaction norms applied to behaviourally-mediated trophic cascades. Panel (a) shows hypothetical reaction norms for per capita consumption rate of an herbivore that is predator 'naïve' (i.e. locally adapted to the absence of the predator, squares) and predator 'experienced' (i.e. locally adapted to the presence of the predator, triangles) in environments where the predator is absent ('no predator', open symbols) and present ('predator', closed symbols). Panel (b) shows how per capita changes in consumption rate driven by plasticity (the slope of the reaction norm), evolution (the elevation of the reaction norm), and the evolution of plasticity (the *change* in the slope of the reaction norm), translate into ecosystem effects on primary productivity. See the text and Box 13.1 for details.

'no predator' environment (Fig. 13.4a, *a*). The introduction of the predator causes an immediate plastic response in the herbivores, which includes a reduction in per capita consumption rate (Fig. 13.4a, *b*). As discussed previously, such a reduction in consumption rate is a typical plastic response of prey individuals when faced with predation risk. This plastic response is captured by the slope of the reaction norm depicting the change in consumption rate from the 'no predator' environment to the 'predator' environment displayed by the 'naïve' herbivore population. Understanding the ecosystem consequences of such a plastic shift in phenotype is the goal of many studies of behaviourally-mediated trophic cascades (Schmitz et al. 2008). Such studies are typically performed using a single herbivore population at one point in time exposed to different environments (i.e. 'no predator' and 'predator').

The immediate plastic shift in behaviour is only part of the story, however. Over time (perhaps just several generations), the herbivore population is expected to become locally adapted to the presence of the predator. This local adaptation is described by the change in the elevation of the reaction norm from the predator 'naïve' population to the predator 'experienced' population, assuming that the herbivore exhibits no plastic response to the predator (Fig. 13.4a, *c*). Understanding the ecosystem consequences of this purely evolutionary shift, by removing the effects of plasticity, is the goal of many eco-evolutionary studies (Matthews et al. 2011). Such studies are typically performed using laboratory-born individuals from different populations (i.e. 'naïve' and 'experienced') exposed to a common environment (typically the absence of predators).

In natural ecosystems that have undergone a predator introduction, however, the now-predator 'experienced' herbivore population exists in an environment with the predator, making it important to consider the evolution of the plastic response to the novel environment. The evolution of plasticity is captured by the change in the slope of the reaction norm, in this case resulting in a greater reduction in *per capita* consumption rate than would be expected if plasticity had not evolved (Fig. 13.4a, *d*). Such an increase in the antipredator plastic response (steeper slope of the reaction norm) is expected as an adaptation to predator introduction if there is a fitness gain associated with an increased expression of the predator-induced behaviour, such as when greater reductions in feeding activity further reduce mortality risk (Ghalambor and Martin 2002; Ghalambor et al. 2010).

As described, reaction norms give the expected trait values, but understanding their ecological consequences requires linking these trait values to ecosystem effects. In some cases, we may expect a simple and predictable relationship between traits and ecosystem effects. For example, a linear relationship between consumption rate and primary productivity means that predation environment can drive evolutionary and plastic changes in herbivore prey phenotypes that translate directly into changes in primary productivity (Fig. 13.4b). Such a relationship underlies a simple understanding of behaviourally-mediated trophic cascades. With knowledge of the reaction norms and the relationship between trait values and ecosystem effects, the relative contributions of plasticity, evolution, and the evolution of plasticity to total ecosystem change can be calculated (Box 13.1).

We have described our reaction norm approach in terms of a timeline of events following a perturbation (in this case, the introduction of a novel predator). However, this approach can also be useful for comparing trait and ecosystem responses of extant populations. Thus, the 'naïve' and 'experienced' herbivore populations depicted in Fig. 13.4a could represent two different populations at one point in time. Both populations may currently exist in a 'no predator' environment, perhaps because the 'experienced' population lost its predator in the very recent past. In this case, predator introduction (or reintroduction) would have dramatically different effects on the ecosystem depending on the recent evolutionary history of the herbivore population

---

Box 13.1 Partitioning ecosystem effects using reaction norms

The effects of plasticity, evolution, and the evolution of plasticity can be partitioned for both trait change and ecosystem change if the reaction norms and the function relating trait change to ecosystem change are both known. Here we provide an example for the hypothetical case depicted in Figure 13.4.

(1) First, total *trait* change must be partitioned into the effects of plasticity, evolution, and the evolution of plasticity.

In Fig. 13.4a, the total trait change (i.e. from the 'naïve' population in the 'no predator' environment to the 'experienced' population' in the 'predator' environment) is given by:

$$a - d \text{ or } 10 - 2 = 8$$

The contribution of plasticity to overall trait change (i.e. the immediate plastic response to the introduction of the predator) is given by:

$$a - b \text{ or } 10 - 9 = 1$$

The contribution of evolution to overall trait change (i.e. the evolutionary effect that would have occurred in the absence of plasticity) is given by:

$$a - c \text{ or } 10 - 8 = 2$$

The contribution of the evolution of plasticity (i.e. the difference in the plastic response from the ancestral to the derived population) is given by:

$$(c - d) - (a - b) \text{ or } (8 - 2) - (10 - 9) = 5$$

Thus, the total trait change partitioned into plasticity, evolution, and the evolution of plasticity is given by:

$$(a - b) + (a - c) + ((c - d) - (a - b)) \text{ or } 1 + 2 + 5 = 8$$

Note that this represents the total shift from *a* to *d*.

(2) Once the partitioning has been performed for trait change, the function relating trait change to ecosystem change can be applied to discern relative ecosystem effects.

Figure 13.4b depicts a linear relationship between consumption rate and primary productivity with a slope of 0.1 (rise/run = 1/10). This means that each unit of trait change (reduced consumption) translates into 10 units of ecosystem change (increased primary production). Therefore, of the total change in primary production (*a'* to *d'* = 80 units), the change due to plasticity represents 12.5%, the change due to evolution represents 25%, and the change due to the evolution of plasticity represents 62.5%. In this example, we have described the relationship between traits and ecosystem effects using a simple linear function. However, more complex functions can also be applied.

(Fig. 13.4b). Specifically, short-term ecosystem change due to plasticity would be expected to be smaller for the 'naïve' population (Fig. 13.4b, a'–b') compared to the 'experienced' population (Fig. 13.4b, c'–d'). Of course, the maladapted behaviour of the 'naïve' herbivore may lead to a higher predator-induced mortality rate. This mortality may strengthen the short-term density-mediated trophic cascade and, over several generations, lead to the evolution of the 'experienced' reaction norm.

We have described how reaction norms can be used to study ecosystem change. But what are the practical benefits of applying such an approach in the context of human disturbance? Much effort has gone into developing theory to understanding how reaction norms for diverse traits, including behaviour, will evolve under different environmental conditions (Ghalambor et al. 2010). One key benefit of applying a reaction norm framework to ecosystems is that this existing theory can be used to make predictions about ecosystem responses. In a rapidly changing world, the ability to make predictions is more than an academic concern. Returning to the example of trophic cascades, humans are adept at 'tinkering' with top predators (Strong and Frank 2010). Human predation on top predators is a novel selective force on these populations, potentially driving evolutionary changes in their behaviour, just as natural predators do to their prey. Humans have eliminated top predators from many ecosystems and added novel top predators to others. When top predators are eliminated or introduced, prey populations evolve in new environments with altered predation risk. The development of robust evolutionary and ecological linkages between natural selection acting on predators, the evolution of behavioural reaction norms, and the resulting ecosystem consequences could potentially go a long way towards predicting the impacts of top predator removals and introductions on the world's ecosystems.

13.5 Conclusions

Animal behaviour is a critical, but underappreciated, link between human activity and ecosystem processes. Behavioural traits related to consumption and nutrient cycling, in particular, can cause widespread effects on ecosystems. Therefore, rapid changes to these traits due to phenotypic plasticity and contemporary evolution may cause important changes to ecosystems. While plasticity effects have been recognized for some time, evolutionary effects are just beginning to emerge as a potentially important driver of ecosystem change. The effects of plasticity and evolution have largely been considered separately, but these factors work in concert to shape behavioural traits. Therefore, new approaches are needed to integrate these mechanisms of trait change and link them to ecosystem effects. We propose a framework for considering the joint effects of evolution and plasticity based on reaction norms. Natural selection provides a predictive theoretical framework for understanding the evolution of reaction norms under different environmental conditions. Linking this theoretical framework to ecosystem processes has the potential to add predictive power to ecosystem ecology. Such information can potentially be used to predict the consequences of human activity for ecosystems.

Acknowledgements

We thank Andrew Hendry, Os Schmitz, Cameron Ghalambor, Nelson Hairston, Mike Kinnison, Ulrika Candolin, Bob Wong, Andrés López-Sepulcre, and an anonymous reviewer for helpful comments and suggestions that improved the manuscript.

References

Ainley, D. G., Ballard, G., and Dugger, K. M. (2006). Competition among penguins and cetaceans reveals trophic cascades in the western Ross Sea, Antarctica. *Ecology*, 87, 2080–93.

Alba-Lynn, C. and Detling, J. K. (2008). Interactive disturbance effects of two disparate ecosystem engineers in North American shortgrass steppe. *Oecologia*, 157, 269–78.

Bassar, R. D., Marshall, M. C., López-Sepulcre, A., Zandonà, E., Auer, S., Travis, J., Pringle, C. M., Flecker, A. S., Thomas, S. A., Fraser, D. F., and Reznick, D. N. (2010). Local adaptation in Trinidadian guppies alters ecosystem processes. *Proceedings of the National Academy of Sciences of the USA*, 107, 3616–21.

Beyer, H. L., Merrill, E. H., Varley, N., and Boyce, M. S. (2007). Willow on Yellowstone's northern range: evidence for a trophic cascade? *Ecological Applications*, 17, 1563–71.

Brooks, J. L. and Dodson, S. I. (1965). Predation, body size, and composition of plankton. *Science*, 150, 28–35.

Bump, J. K., Peterson, R. O. and Vucetich, J. A. (2009). Wolves modulate soil nutrient heterogeneity and foliar nitrogen by configuring the distribution of ungulate carcasses. *Ecology*, 90, 3159–67.

Calcagno, V., Sun, C., Schmitz, O. J., and Loreau, M. (2011). Keystone predation and plant species coexistence: the role of carnivore hunting mode. *American Naturalist*, 177, E1–E13.

Carey, M. P. and Wahl, D. H. (2010). Interactions of multiple predators with different foraging modes in an aquatic food web. *Oecologia*, 162, 443–52.

Christie, K. S., Hocking, M. D., and Reimchen, T. E. (2008). Tracing salmon nutrients in riparian food webs: isotopic evidence in a ground-foraging passerine. *Canadian Journal of Zoology-Revue Canadienne De Zoologie*, 86, 1317–23.

Ellner, S. P., Geber, M. A., and Hairston, N. G., Jr. (2011). Does rapid evolution matter? Measuring the rate of contemporary evolution and its impacts on ecological dynamics. *Ecology Letters*, 14, 603–14.

Estes, J. A., Tinker, M. T., Williams, T. M., and Doak, D. F. (1998). Killer whale predation on sea otters linking oceanic and nearshore ecosystems. *Science*, 282, 473–6.

Flecker, A. S., McIntyre, P. B., Moore, J. W., Anderson, J. T., Taylor, B. W., and Hall, R. O. (2010). Migratory fishes as material and process subsidies in riverine ecosystems. In: Gido, K. B. and Jackson, D. (eds) *Community Ecology of Stream Fishes: Concepts, Approaches, and Techniques*. Bethesda, Maryland, American Fisheries Society Symposium.

Fortin, D., Beyer, H. L., Boyce, M. S., Smith, D. W., Duchesne, T., and Mao, J. S. (2005). Wolves influence elk movements: behavior shapes a trophic cascade in Yellowstone National Park. *Ecology*, 86, 1320–30.

Geddes, P. and Trexler, J. C. (2003). Uncoupling of omnivore-mediated positive and negative effects on periphyton mats. *Oecologia*, 136, 585–95.

Ghalambor, C. K., Angeloni, L., and Carroll, S. P. (2010). Behavior as phenotypic plasticity. In: Fox, C. and Westneat, D. (eds) *Evolutionary Behavioral Ecology*. New York, Oxford University Press.

Ghalambor, C. K. and Martin, T. E. (2002). Comparative manipulation of predation risk in incubating birds reveals variability in the plasticity of responses. *Behavioral Ecology*, 13, 101–8.

Ghalambor, C. K., McKay, J. K., Carroll, S. P., and Reznick, D. N. (2007). Adaptive versus non-adaptive phenotypic plasticity and the potential for contemporary adaptation in new environments. *Functional Ecology*, 21, 394–407.

Glaholt, S. P. and Vanni, M. J. (2005). Ecological responses to simulated benthic-derived nutrient subsidies mediated by omnivorous fish. *Freshwater Biology*, 50, 1864–81.

Greenwood, J. L., Rosemond, A. D., Wallace, J. B., Cross, W. F., and Weyers, H. S. (2007). Nutrients stimulate leaf breakdown rates and detritivore biomass: bottom-up effects via heterotrophic pathways. *Oecologia*, 151, 637–49.

Hairston, N. G., Sr., Smith, F. E., and Slobodkin, L. B. (1960). Community structure, population control, and competition. *American Naturalist*, 94, 421–5.

Halaj, J. and Wise, D. H. (2001). Terrestrial trophic cascades: how much do they trickle? *American Naturalist*, 157, 262–81.

Hawlena, D. and Schmitz, O. J. (2010a). Herbivore physiological response to predation risk and implications for ecosystem nutrient dynamics. *Proceedings of the National Academy of Sciences of the USA*, 107, 15503–7.

Hawlena, D. and Schmitz, O. J. (2010b). Physiological stress as a fundamental mechanism linking predation to ecosystem functioning. *American Naturalist*, 176, 537–56.

Hendry, A. P., Bohlin, T., Jonsson, B., and Berg, O. K. (2004). To sea or not to sea? Anadromy versus non-anadromy in Salmonids. In: Hendry, A. P. and Stearns, S. C. (eds) *Evolution Illuminated Salmon and their Relatives*. Oxford, Oxford University Press.

Henschel, J. R., Mahsberg, D., and Stumpf, H. (2001). Allochthonous aquatic insects increase predation and decrease herbivory in river shore food webs. *Oikos*, 93, 429–38.

Holtgrieve, G. W., Schindler, D. E., and Jewett, P. K. (2009). Large predators and biogeochemical hotspots: brown bear (*Ursus arctos*) predation on salmon alters nitrogen cycling in riparian soils. *Ecological Research*, 24, 1125–35.

Hooper, D. U., Chapin, F. S., Ewel, J. J., Hector, A., Inchausti, P., Lavorel, S., Lawton, J. H., Lodge, D. M., Loreau, M., Naeem, S., Schmid, B., Setala, H., Symstad, A. J., Vandermeer, J., and Wardle, D. A. (2005). Effects of biodiversity on ecosystem functioning: a consensus of current knowledge. *Ecological Monographs*, 75, 3–35.

Houston, A. I. (2010). Evolutionary models of metabolism, behaviour and personality. *Philosophical Transactions of the Royal Society B: Biological Sciences*, 365, 3969–75.

Iwasa, Y., Higashi, M., and Yamamura, N. (1981). Prey distribution as a factor determining the choice of optimal foraging strategy. *American Naturalist*, 117, 710–23.

Jiang, L. and Morin, P. J. (2005). Predator diet breadth influences the relative importance of bottom-up and top-down control of prey biomass and diversity. *American Naturalist*, 165, 350–63.

Knoll, L. B., McIntyre, P. B., Vanni, M. J., and Flecker, A. S. (2009). Feedbacks of consumer nutrient recycling on producer biomass and stoichiometry: separating direct and indirect effects. *Oikos*, 118, 1732–42.

Lazzaro, X., Lacroix, G., Gauzens, B., Gignoux, J., and Legendre, S. (2009). Predator foraging behaviour drives food-web topological structure. *Journal of Animal Ecology*, 78, 1307–17.

López-Sepulcre, A. (2011). The many ecologies of behavior. *Behavioral Ecology*, 22, 232–3.

Matthews, B., Narwani, A., Hausch, S., Nonaka, E., Peter, H., Yamamichi, M., Sullam, K. E., Bird, K. C., Thomas, M. K., C., H. T., and Turner, C. B. (2011). Toward an integration of evolutionary biology and ecosystem science. *Ecology Letters*, 14, 690–701.

Mattos, K. J. and Orrock, J. L. (2010). Behavioral consequences of plant invasion: an invasive plant alters rodent antipredator behavior. *Behavioral Ecology*, 21, 556–61.

McIntyre, P. B., Flecker, A. S., Vanni, M. J., Hood, J. M., Taylor, B. W., and Thomas, S. A. (2008). Fish distributions and nutrient cycling in streams: can fish create biogeochemical hotspots. *Ecology*, 89, 2335–46.

Meyer, J. L. and Schultz, E. T. (1985). Migrating Haemulid fishes as a source of nutrients and organic-matter on coral reefs. *Limnology and Oceanography*, 30, 146–56.

Miner, B. G., Sultan, S. E., Morgan, S. G., Padilla, D. K., and Relyea, R. A. (2005). Ecological consequences of phenotypic plasticity. *Trends in Ecology and Evolution*, 20, 685–92.

O'Steen, S., Cullum, A. J., and Bennett, A. F. (2002). Rapid evolution of escape ability in Trinidadian guppies (*Poecilia reticulata*). *Evolution*, 56, 776–84.

Pace, M. L., Cole, J. L., Carpenter, S. R., and Kitchell, J. F. (1999). Trophic cascades revealed in diverse ecosystems. *Trends in Ecology and Evolution*, 14, 483–8.

Paine, R. T. (1966). Food web complexity and species diversity. *American Naturalist*, 100, 65–75.

Paine, R. T. (1969). A note on trophic complexity and community stability. *American Naturalist*, 103, 91–93.

Palkovacs, E. P., Dion, K. B., Post, D. M., and Caccone, A. (2008). Independent evolutionary origins of landlocked alewife populations and rapid parallel evolution of phenotypic traits. *Molecular Ecology*, 17, 582–97.

Palkovacs, E. P., Marshall, M. C., Lamphere, B. A., Lynch, B. R., Weese, D. J., Fraser, D. F., Reznick, D. N., Pringle, C. M., and Kinnison, M. T. (2009). Experimental evaluation of evolution and coevolution as agents of ecosystem change in Trinidadian streams. *Philosophical Transactions of the Royal Society B: Biological Sciences*, 364, 1617–28.

Palkovacs, E. P. and Post, D. M. (2008). Eco-evolutionary interactions between predators and prey: can predator-induced changes to prey communities feed back to shape predator foraging traits? *Evolutionary Ecology Research*, 10, 699–720.

Palkovacs, E. P. and Post, D. M. (2009). Experimental evidence that phenotypic divergence in predators drives community divergence in prey. *Ecology*, 90, 300–5.

Palkovacs, E. P., Wasserman, B. A., and Kinnison, M. T. (2011). Eco-evolutionary trophic dynamics: loss of top predators drives trophic evolution and ecology of prey. *PLOS ONE*, 6, e18879.

Post, D. M. and Palkovacs, E. P. (2009). Eco-evolutionary feedbacks in community and ecosystem ecology: interactions between the ecological theatre and the evolutionary play. *Philosophical Transactions of the Royal Society B: Biological Sciences*, 364, 1629–40.

Post, D. M., Palkovacs, E. P., Schielke, E. G., and Dodson, S. I. (2008). Intraspecific variation in a predator affects community structure and cacading trophic interactions. *Ecology*, 89, 2019–32.

Post, D. M., Taylor, J. P., Kitchell, J. F., Olson, M. H., Schindler, D. E., and Herwig, B. R. (1998). The role of migratory waterfowl as nutrient vectors in a managed wetland. *Conservation Biology*, 12, 910–20.

Power, M. E., Tilman, D. E., Estes, J. A., Menge, B. A., Bond, W. J., Mills, L. S., Daily, G. S., Castilla, J. C., Lubchencho, J., and Paine, R. T. (1996). Challenges in the quest for keystones. *Bioscience*, 46, 609–20.

Preisser, E. L., Orrock, J. L., and Schmitz, O. J. (2007). Predator hunting mode and habitat domain alter non-consumptive effects in predator-prey interactions. *Ecology*, 88, 2744–51.

Reznick, D. N., Ghalambor, C. K., and Crooks, K. (2008). Experimental studies of evolution in guppies: a model for understanding the evolutionary consequences of predator removal in natural communities. *Molecular Ecology*, 17, 97–107.

Saba, G. K., Steinberg, D. K., and Bronk, D. A. (2009). Effects of diet on release of dissolved organic and inorganic nutrients by the copepod *Acartia tonsa*. *Marine Ecology-Progress Series*, 386, 147–61.

Schindler, D. E., Scheuerell, M. D., Moore, J. W., Gende, S. M., Francis, T. B., and Palen, W. J. (2003). Pacific salmon and the ecology of coastal ecosystems. *Frontiers in Ecology and the Environment*, 1, 31–7.

Schmitz, O. J. (1998). Direct and indirect effects of predation and predation risk in old-field interaction webs. *American Naturalist*, 151, 327–42.

Schmitz, O. J. (2008). Effects of predator hunting mode on grassland ecosystem function. *Science*, 319, 952–4.

Schmitz, O. J., Beckerman, A. P., and O'Brien, K. M. (1997). Behaviorally mediated trophic cascades: effects of predation risk on food web interactions. *Ecology*, 78, 1388–99.

Schmitz, O. J., Grabowski, J. H., Peckarsky, B. L., Preisser, E. L., Trussell, G. C., and Vonesh, J. R. (2008). From individuals to ecosystem function: toward an integration of evolutionary and ecosystem ecology. *Ecology*, 89, 2436–45.

Schmitz, O. J., Hawlena, D., and Trussell, G. C. (2010). Predator control of ecosystem nutrient dynamics. *Ecology Letters*, 13, 1199–209.

Schmitz, O. J., Krivan, V., and Ovadia, O. (2004). Trophic cascades: the primacy of trait-mediated indirect interactions. *Ecology Letters*, 7, 153–63.

Schoener, T. W. (2011). The newest synthesis: understanding the interplay of evolutionary and ecological dynamics. *Science*, 331, 426–9.

Sih, A. (1980). Optimal behavior—can foragers balance two conflicting demands? *Science*, 210, 1041–3.

Silliman, B. R. and Bertness, M. D. (2002). A trophic cascade regulates salt marsh primary production. *Proceedings of the National Academy of Sciences of the USA*, 99, 10500–5.

Springer, A. M., Estes, J. A., van Vliet, G. B., Williams, T. M., Doak, D. F., Danner, E. M., and Pfister, B. (2008). Mammal-eating killer whales, industrial whaling, and the sequential megafaunal collapse in the North Pacific Ocean: a reply to critics of Springer et al. 2003. *Marine Mammal Science*, 24, 414–42.

Statzner, B., Fievet, E., Champagne, J. Y., Morel, R., and Herouin, E. (2000). Crayfish as geomorphic agents and ecosystem engineers: biological behavior affects sand and gravel erosion in experimental streams. *Limnology and Oceanography*, 45, 1030–40.

Strauss, R. E. (1979). Reliability estimates for Ivlev electivity index, the forage ratio, and a proposed linear index of food selection. *Transactions of the American Fisheries Society*, 108, 344–52.

Strong, D. R. and Frank, K. T. (2010). Human involvement in food webs. *Annual Review of Environment and Resources*, 35, 1–23.

Terborgh, J., Holt, R. D., and Estes, J. A. (2010). Trophic cascades: what they are, how they work, and why they matter. In: Terborgh, J. and Estes, J. A. (eds) *Trophic Cascades: Predators, Prey, and the Changing Dynamics of Nature*. Washington, DC, Island Press.

Thomas, F., Renaud, F., de Meeus, T., and Poulin, R. (1998). Manipulation of host behaviour by parasites: ecosystem engineering in the intertidal zone? *Proceedings of the Royal Society of London B, Biological Sciences*, 265, 1091–6.

Tiegs, S. D., Campbell, E. Y., Levi, P. S., Ruegg, J., Benbow, M. E., Chaloner, D. T., Merritt, R. W., Tank, J. L., and Lamberti, G. A. (2009). Separating physical disturbance and nutrient enrichment caused by Pacific salmon in stream ecosystems. *Freshwater Biology*, 54, 1864–75.

Trussell, G. C., Ewanchuk, P. J., and Matassa, C. M. (2006). Habitat effects on the relative importance of trait- and density-mediated indirect interactions. *Ecology Letters*, 9, 1245–52.

Uliano, E., Cataldi, M., Carella, F., Migliaccio, O., Iaccarino, D., and Agnisola, C. (2010). Effects of acute changes in salinity and temperature on routine metabolism and nitrogen excretion in gambusia (*Gambusia affinis*) and zebrafish (*Danio rerio*). *Comparative Biochemistry and Physiology a-Molecular and Integrative Physiology*, 157, 283–90.

Vanni, M. J. (2002). Nutrient cycling by animals in freshwater ecosystems. *Annual Review of Ecology and Systematics*, 33, 341–70.

Vanni, M. J., Arend, K. K., Bremigan, M. T., Bunnell, D. B., Garvey, J. E., Gonzalez, M. J., Renwick, W. H., Soranno, P. A., and Stein, R. A. (2005). Linking landscapes and food webs: effects of omnivorous fish and watersheds on reservoir ecosystems. *Bioscience*, 55, 155–67.

Vitousek, P. M. and Howarth, R. W. (1991). Nitrogen limitation on land and in the sea—how can it occur? *Biogeochemistry*, 13, 87–115.

Walters, A. W., Barnes, R. T., and Post, D. M. (2009). Anadromous alewives (*Alosa pseudoharengus*) contribute marine-derived nutrients to coastal stream food webs. *Canadian Journal of Fisheries and Aquatic Sciences*, 66, 439–48.

Wirsing, A. J., Cameron, K. E., and Heithaus, M. R. (2010). Spatial responses to predators vary with prey escape mode. *Animal Behaviour*, 79, 531–7.

Yako, L. A., Dettmers, J. M., and Stein, R. A. (1996). Feeding preferences of omnivorous gizzard shad as influenced by fish size and zooplankton density. *Transactions of the American Fisheries Society*, 125, 753–9.

Zandonà, E., Auer, S. K., Kilham, S. S., Howard, J. L., López-Sepulcre, A., O'Connor, M. P., Bassar, R. D., Osorio, A., Pringle, C. M., and Reznick, D. N. (2011). Diet quality and prey selectivity correlate with life histories and predation regime in Trinidadian guppies. *Functional Ecology*, 25, 964–73.

CHAPTER 14

The role of behavioural variation in the invasion of new areas

Ben L. Phillips and Andrew V. Suarez

⊃ Overview

Behaviour determines the rate at which invasive species spread, as well as the impact they have on natives. When behaviour varies between individuals (as it almost always does), then the mean behaviour is often less important than the extremes of behaviour. The rate at which a species spreads, for example, is governed primarily by the most extreme dispersers. Similarly, individuals of native species that are extreme in their behaviour may be more, or less, likely to suffer impact from a given invasive species. Thus, we argue, an understanding of behavioural variation is critical if we are to understand the long-term impacts of invasive species in a changing world.

14.1 Introduction

Biological invasions are now widely considered to be a form of global change (Vitousek et al. 1996). In addition to severe economic consequences (Pimental et al. 2000), the introduction of species into new environments can have devastating ecological impacts (Mack et al. 2000; Parker et al. 1999). It is something of a relief then that we don't have a lot more invasive species, particularly given that every year probably hundreds to thousands of species are introduced outside their natural range. Luckily, very few of these go on to establish, spread, and have large ecological impacts. Why is this so? It turns out that the different stages of the invasion process—transportation across dispersal barriers, establishment of a population at the new location, and post-establishment spread—act as strong phenotypic filters, winnowing out the majority of species. Thus, a non-random subset of all taxa are transported, a non-random subset of these are then capable of establishment, and then a non-random subset of these become spreading invaders (Blackburn and Duncan 2001; Simons 2003; Tingley et al. 2010). Very few species pass easily through all three filters.

Importantly, the same filtering process that applies to taxa also applies to individuals within an invasive taxon (in fact, it applies to the individuals first, and the taxon only as an after-effect). The individuals that are transported, that form established populations, and that become invasive are all potentially non-random subsets of the population from which they originated (Phillips et al. 2010a). Thus, population-level variation in traits associated with invasion success is a key determinant of the success, or otherwise, of a particular invasion.

Invasive species do not, however, occur in a vacuum. As an invader spreads, its interactions with resident species become more widespread and the success of the invader, as well as that of the invaded, may well depend on these interactions. Variation is important here too. Introduced species often interact negatively with native species (either directly, as predators, competitors, and parasites; or indirectly, through habitat modification). So the variation in

Behavioural Responses to a Changing World. First Edition. Edited by Ulrika Candolin and Bob B.M. Wong.
© 2012 Oxford University Press. Published 2012 by Oxford University Press.

traits mediating these interactions will determine which native individuals survive and which do not: natural selection in action.

Many of the traits associated with invasion and interspecific interaction are behavioural. In this chapter, we focus primarily on behaviour during the post-establishment stage of invasion: spread. We discuss the processes that lead to population spread, and some of the behaviours that facilitate those processes. We also discuss the role of behaviour in determining the outcomes of interspecific interactions between introduced and resident species. In both cases we focus not just on behaviours, but on the effect of behavioural variation. We will see that behavioural variation not only matters, but that it is central to understanding both the process of invasion, and the impact of invaders on natives.

14.2 Behaviours influencing the process of spread

14.2.1 The mechanics of spread

Populations spread through space as a function of two fundamental processes: dispersal and population growth (Skellam 1951). Dispersal moves individuals through space (and into new space), and population growth fills space (including newly colonized space) with individuals. The rate at which a population spreads, then, is determined almost entirely by these two fundamental processes. This said, however, a surprising array of spread dynamics are possible (including lags, accelerating, pulsed, and constant spread rates) depending upon how these fundamental processes are executed (Hastings et al. 2005; Johnson et al. 2006).

If there is variation in a population for traits affecting dispersal or population growth rate, things get more complex. In addition to the two fundamental processes of spread (dispersal and population growth), two unavoidable consequences of spread allow trait variation to come into play (Phillips et al. 2010a). The first unavoidable consequence is that, on the invasion front, individuals are spatially assorted by dispersal ability. This is known as the Olympic village effect, or spatial sorting (Phillips et al. 2008a; Shine et al. 2011), and leads to assortative mating by dispersal ability and runaway selection for increased dispersal on the invasion front (Travis and Dytham 2002, Hughes et al. 2007; Fig. 14.1). The second unavoidable consequence of range advance is that individuals on the expanding

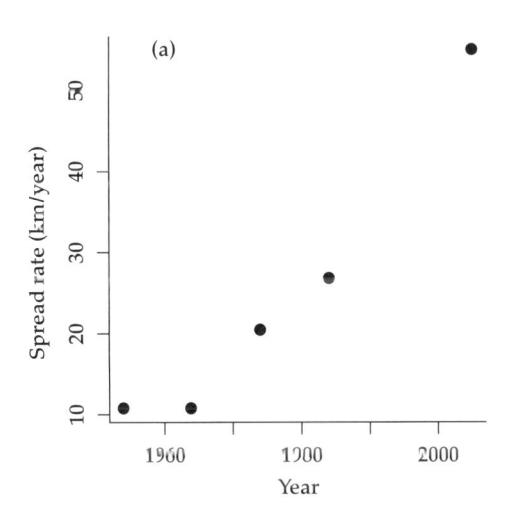

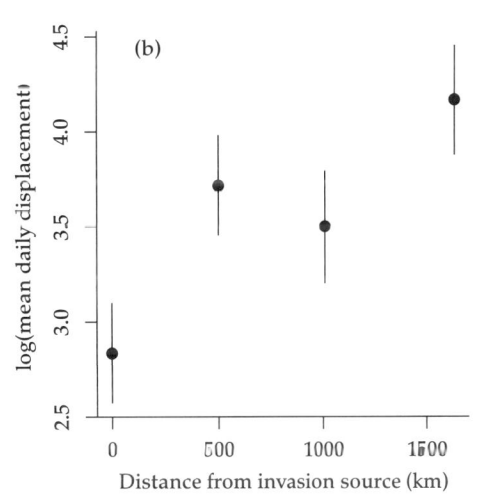

Figure 14.1 The accelerating rate of invasion of cane toads *Rhinella marina* across northern Australia (a), is associated with evolutionarily increased dispersal rates (b) shown here as an increase in mean daily displacement of animals collected from populations spanning the earliest introduction point (zero on the x-axis) through to the still-expanding invasion front. This is a pattern increasingly seen in the rate of spread of many invasive species. Error bars are one standard error. Data redrawn from Phillips et al. 2006, 2008a.

edge are in an environment with a very low density of conspecifics relative to established populations. In such an environment, traits that increase reproductive rate or decrease Allee effects are favoured (Burton et al. 2010).

With these two fundamental processes and two unavoidable consequences, we now have a framework with which to examine the influence of behaviour, and behavioural variation, on spread.

14.2.2 Dispersal behaviour during spread

Dispersal, the permanent movement of an animal from its birth place to its place of first reproduction (*sensu* Howard, 1960), carries strong costs, but despite this, is almost ubiquitous across all living organisms (Clobert et al. 2001). Dispersal, therefore, must also carry strong rewards, and these include the avoidance of inbreeding and kin competition, escape from parasites and pathogens, as well as the subtler advantages of colonizing newly available patches in an extinction prone landscape (e.g. Hamilton and May 1977; Van Valen 1971; Gandon and Michalakis 2001; Weisser et al. 2001). Thus, most organisms disperse, but getting from A to B can be done in an astonishing number of ways. It can be done passively, such as a windborne seed, actively, such as a kangaroo hopping, or in an amalgam of the two.

The most successful invaders are those that spread over large areas very quickly. Invasive species will therefore tend to have behaviours that generate long-distance dispersal relatively cheaply (Bartoń et al. 2009). Typically, such dispersal strategies involve some form of passive or active/passive combination. In terrestrial realms, flight is perhaps the most obvious amalgam of active and passive dispersal: although getting airborne might be energetically expensive, the ability to harness the movement of air once aloft can disperse airborne animals over vast distances. Thus, many of the most rapid and famous invasions have involved flying species, such as the house finch *Carpodacus mexicanus*, house sparrow *Passer domesticus*, the European starling *Sturnus vulgaris*, the Eurasian collared dove *Streptopelia decaocto*, and the gypsy moth *Lymantria dispar* (Elton 1958; Veit and Lewis 1996).

Another cheap technique is to hitch a lift. Small species that allow other, typically much larger, species to effect their movement may get around very cheaply indeed. Some of the best examples of such facilitated dispersal come from epidemiology, where diseases are spread rapidly across an area by their animal vectors (e.g. spread of West Nile virus, Loss et al. 2009; Seidowski et al. 2010). Of course humans also make excellent targets for facilitated dispersal. Human-mediated jump dispersal is not only responsible for creating new foci of invasion well away from established populations, but may also be the primary means by which some species spread at the landscape level (e.g. Argentine ants *Linepithema humile*, Suarez et al. 2001; snails *Xeropicta derbentina*, Aubry et al., 2006).

Finally, in cases where dispersal is active, increased efficiency can be gained by straightening the movement path (Bartoń et al. 2009): individuals that disperse in a relatively straight line will, ultimately, displace much further than individuals following a more tortuous path. Indeed, a tendency to follow straight paths is a clear attribute of the most rapidly spreading populations of invasive cane toads *Rhinella marina* in northern Australia (Alford et al. 2009; Phillips et al., 2008a).

14.2.3 Behaviour and population growth during spread

In broad terms, population growth is the net result of the number of births in a population minus the number of deaths. The number of births and deaths, however, is a gross summary of the myriad behaviours, morphologies, and physiologies that allow individuals in a population to survive, grow, and reproduce. In this light, then, almost any behaviour has some influence on population growth. For logistical reasons then, we focus here on behaviours that have an obvious bearing on the colonization of new areas: behaviours that allow individuals to survive and quickly capture resources for growth and reproduction in a new environment. Although there has been no definitive 'laundry list' of behavioural characteristics that convey invasion success, it has been suggested that some degree of sociality along with high levels of aggression, activity (foraging/search-

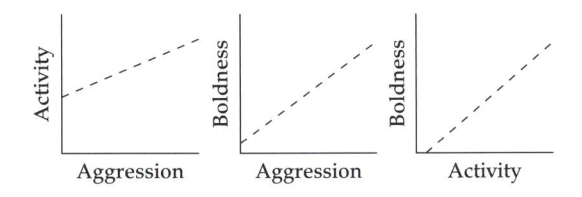

Figure 14.2 Relationship between foraging activity, intraspecific aggression, and boldness in the invasive signal crayfish *Pacifastacus leniusculus*. All correlations are significant (p <0.05) and are based upon comparisons of population means. Data redrawn from Pintor et al. 2008.

ing behaviour), and vigilance (boldness/antipredator behaviour) may be important (Cote et al. 2010; Holway and Suarez 1999; Kolar and Lodge 2001).

Over the past decade, there has been a growing literature on consistent individual differences in behaviour, frequently termed behavioural syndromes or animal personality (Sih et al. 2004a). Some common behavioural traits, such as exploratory behaviour, aggression towards conspecifics, and boldness in risky situations, may be both consistent and positively correlated within individuals (Reale et al. 2007) and could provide an advantage to colonizing species (Sih et al. 2004a; Fogarty et al. 2011). For example, using the framework of behavioural syndromes, Pintor and colleagues (2008) found that foraging activity, intraspecific aggression and boldness were highly correlated across populations of signal crayfish *Pacifastacus leniusculus* (Fig. 14.2). Crayfish were most aggressive, active and bold in introduced areas with low prey biomass and an absence of interspecific competitors, suggesting this correlated suite of behaviours may be beneficial for establishing new populations where resources are scarce (Pintor et al. 2008).

14.3 The effect of behavioural variation on spread

The unavoidable consequences of invasion—reduced conspecific density and assortative mating by dispersal ability on the invasion front can potentially drive phenotypic changes. These shifts might be simple plastic responses to the new environment, or might reflect evolved shifts in response to directional selection during spread. Irrespective, these phenotypic changes can, in turn, feed back into the spread process. Importantly, plastic responses will have an almost instantaneous effect on spread rate (as they occur as a direct response to the environment), whereas evolved responses will take longer (several to many generations) to play out. Plastic responses may or may not increase spread rate, whereas evolved responses will likely always increase spread rate.

14.3.1 Plastic responses

Behavioural flexibility, particularly in relation to novel stimuli and introduction into novel environments, has been suggested as a possible explanation for why some species become successful invaders (Sol et al. 2002; Wright et al. 2010). For example, some individuals might quickly notice a new prey type and alter their foraging tactics, whereas others might adopt less plastic strategies (Sih et al. 2010). The more behaviourally flexible individuals would be more likely to prosper in newly-invaded areas; they would be more likely to have positive population growth and less likely to suffer Allee effects. Using brain size as a surrogate for behavioural flexibility, Sol and colleagues found that relative brain size was a strong predictor of invasion success for birds and mammals (Sol and Lefebvre 2000; Sol et al. 2008). A similar pattern has since been observed for invasive amphibians (Amiel et al. 2011), so brain size is shaping up to be a strong predictor of invasion success generally. Although bigger brains might provide greater behavioural flexibility, it is important to note that brain size (or behavioural flexibility) itself may be under selection during the course of an invasion: invasion may be selecting for behavioural flexibility. Whether brain size changes during the course of invasion, or influences spread rate has, however, yet to be adequately tested.

Other brain-size related plastic responses might include learning or social cues. In a world where conspecifics are rare (such as an invasion front), individuals may need to invest greater effort into locating those conspecifics. For example starlings from newly invaded areas are more responsive to the calls of conspecifics than are starlings from long-established areas (Rodriguez et al. 2010). Again, whether this results purely from growing up lonely, or as a consequence of longer term selection for individuals that are better at finding mates in situations of low conspecific density, is unknown.

Plastic responses might also occur as accidents of a changed environment during invasion. For example, many frog species alter both their behaviour and developmental rate as tadpoles in response to the presence of aquatic predators (Kiesecker and Blaustein 1997; Relyea 2001). One of the consequences of low conspecific density and assortment by dispersal on the edge of an invading population is that coevolved predators often get left behind (Phillips et al. 2010d). If this happens, then the cues driving the typical predator response may be absent, and a morphologically and behaviourally 'predator free' phenotype will emerge (even if predators, albeit unfamiliar ones, are still present). Typically, tadpoles in the absence of predators spend more time active, and have higher growth rates than conspecifics in the presence of predators (e.g. Thiemann and Wassersug 2000; Relyea 2001). Higher individual growth rates resulting from tadpoles behaving in a predator free manner could, conceivably, lead to higher population growth rates and faster spread, although this net effect would be an accidental byproduct of an animal in an unfamiliar environment. Alternatively, of course, increased predation rates driven by inappropriate responses to unfamiliar predators could depress population growth and slow spread. The effects of plasticity could, conceivably, either accelerate or decelerate an invasion.

14.3.2 Evolved responses

Traditionally, we have thought about evolution in invasive species as being about adaptation to their new environment (Mack et al. 2000). While this standard adaptive process must indeed be happening after an invasive population has become established, it is increasingly apparent that evolutionary processes on the invasion front itself are driven by additional forces (Phillips et al. 2010a). Invasion fronts can capture rare, and even deleterious, mutations, drive them to high frequency, and then spread them across large areas as the invasion front moves through space (Travis et al. 2007). Coupled with this are the processes of assortative mating by dispersal ability, and *r*-selection (driven by low conspecific density), which drive selection for increased dispersal and reproductive rates (Burton et al. 2010). Thus, during invasion, behaviours will evolve to increase dispersal and, if possible given trade-offs with dispersal, increase reproductive rates. Our brief survey of behavioural traits that facilitate invasion (above) gives us an idea as to the potential suite of traits that might evolve in this way during invasion.

Empirical evidence for changed dispersal behaviour during invasion is rapidly accumulating. Insects on expanding range edges have been shown to invest more in dispersal than their conspecifics in older, established populations (Hughes et al. 2007; Simmons and Thomas 2004; Léotard et al. 2009), and invasive cane toads in northern Australia have evolved increased dispersal rates on the invasion front relative to their conspecifics from older, established populations (Phillips et al. 2010b; Phillips et al. 2008a).

Empirical evidence for the evolution of behaviours facilitating population growth is, however, weaker. Higher growth rates have been found in invasion front populations of cane toads (Phillips 2009), but the behavioural correlates of this higher growth rate, if any, remain to be elucidated. The increased sensitivity of invasion front starlings to the calls of their conspecifics (Rodriguez et al. 2010) might be plastic or evolved (or both), we simply do not know yet. But as an adaptation to reducing Allee effects and increasing population growth rate on an invasion front, we might expect it to be an evolved shift. Generally, the importance of comparing newly colonized and older established populations inside the invasive range is only recently being appreciated. So the field is open for behavioural ecologists to start investigating behavioural shift in response to range shift.

Another intriguing aspect of evolution during invasive spread is that it may bring together multiple behavioural traits into sets. Being such a strong directed selective pressure, and being directed at dispersal and reproductive rates (complex traits typically driven by multiple behaviours), we would expect range shift to start accumulating individuals with multiple behaviours that increase dispersal and reproductive rates, whilst minimizing Allee effects (Travis et al. 2010; Shine et al. 2011). In Australian cane toads, individuals from the invasion front grow faster, move more often, move further when they do move, and follow straighter paths than their trailing conspecifics, all of which conspires to increase their net dispersal and reproductive rates (Phillips et al. 2008a; Alford et al. 2009; Phillips 2009). Similarly, we might expect all traits that influence dispersal and reproductive rates to become associated in invasion front populations. Given that many of the traits associated with animal personality (e.g. aggression and boldness: Duckworth 2008; Duckworth and Badyaev 2007; Sih et al. 2004b) may also be associated with dispersal and/or reproduction, range shifts can, over time, assemble populations with distinct personalities (e.g. aggressive, risk taking, and highly active). Why animals exhibit distinct 'personalities' is a current focus amongst behavioural ecologists (Dingemanse et al. 2010; Sih et al. 2004b). It may be that selection for dispersal (Cote et al. 2010) during invasion is a currently under-appreciated way for animal personalities to be assembled. Future work exploring this possibility would likely be rewarding. Such work would focus strongly on behavioural variation, and how it varies through invasion history.

14.4 Behavioural variation and the impacts of invasive species on natives

Invasive species are typically of broad interest because of their perceived impacts on native species. Almost all the impacts of invasive species on natives are likely mediated by behaviour. Both the behaviour of the invader, and that of the native, determine the outcome. Where the outcome of an interaction is negative (either for the native or the invader, or both), we might expect to see rapid evolution of behaviour. Thus, species invasions offer immense opportunities for behavioural ecologists interested in the evolution (and coevolution) of behaviour. Invasive species often represent a novel selective force in the environment that native species need to respond to through behavioural means (Ashley et al. 2003).

Many of the biggest impacts from invasive animals occur when the invader is a predator. More than 22 species of native mammal were lost from the Australian continent following the arrival of the invasive cat and fox (McKenzie et al. 2007). Similarly, in Lake Victoria, the introduction of the Nile perch *Lates niloticus* led to the extinction of around 200 species of native cichlid fish (Ogutu-Ohwayo 1999). In New Zealand, the introduction of Pacific rats *Rattus exulans* led to the extinction from the mainland of many species of birds and lizards (Towns et al. 2007). These examples of extinction point to situations where native species simply didn't have appropriate behavioural variation (in both evolved and plastic senses) to cope with the arrival of a new predator. Potentially, however, there may be many more (unreported) instances where natives can modify behaviours in response to predation, and so persist. For example, Hoare et al. (2007) show that, on New Zealand islands colonized by Pacific rats, native geckoes *Hoplodactylus duvaucelii* rapidly shift their habitat preferences to avoid interactions with the predatory rats. Unfortunately, these subtler examples (where native species are not extirpated, but survive through behavioural shifts) are understudied.

The impacts of introduced species are also likely to be density dependent (Yokomizo et al. 2009). Therefore, behaviours that influence the density of species in introduced populations may disproportionately influence their impact. For example, intraspecific competition is often cited as a primary mechanism that regulates relative abundance (e.g. Ryti and Case 1988; Comita et al. 2010), so highly territorial species that exclude conspecifics from large areas are not likely to become 'high-impact' invasives. In contrast, social species that do not exhibit discrete or exclusive territories may attain higher densities and may also avoid Allee effects that inhibit population spread. While an association between intraspecific aggression and impact has not been explored across many taxa, it has been suggested as a mechanism for the success of invasive ants (Holway and Suarez 1999). Many introduced populations of highly invasive ants are 'unicolonial'—spatially separate nests behave as 'one colony' across the entire population (Holway et al. 2002); a situation that is not always the case in their native range. In unicolonial populations where intraspecific territorial behaviour is absent or reduced, colonies often exhibit higher resource retrieval rates, and greater brood and worker production relative to native species that exhibit territorial behaviour (Holway et al. 1998). The origin and maintenance of unicoloniality in ants may provide insight into the role of behavioural variation in the success and impact of invasive species more generally, and is currently the subject of intense debate and study (Helantera et al. 2009).

Behavioural variation across environmental contexts may also be important in influencing the outcomes of competition with invasive species. For example, by hanging lights in abandoned Second World War aircraft hangers in Hawaii, Petren and colleagues (1993) were able to experimentally demonstrate how increasing the local abundance of insects influenced competitive interactions between two species of gecko: the invasive *Hemidactylus frenatus* and the resident *Lepidodactylus lugubris*. The larger, more aggressive *H. frenatus* was able to forage more effectively and monopolize clumped resources in open areas relative to the less aggressive *L. lugubris*. However, the effects of competition between the species disappeared if resources were not clumped or if complex topography was added around the light sources to provide safe refuges for the less aggressive species to hide in (Petren and Case 1998; Fig. 14.3).

Another well-studied example of behavioural mediation of impacts concerns the arrival, in Australia, of the toxic cane toad. In Australia, toads occur in very high densities, and show no territoriality. They are a predator, but their major impact appears to be on native predators that mistake toads as palatable prey. Numerous native predators, which were naïve to the toxins in the toads' skin, have been fatally poisoned following the arrival of toads. Despite the simplicity of the impact mode—poisoning—impacts have varied tremendously, both within populations as well as between closely related species (Shine 2010). Much of this variation in impact is driven by subtle differences in the natives' prey handling behaviour. For example, death adders *Acanthophis praelongus* in northern Australia handle different prey species (all of which are frogs) differently, depending upon the prey species' toxicity (Phillips and Shine 2007). Non-toxic frogs are simply grabbed and swallowed, whereas frogs which secrete a sticky glue or which are neurotoxic are envenomated and released to die, after which the snake then tracks the meal down and ingests it. Snakes wait around ten minutes between bite and consumption for the gluey frogs but wait much longer (typically around 40 minutes) before tasting and, ultimately, consuming the neurotoxic species. When these snakes first encounter a toad, the response of individuals varied: some snakes treat the toad as if it was non-toxic, while others treat them as if they were gluey, neurotoxic, or something that is ultimately inedible (Hagman et al. 2009). Almost all death adders that consume a toad die, thus, a snake's classification of this new prey item (and the subsequent behaviours that flow from that classification), strongly defines the impact that the arrival of toads will have on that individual. This correlation between behaviour and impact suggests strong selection operating on prey preference in this species (Phillips et al. 2010c), and indeed at least one Australian snake species appears to have evolved behavioural avoidance of toads as prey due to this strong selection pressure (Phillips and Shine 2006).

These examples show the critical importance of the behaviour of both native and invasive species in mediating the invasive's impact. Importantly, however, they also demonstrate the critical importance of behavioural variability (both plastic and evolved) in determining the long-term outcome of that impact. Behavioural plasticity or rapid evolved responses may be integral for native species to persist in invaded landscapes.

14.5 Conclusion and future directions

Through both theory and example, we have demonstrated the importance of behaviour to both the spread and impact of invasive species. Spread rate (governed by dispersal and population growth) is, ultimately, an outcome of the behaviour of the invader. Similarly, the impact of the invader depends on the interaction between its behaviour and that of nearby native species. The central importance of behaviour in these cases is not particularly surpris-

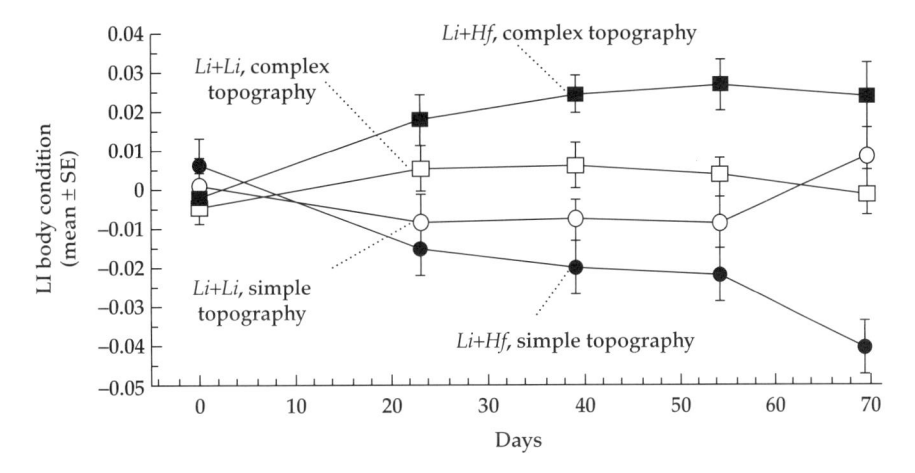

Figure 14.3 The distribution of resources (clumped versus dispersed) and the degree of habitat complexity influences the outcome of interspecific competition between invasive *Hemidactylus frenatus Hf*, and resident *Lepidodactylus lugubris Li*, geckos. When insect resources are clumped at light sources, and the space around the lights is open, the larger more aggressive *Hemidactylus* can monopolize access to insects. However, if resources are not clumped and/or baffles are placed around the lights creating refuges for the geckos, *Lepidodactylus* are able to forage more effectively, evidenced by an increase in body condition. Figure reprinted from Petren and Case 1998, with permission. Copyright (1998) National Academy of Sciences, U.S.A.

ing, but consideration of behavioural variation (either plastic or evolved) can lead to surprising results. For example, early predictions of the spread rate of cane toads across northern Australia were rendered wildly inaccurate by the rapid evolution of increased dispersal behaviour in this species (Phillips et al. 2008b). Similarly, many predictions of deleterious impacts of invasive species on natives may prove to be fleeting as native species exhibit behavioural flexibility in response to the invader (e.g. Langkilde 2009).

The future, thus, should be about variation. When an invader arrives, the average behaviour often turns out to be much less important than extremes of behaviour. Extreme dispersers come to dominate the vanguard of an invasive population, and extreme behaviours may be associated with individuals most or least at risk from the presence of an invader. The challenge for behavioural ecologists is thus clear. We need to understand behavioural variability: within and between individuals and, within and between populations. We need to understand both the genesis of such variation as well as its environmental correlates. Only by understanding this variation can we hope to predict behavioural responses into the future, in a changing world.

References

Alford, R. A., Brown, G. P., Schwarzkopf, L., Phillips, B. L., and Shine, R. (2009). Comparisons through time and space suggest rapid evolution of dispersal behaviour in an invasive species. *Wildlife Research*, 36, 23–8.

Amiel, J. J., Tingley, R., and Shine, R. (2011). Effects of relative brain size on establishment success of invasive amphibians and reptiles. *PLOS ONE* 6: e18277. doi:10.1371/journal.pone.0018277.

Ashley, M. V., Willson, M. F., Pergams, O. R. W., O'Dowd, D. J., Gende, S. M., and Brown, J. S. (2003). Evolutionarily enlightened management. *Biological Conservation*, 111, 115–23.

Aubry, S., Labaune, C., Magnin, F., Roche, P., and Kiss, L. (2006). Active and passive dispersal of an invading land snail in Mediterranean France. *Journal of Animal Ecology*, 75, 802–13.

Barton, K. A., Phillips, B. L., Morales, J. M., and Travis, J. M. J. (2009). The evolution of an 'intelligent' dispersal strategy: biased, correlated random walks on patchy landscapes. *Oikos*, 118, 309–19.

Blackburn, T. M. and Duncan, R. P. (2001). Establishment patterns of exotic birds are constrained by non-random patterns in introduction. *Journal of Biogeography*, 28, 927–39.

Burton, O. J., Travis, J. M. J., and Phillips, B. L. (2010). Trade-offs and the evolution of life-histories during range expansion. *Ecology Letters*, 13, 1210–20.

Clobert, J., Danchin, E., Dhondt, A. A., and Nichols, J. D. (eds) (2001). *Dispersal.* Oxford, Oxford University Press.

Comita, L. S., Muller-Landau, H. C., Aguillar, S., and Hubbell, S. P. (2010). Asymmetric density dependence shapes species abundances in a tropical tree community. *Science*, 329, 330–2.

Cote, J., Clobert, J., Brodin, T., Fogarty, S., and Sih, A. (2010). Personality-dependent dispersal: characterization, ontogeny and consequences for spatially structured populations. *Philosophical Transactions of the Royal Society of London Series B-Biological Sciences*, 365, 4065–76.

Dingemanse, N. J., Kazem, A. J. N., Reale, D., and Wright, J. (2010). Behavioural reaction norms: animal personality meets individual plasticity. *Trends in Ecology and Evolution*, 24, 81–9.

Duckworth, R. (2008). Adaptive dispersal strategies and the dynamics of range expansion. *The American Naturalist*, 172, S4–S17.

Duckworth, R. A. and Badyaev, A. V. (2007). Coupling of dispersal and aggression facilitates the rapid range expansion of a passerine bird. *Proceedings of the National Academy of Sciences of the USA*, 104, 15017–22.

Elton, C. S. (1958). *The Ecology of Invasions By Animals and Plants.* London, Methuen.

Fogarty, S., Cote, J., and Sih, A. (2011). Social personality polymorphism and the spread of invasive species: a model. *The American Naturalist*, 177, 273–87.

Gandon, S. and Michalakis, Y. (2001). Multiple causes of the evolution of dispersal. In: Clobert, J., Danchin, E., Dhondt, A. A., and Nichols, J. (eds) *Dispersal.* Oxford, Oxford University Press.

Hagman, M., Phillips, B. L., and Shine, R. (2009). Fatal attraction: adaptations to prey on native frogs imperil snakes after invasion of toxic prey. *Proceedings of the Royal Society of London B, Biological Sciences*, 276, 2813–18.

Hamilton, W. D. and May, R. M. (1977). Dispersal in stable habitats. *Nature*, 269, 578–81.

Hastings, A., Cuddington, K., Davies, K. F., Dugaw, C. J., Elmendorf, S., Freestone, A., Harrison, S., Holland, M., Lambrinos, J., Malvadkar, U., Melbourne, B. A., Moore, K., Taylor, C., and Thomson, D. (2005). The spatial spread of invasions: new developments in theory and evidence. *Ecology Letters*, 8, 91–101.

Helantera, H., Strassman, J. E., Carrillo, J., and Queller, D. C. (2009). Unicolonial ants: where do they come from, what are they and where are they going? *Trends in Ecology and Evolution*, 24, 341–9.

Hoare, J. M., Shirley, P., Nelson, N. J., and Daugherty, C. H. (2007). Avoiding aliens: Behavioural plasticity in habitat use enables large, nocturnal geckos to survive Pacific rat invasions. *Biological Conservation*, 136, 510–19.

Holway, D. A., Lach, L., Suarez, A. V., Tsutsui, N. D. and Case, T. J. (2002). The causes and consequences of ant invasions. *Annual Review of Ecology and Systematics*, 33, 181–233.

Holway, D. A. and Suarez, A. V. (1999). Animal behaviour: an essential component of invasion biology. *Trends in Ecology and Evolution*, 14, 328–30.

Holway, D. A., Suarez, A. V., and Case, T. J. (1998). Loss of intraspecific aggression in the success of a widespread invasive social insect. *Science*, 282, 949–52.

Howard, W. E. (1960). Innate and environmental dispersal of individual vertebrates. *American Midland Naturalist*, 63, 152–61.

Hughes, C. L., Dytham, C., and Hill, J. K. (2007). Modelling and analysing evolution of dispersal in populations at expanding range boundaries. *Ecological Entomology*, 32, 437–45.

Johnson, D. M., Liebhold, A. M., Tobin, P. C., and Bjornstad, O. N. (2006). Allee effects and pulsed invasion by the gypsy moth. *Nature*, 444, 361–3.

Kiesecker, J. M. and Blaustein, A. M. (1997). Population differences in responses of red-legged frogs (*Rana aurora*) to introduced bullfrogs. *Ecology*, 78, 1752–60.

Kolar, C. S. and Lodge, D. M. (2001). Progress in invasion biology: predicting invaders. *Trends in Ecology and Evolution*, 16, 199–204.

Langkilde, T. (2009). Invasive fire ants alter behavior and morphology of native lizards. *Ecology*, 90, 208–17.

Léotard, G., Debout, G., Dalecky, A., Guillot, S., Gaume, L., Mckey, D., and Kjellberg, F. (2009). range expansion drives dispersal evolution in an equatorial three-species symbiosis. *PLOS One*, 4, e5377. doi:10.1371/journal.pone.0005377.

Loss, S. R., Hamer, G. L., Waljer, E. D., Ruiz, M. O., Goldberg, T. L., Kitron, U. D., and Brawn, J. D. (2009). Avian host community structure and prevalence of West Nile Virus in Chicago, Illinois. *Oecologia*, 159, 415–24.

Mack, R. N., Simberloff, D., Lonsdale, W. M., Evans, H., Clout, M., and Bazzaz, F. (2000). Biotic invasions: Causes, epidemiology, global consequences and control. *Ecological Applications*, 10, 689–710.

Mckenzie, N. L., Burbidge, A. A., Baynes, A., Brereton, R. N., Dickman, C. R., Gordon, G., Gibson, L. A., Menkhorst, P. W., Robinson, A. C., Williams, M. R., and Woinarski, J. C. Z. (2007). Analysis of factors implicated in the recent decline of Australia's mammal fauna. *Journal of Biogeography*, 34, 597–611.

Ogutu-Ohwayo, R. (1999). Nile perch in Lake Victoria: the balance between benefits and negative impacts of aliens.

In: Sandlund, O. T., Schei, P. J., and Viken, A. (eds) *Invasive Species and Biodiversity Management*. Boston, Kluwer Academic.

Parker, I. M., Simberloff, D., Lonsdale, W. M., Goodell, K., Wonham, M., Kareiva, P. M., Williamson, M. H., Von Holle, B., Moyle, P. B., Byers, J. E., and Goldwasser, L. (1999). Impact: Toward a framework for understanding the ecological effects of invaders. *Biological Invasions*, 1, 3–19.

Petren, K., Bolger, D. T., and Case, T. J. (1993). Mechanisms in the competitive success of an invading sexual gecko over an asexual native. *Science*, 259, 354–8.

Petren, K. and Case, T. J. (1998). Habitat structure determines competition intensity and invasion success in gecko lizards. *Proceedings of the National Academy of Sciences of the USA*, 95, 11739–44.

Phillips, B. L. (2009). The evolution of growth rates on an expanding range edge. *Biology Letters*, 5, 802–4.

Phillips, B. L., Brown, G. P., and Shine, R. (2010a). The evolution of life-histories during range-advance. *Ecology*, 91, 1617–27.

Phillips, B. L., Brown, G. P., and Shine, R. (2010b). Evolutionarily accelerated invasions: the rate of dispersal evolves upwards during range advance of cane toads. *Journal of Evolutionary Biology*, 23, 2595–601.

Phillips, B. L., Brown, G. P., Travis, J. M. J., and Shine, R. (2008a). Reid's paradox revisited: the evolution of dispersal in range-shifting populations. *The American Naturalist*, 172, S34–S48.

Phillips, B. L., Chipperfield, J. D., and Kearney, M. R. (2008b). The toad ahead: challenges of modelling the range and spread of an invasive species. *Wildlife Research*, 35, 222–34.

Phillips, B. L., Greenlees, M. J., Brown, G. P., and Shine, R. (2010c). Predator behaviour and morphology mediates the impact of an invasive species: cane toads and death adders in Australia. *Animal Conservation*, 13, 53–9.

Phillips, B. L., Kelehear, C., Pizzatto, L., Brown, G. P., Barton, D., and Shine, R. (2010d). Parasites and pathogens lag behind their host during periods of host range-advance. *Ecology*, 91, 872–81.

Phillips, B. L. and Shine, R. (2006). An invasive species induces rapid adaptive change in a native predator: cane toads and black snakes in Australia. *Proceedings of the Royal Society of London B, Biological Sciences*, 273, 1545–50.

Phillips, B. L. and Shine, R. (2007). When dinner is dangerous: toxic frogs elicit species-specific responses from a generalist snake predator. *The American Naturalist*, 170, 936–42.

Pimental, D., Lach, L., Zuniga, R., and Morrison, D. (2000). Environmental and economic costs of nonindigenous species in the United States. *BioScience*, 50, 53–64.

Pintor, L. M., Sih, A., and Bauer, M. L. (2008). Differences in aggression, activity and boldness between native and introduced populations of an invasive crayfish. *Oikos*, 117, 1629–36.

Reale, D., Reader, S. M., Sol, D., Mcdougall, P. T., and Dingemanse, N. J. (2007). Integrating animal temperament within ecology and evolution. *Biological Reviews of the Cambridge Philosophical Society*, 82, 291–318.

Relyea, R. A. (2001). The lasting effects of adaptive plasticity: predator-induced tadpoles become long-legged frogs. *Ecology*, 82, 1947–55.

Rodriguez, A., Hausberger, M., and Clergeau, P. (2010). Flexibility in European starlings' use of social information: experiments with decoys in different populations. *Animal Behaviour*, 80, 965–73.

Ryti, R. R. and Case, T. J. (1988). Field experiments on desert ants: testing for competition between colonies. *Ecology*, 69, 1993–2003.

Seidowski, D., Ziegler, U., Von Rönn, J. A. C., Müller, K., Hüppop, K., Müller, T., Freuling, C., Mühle, R.-U., Nowotny, N., Ulrich, R. G., Niedrig, M., and Groschup, M. H. (2010). West Nile Virus monitoring of migratory and resident birds in Germany. *Vector-Borne and Zoonotic Diseases*, 10, 639–47.

Shine, R. (2010). The ecological impact of invasive cane toads (Bufo marinus) in Australia. *Quarterly Review of Biology*, 85, 253–91.

Shine, R., Brown, G. P., and Phillips, B. L. (2011). An evolutionary process that assembles phenotypes through space rather than through time. *Proceedings of the National Academy of Sciences of the USA*, 108, 5708–11.

Sih, A., Bell, A., and Chadwick Johnson, J. (2004a). Behavioural syndromes: an ecological and evolutionary overview. *Trends in Ecology and Evolution*, 19, 372–8.

Sih, A., Bell, A. M., Johnson, J. C., and Ziemba, R. E. (2004b). Behavioral syndromes: an integrative overview. *The Quarterly Review of Biology*, 79, 241–77.

Sih, A., Bolnick, D. I., Luttbeg, B., Orrock, J. L., Peacor, S. D., Pintor, L. M., Preisser, E., Rehage, J. S., and Vonesh, J. R. (2010). Predator–prey naïveté, antipredator behavior, and the ecology of predator invasions. *Oikos*, 119, 610–21.

Simmons, A. D. and Thomas, C. D. (2004). Changes in dispersal during species' range expansions. *American Naturalist*, 164, 378–95.

Simons, A. M. (2003). Invasive aliens and sampling bias. *Ecology Letters*, 6, 278–80.

Skellam, J. G. (1951). Random dispersal in theoretical populations. *Biometrika*, 38, 196–218.

Sol, D., Bacher, S., Simon, M. R., and Lefebvre, L. (2008). Brain size predicts the success of mammal species introduced into novel environments. *The American Naturalist*, 172, S63–S71.

Sol, D. and Lefebvre, L. (2000). Behavioural flexibility predicts invasion success in birds introduced to New Zealand. *Oikos*, 90, 599–605.

Sol, D., Timmermans, S., and Lefebvre, L. (2002). Behavioural flexibility and invasion success in birds. *Animal Behaviour*, 63, 495–502.

Suarez, A. V., Holway, D. A., and Case, T. J. (2001). Patterns of spread in biological invasions dominated by long-distance jump dispersal: Insights from Argentine ants. *Proceedings of the National Academy of Sciences of the USA*, 98, 1095–100.

Thiemann, G. W. and Wassersug, R. J. (2000). Patterns and consequences of behavioural responses to predators and parasites in Rana tadpoles. *Biological Journal of the Linnean Society*, 71, 513–28.

Tingley, R., Romagosa, C. M., Kraus, F., Bickford, D., Phillips, B. L., and Shine, R. (2010). The frog filter: amphibian introduction bias driven by taxonomy, body size, and biogeography. *Global Ecology and Biogeography*, 19, 496–503.

Towns, D. R., Parrish, G. R., Tyrrell, C. L., Ussher, G. T., Cree, A., Newman, D. G., Whitaker, A. H., and Westbrooke, I. (2007). Responses of tuatara (*Sphenodon punctatus*) to removal of introduced Pacific rats from islands. *Conservation Biology*, 21, 1021–31.

Travis, J. M. J. and Dytham, C. (2002). Dispersal evolution during invasions. *Evolutionary Ecology Research*, 4, 1119–29.

Travis, J. M. J., Münkemüller, T., and Burton, O. J. (2010). Mutation surfing and the evolution of dispersal during range expansions. *Journal of Evolutionary Biology*, 23, 2656–67.

Travis, J. M. J., Münkemüller, T., Burton, O. J., Best, A., Dytham, C., and Johst, K. (2007). Deleterious mutations can surf to high densities on the wave front of an expanding population. *Molecular Biology and Evolution*, 24, 2334–43.

Van Valen, L. (1971). Group selection and the evolution of dispersal. *Evolution*, 25, 591–8.

Veit, R. R. and Lewis, M. A. (1996). Dispersal, population growth and the allee effect: dynamics of the house finch invasion of eastern North America. *The American Naturalist*, 148, 255–74.

Vitousek, P. M., D'Antonio, C. M., Loope, L. L., and Westbrooks, R. (1996). Biological Invasions as global environmental change. *American Scientist*, 84, 469–78.

Weisser, W. W., Mccoy, K. D., and Boulinier, T. (2001). Parasitism and predation as causes of dispersal. In: Clobert, J., Danchin, E., Dhondt, A. A., and Nichols, J. (eds) *Dispersal*. Oxford, Oxford University Press.

Wright, T. F., Eberhard, J. R., Hobson, E. A., Avery, M. L., and Russello, M. A. (2010). Behavioural flexibility and species invasions: the adaptive flexibility hypothesis. *Ethology, Ecology and Evolution*, 22, 393–404.

Yokomizo, H., Possingham, H. P., Thomas, M. B., and Buckley, Y. M. (2009). Managing the impact of invasive species: the value of knowing the density-impact curve. *Ecological Applications*, 19, 376–86.

CHAPTER 15

Sexual selection in changing environments: consequences for individuals and populations

Ulrika Candolin and Bob Wong

⊃ Overview

Securing suitable mates is rapidly becoming a lot more challenging in an increasingly human-dominated world. Environmental changes are having dramatic effects on reproductive behaviour: male beetles are copulating with discarded beer bottles instead of actual females; cichlid fishes are mating with members of the wrong species because of reduced visibility from algal blooms; and frog calls are being drowned out by the noise of urban traffic. In this chapter, we explore the impacts of human-induced environmental change on sexually selected behaviours. We begin with an overview of the mechanisms of sexual selection and explain how they depend on the environment. We then examine the effects of environmental disturbances on sexually selected behaviours, and what consequences these might have for the fate of individuals, populations, and species. Finally, we discuss the degree to which animals can adjust their sexually selected behaviour to environmental change, and whether these behavioural responses can help populations to persist under human-altered conditions.

15.1 The importance of sexual selection

Sexual selection is a potent evolutionary force (Darwin 1871; Andersson 1994). It favours the elaboration of traits that increase success in the competition for fertilizations or, more precisely, access to opposite-sex gametes, and is responsible for some of the most spectacular and complex behaviours found in nature—from ostentatious courtship displays to highly ritualized fights over mates. Spectacular and showy these traits may be, but sexual traits are also costly. Courtship displays, for example, are often time consuming and energetically taxing to perform (Hunt et al. 2004). They can also draw unwanted attention from would-be predators and parasitoids, and increase an individual's risk of injury or even death (Zuk and Kolluru 1998; Woods et al. 2007). Thus, in order for sexually selected traits to evolve and be maintained within a population, the fitness benefits conferred by such traits have to exceed the fitness costs. In other words, sexual traits have to increase the lifetime mating or fertilization success of an individual relative to those of other individuals—but may not necessarily increase the fitness, or viability, of the population.

Sexual selection is generally ascribed to the operation of two key mechanisms: competition over, and choice of, opposite-sex gametes. Both mechanisms can favour the evolution of costly traits. The cost of competition can be quite extreme in the case of injury or death from fighting; but more subtle costs of competition can also exist, such as those associated with the production and expenditure of

Behavioural Responses to a Changing World. First Edition. Edited by Ulrika Candolin and Bob B.M. Wong.
© 2012 Oxford University Press. Published 2012 by Oxford University Press.

ejaculates. Cost of choice, by contrast, is typically associated with mate search and evaluation, and the rejection of unwanted, but persistent, suitors (Real 1990; Booksmythe et al. 2008). A central tenet in sexual selection theory is that members of the choosing sex can use certain traits to assess potential mates (see also Chapter 2 on communication). Mate choice traits, in this regard, are widely purported to reflect a whole suite of direct (i.e. material) and/or indirect (i.e. genetic) fitness benefits, from superior parental care to increased offspring viability (Jennions and Petrie 2000; Kirkpatrick and Ryan 1991). Theory suggests that the reliability of these traits is safeguarded by constraints (e.g. dependency on body size), by the cost of producing or maintaining the traits, or by the cost of cheating (Searcy and Nowicki 2005). Based on this argument, only individuals well adapted to the environment should be able to display the most exaggerated traits, thus ensuring that traits accurately reflect fitness benefits under prevailing environmental conditions.

15.1.1 Population-level consequences

Because sexually selected traits influence the survival and reproductive output of individuals, sexual selection can have both positive and negative consequences for the demography of populations. Sexual selection will impact negatively on population growth if investment into such traits reduces female fecundity, the number of females reproducing, or the fitness of offspring. Behaviours that reduce female fecundity are particularly influential. For instance, intersexual conflict, where males hinder the reproduction of other males by preventing or reducing female reproduction, can limit the potential for population growth (Rankin and Arnqvist 2008). Similarly, intralocus sexual conflict, where alleles that are favourable in one sex are maladaptive in the other, can inflict costs at the population level (Cox and Calsbeek 2009). Thus, a discrepancy can exist between what is best for the individual and what is best for the population. Such a discrepancy arises because individuals behave in ways that help them to maximize their own reproductive success relative to those of other individuals within the population. This can occur even if it results in a decrease in the reproductive success of the population as a whole ('tragedy of the commons'; sensu Rankin et al. 2011).

But sexual selection also has another important and potentially countering effect: it can increase the frequency of favourable alleles within a population and, in so doing, enhance population viability (Houle and Kondrashov 2002; Møller and Alatalo 1999; Jarzebowska and Radwan 2010). This occurs when success in mate competition is positively related to the genetic quality of the individuals that are vying to reproduce, that is, when mating and fertilization success correlate with how well individuals are adapted to their environment. Moreover, costs of sexually selected traits can remove poorly adapted individuals from the mating pool. This can have the positive effect of increasing the resources available for those remaining in the population. Interestingly, if the excluded individuals are mostly male, we might even expect an increase in the reproductive output of the population—since reproductive output is determined by female fecundity (Clutton-Brock et al. 2002). On the other hand, a reduction in effective population size (N_e) can reduce genetic diversity and, hence, the potential to adapt to new or changing conditions. Thus, the evolution of sexually selected traits can impose both costs and benefits at the population level. What consequences this might have for population viability, however, remain highly contentious (Candolin and Heuschele 2008).

15.2 Sexual selection and the environment

The costs and benefits of sexually selected behaviours depend fundamentally on the environment (Fig. 15.1). For instance, the cost of searching for, and evaluating, potential mates varies with predation risk and the distribution of individuals within the population. This is seen in female fiddler crabs *Uca uruguayensis*, which have a lower probability of encountering males when population density is low, resulting in a high net cost of mate searching (Ribeiro et al. 2010). Courtship behaviours are similarly sensitive to environmental conditions. Male

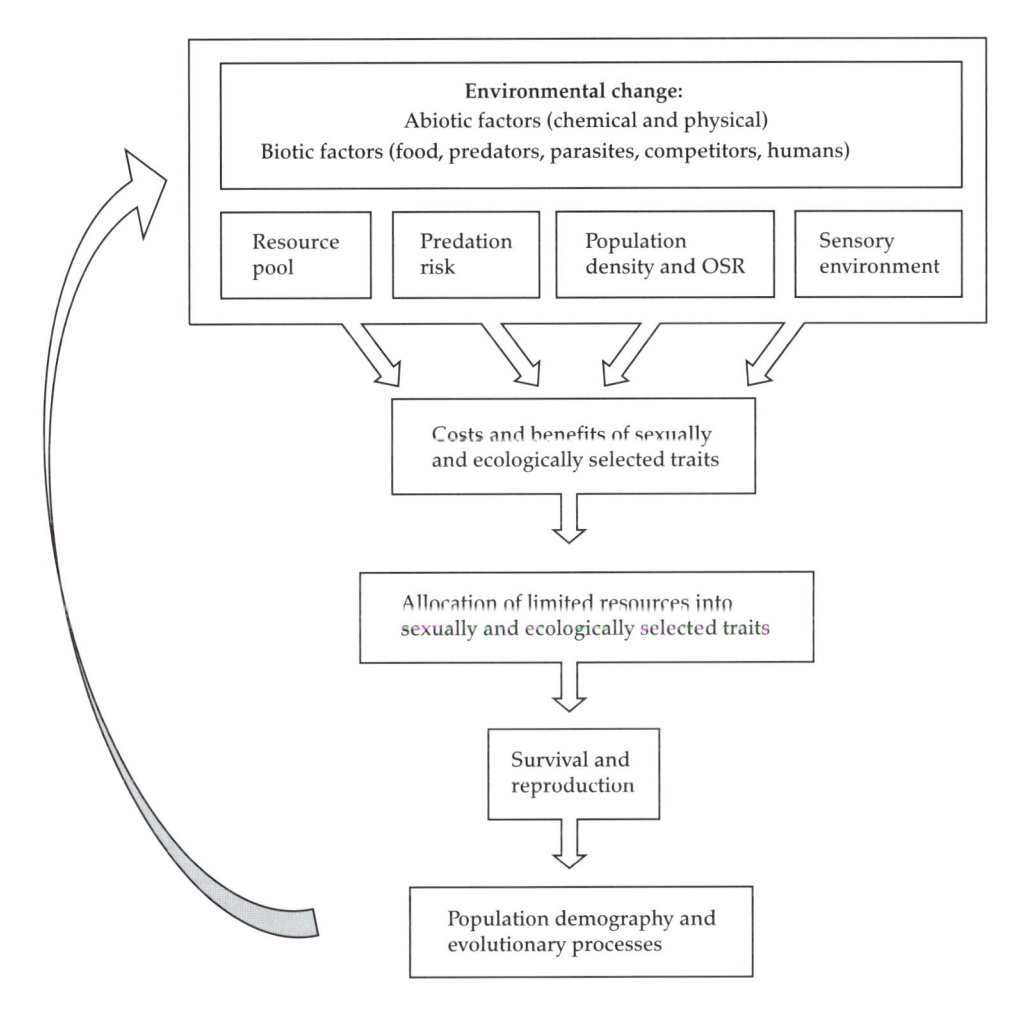

Figure 15.1 A simplified graph of the feedback between environmental change, sexually selected traits, population demography, and evolutionary processes. Changed environmental conditions—such as resource availability, predation risk, population density, operational sex ratio (OSR), and sensory environment—influence the costs and benefits of sexually and ecologically selected traits, and, thus, how limited resources are allocated among the various traits. This, in turn, influences individual survival and reproduction. Environmental change can result in an ecosystem-wide feedback system, where sexually selected traits are just one component determining the persistence of the population.

Alpine newts *Mesotriton alpestris*, for example, benefit less from visual displays when light conditions are poor and rely, instead, on tactile and chemical signals to attract females for mating (Denoel and Doellen 2010).

Environmental dependence can help ensure the reliability of sexually selected displays. A positive correlation is expected to develop between the trait and the quality it is reflecting when both are influenced by the environment. When this occurs, the trait can provide reliable information on the direct or indirect fitness benefits being offered by the bearer. Thus, sexually selected displays are often condition-dependent, where condition is the pool of resources (energy, nutrients and time) available to an individual for allocating into fitness enhancing traits (Tomkins et al. 2004).

In a deterministic environment, where conditions are stable or where changes are predictable, sexually selected behaviours are expected to evolve to an optimum state at the individual level. Over-expression is selected against because of high mortality and low reproductive success. This prevents the cost from becoming so untenable that it threatens population persistence (Kokko and Rankin 2006). However, population viability, even under stable conditions, can potentially be undermined if individuals other than those who are performing the behaviour have to bear some of the cost. For example, males attempting to increase their own reproductive success may try to prevent females from mating with other males. This can reduce female fecundity and, as a result, impact negatively on population output (Rankin et al. 2011). Over time, selection at the population level is expected to eliminate populations with a high cost of sexual selection.

15.3 Consequences of environmental change

Unpredictable environmental change can cause drastic shifts in the costs and benefits of sexually selected behaviours. This can move the optimum value of the behaviours to either lower or higher values. The ability of individuals to adjust their behaviour to the new optimum will then determine their fitness. In this regard, fitness is expected to decline if the optimum value decreases and individuals do not adjust their behaviour to the new optimum. For instance, if increased predation risk favours less conspicuous courtship behaviours, individuals that do not decrease their conspicuousness will suffer from lower survival probability. On the other hand, if the optimal conspicuousness of courtship increases, as may occur when predation pressure declines, then individuals that continue to court as before may have a lower fitness compared to those that increase the conspicuousness of their displays.

Changes in the optimum state of a behaviour can be difficult to predict. Increased predation pressure, for instance, can increase the mortality cost of conspicuous courtship whilst leaving the benefit—in terms of female attraction—unchanged. At the same time, increased predation on females can shift the operational sex ratio towards males and increase the benefit of courtship (because of intensified male competition) without influencing the cost. Thus, the ultimate effect of environmental change on the cost-benefit ratio of behaviours hinges on how the environment affects all fitness influencing factors, including the resource pool and resource needs of individuals and interactions among ecologically and sexually selected traits (Fig 15.1).

15.3.1 Resource allocation and trade-offs

The influence of environmental change on the costs and benefits of sexually selected behaviours is closely linked with resource availability. The pool of resources accessible to individuals, as well as the allocation of these resources in competition with other traits (i.e. ecologically selected traits), determine how much can be invested into sexually selected traits. For example, a decline in food availability can elevate the cost of searching for both mates and food, forcing individuals to redistribute their resources among the different behaviours. Female Galápagos marine iguanas *Amblyrhynchus cristatus*, for instance, reduced their investment in mate choice following an El Niño event when food availability declined and the relative cost of mate searching increased (Fig. 15.2; Vitousek 2009). Likewise, increased predation risk can cause individuals to allocate resources from courtship displays into ecologically selected vigilance, as demonstrated, for example, in wolf spiders *Schizocosa ocreata* (Lohrey et al. 2009).

Changes in environmental conditions that alter the allocation of resources into sexually selected behaviours can also affect signal reliability by influencing the relationship between the signal and the quality being advertised. For instance, when food is scarce and competition for limited resources is intense, only males that are well adapted to the environment may be able to display the most exaggerated courtship behaviours. In favourable environments, on the other hand, a wider range of males may be able to invest in these displays without compromising their fitness. This implies that resource scarcity could facilitate mate choice and strengthen sexual selection (Cothran and Jeyasingh 2010).

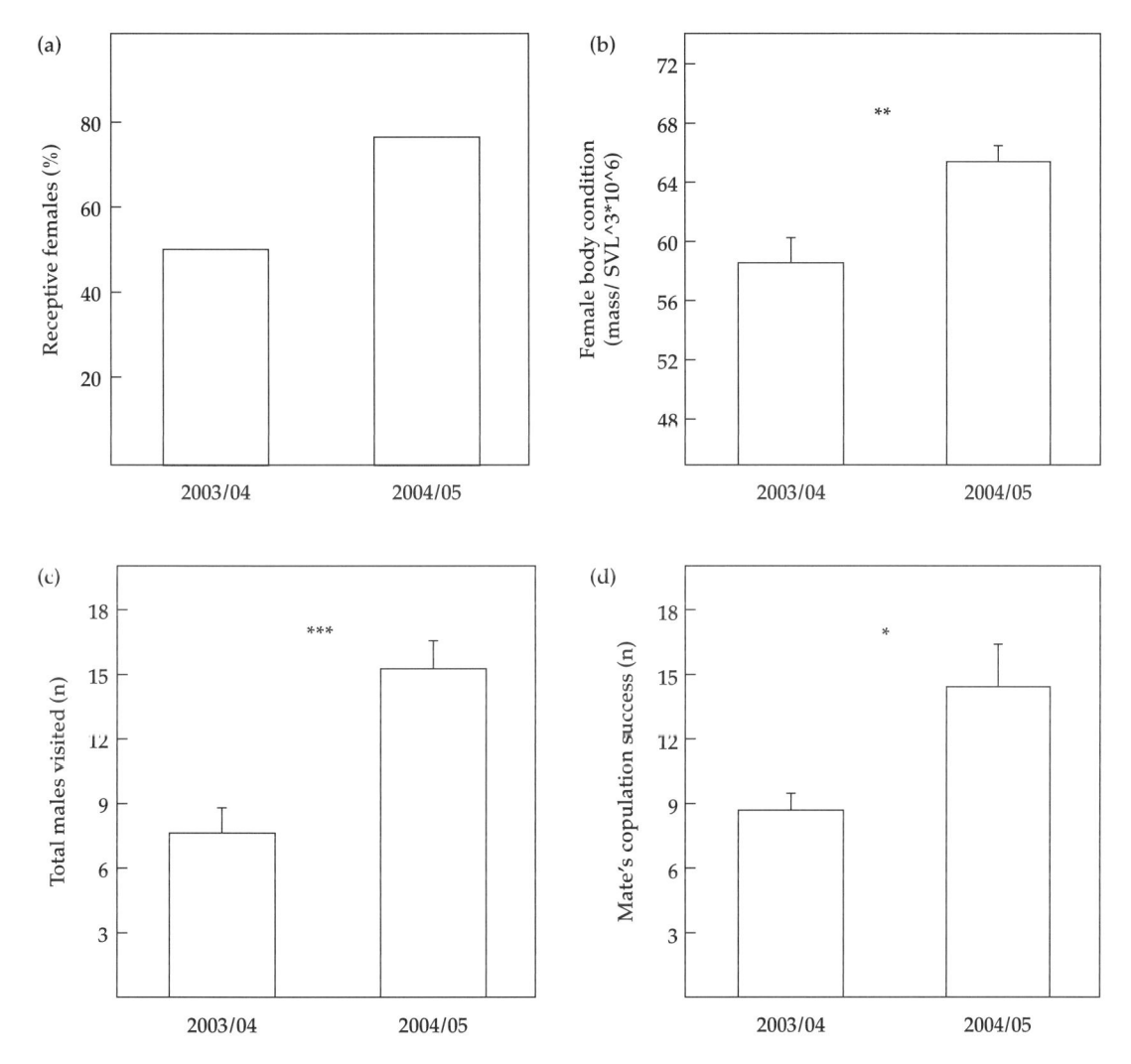

Figure 15.2 Female Galápagos marine iguanas *Amblyrhynchus cristatus* invested less in mate choice when resources were limited after a moderate El Niño event (years 2003/04) than when resources levels were higher in later years (2004/05). In years with low resource levels, (a) the percentage of receptive females was not significantly altered, but the females were (b) in poorer condition, (c) assessed fewer territorial males prior to mating, and (d) mated with a male that received fewer total copulations. Mean ± SE are presented. *P<0.05; **P<0.01; ***P<0.001. Adapted from Vitousek (2009) with permission from Springer Science + Business Media.

15.3.2 Interactions among sexually selected traits

Mate competition and mate choice typically involve multiple traits (Candolin 2003). Many birds, for instance, use bright colours, elaborate songs, and conspicuous courtship behaviours to advertise their quality and attract mates. Due to the myriad of ways in which multiple traits can interact, changes affecting one trait can potentially influence other traits, even if these are not directly affected by the change. For example, increased investment in bright colours can amplify the predation cost of elaborate courtship displays. Yet, much of what we know about the effects of rapid environmental change on trait expression has come from studies focusing on single traits. By contrast, far less is known about the impact of environmental change on multiple traits and, more specifically, what consequences this might have for individual fitness, population viability, and evolutionary processes.

A few recent studies suggest that complex changes in integrated traits are common. For instance, increased water turbidity associated with eutrophication causes female three-spined sticklebacks *Gasterosteus aculeatus* to switch from visual to olfactory cues in mate choice (Heuschele et al. 2009). Such a switch could, in turn, favour changes in male sexual signals to match female choice behaviour. Similarly, increased background noise that hampers the transmission of vocal signals in many bird species could favour a compensatory increase in the use of traits across other sensory modalities (Slabbekoorn and Ripmeester 2008). Such interactions and synergies make it challenging to predict the effect of environmental changes on sexually selected behaviours (see also Chapter 2 on communication).

15.3.3 Honesty of behavioural displays

Environmental changes can profoundly affect the reliability of sexual displays. This is because displays often depend on body condition, which usually develops prior to reproduction. If individuals that have experienced different environments come together to reproduce, a high-fitness genotype that has experienced harsh conditions can appear inferior to a genotype that is less fit but has experienced favourable conditions (Fig. 15.3 A). In addition, the most superior genotype can vary across environments because of genotype-by-environment interactions (Fig. 15.3 B; Ingleby et al. 2010). For instance, male bank voles *Myodes glareolus* advertise their dominance to females, but females mating with dominant males do not necessarily gain dominant sons if the environment changes from father to son (Mills et al. 2007). This is because genotype-by-environment interactions remove the heritability of dominance. Similarly, the ability to provide direct benefits, such as superior parental care, can change from one environment to another. A male in good condition because of a high foraging rate in past environments can have a low foraging rate in the present environment and, hence, may not be the 'best' male in terms of offspring provisioning.

In the above scenarios, sexual behaviours reflect individual qualities under a set of environmental conditions that are no longer reflective of current or future conditions. Another possibility is that the correlation between behaviour and quality is disrupted because of changes to the very mechanisms that help guarantee signal reliability (Fig.15.3 C). For instance, signal honesty is often under social control, but changes to the environment can alter social interactions and, thus, signal reliability. In the

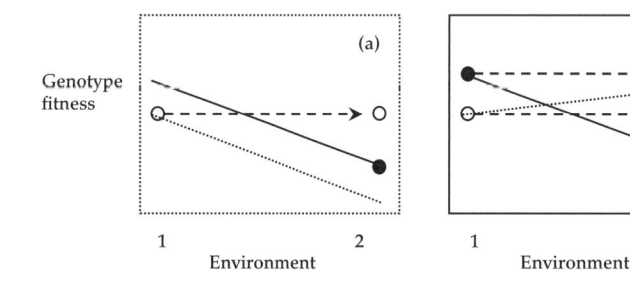

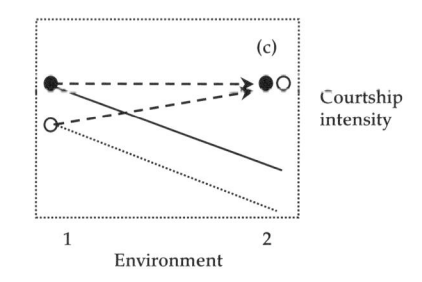

Figure 15.3 Causes of distorted correlations between the fitness of genotypes (lines) and their courtship intensity (dots) when the environment changes from 1 to 2, for genotype a (solid line) and genotype b (dotted line). (a) The fitness rank of the genotypes is constant across environments (parallel lines), while courtship intensity depends on past environment (the dots). Genotype b moves from environment 1 to the harsher environment 2, where genotype a resides. Genotype b is inferior to genotype a, but expresses more intense courtship because of the favourable conditions experienced in environment 1 (broken arrow). (b) The fitness rank of the genotypes varies between the two environments (crossing lines) because of a genotype-by-environment interaction. Courtship intensity is determined in the current environment (i.e. environment 1) and reflects fitness in this environment but not in the future environment 2 (broken arrows). (c) The fitness rank of the genotypes is constant across environments (parallel lines), and courtship reflects fitness in environment 1 but not in environment 2 (broken arrows).

three-spined stickleback, male–male competition ensures that courtship behaviour accurately reflects male condition and parenting ability, because individuals run the risk of costly fights if they dishonestly signal their condition. However, algal-induced water turbidity, by diminishing visibility, reduces the intensity of male–male competition. By relaxing social costs, males in poor condition are able to increase their courtship activity under turbid conditions, thereby reducing the honesty of their displays (Wong et al. 2007). For sexual signalling systems to continue to be adaptive in changing environments, mate preferences—and associated advertising traits—have to be adjusted to the new environment, either through plastic adjustments or evolutionary changes, to ensure signal reliability and beneficial mate preferences.

15.3.4 Impacts on population dynamics and selection processes

As detailed in the previous sections, the consequence of environmental change for population dynamics depends on its effects on current and future offspring production. In Fig. 15.1, we illustrated some of the potential pathways as well as their population-level consequences. Changes in environmental factors, such as resource availability or predation risk, have the capacity to affect the costs and benefits of both sexually and ecologically selected traits. This, in turn, can influence how finite resources are allocated among the various fitness influencing traits and, hence, the survival and reproductive output of individuals. Changes in population demography and evolutionary processes are expected to feed back on environmental factors, resulting in an ecosystem-wide feedback system (Fig. 15.1).

Environmental changes that increase the cost that sexually selected behaviours impose on females (e.g. costly mate searching behaviours) can potentially reduce female fecundity and, hence, population growth. By contrast, changes in the cost to males may not necessarily have the same effect, particularly in polygynous mating systems where a reduction in the number of available males does not limit female fecundity. Correspondingly, environmental changes that increase the benefit of sexually selected behaviours to females can potentially improve offspring production, while the same is not necessarily true for benefits to males. For instance, an increased benefit of Fisherian traits to males may not confer any benefits at the population level, as the traits only indicate the attractiveness of the males rather than their viability (Fisher 1930). Instead, a greater investment into Fisherian traits at the expense of paternal care could actually hamper population growth. Similarly, an increased benefit to males of behaviours that harm females can cause population decline (Rankin et al. 2011). This, in turn, can boost inbreeding and increase female mate search costs, which may further undermine population persistence (Charlesworth and Charlesworth 1987).

It is important to realize, however, that population persistence depends not only on offspring number, but also on offspring fitness. If individuals that are well adapted to changed conditions experience improved reproductive success when population size declines—because of reduced competition for resources—then environmental change could promote evolutionary change and, eventually, facilitate population growth. For example, increased industrial background noise reduces the ability of male ovenbirds *Seiurus aurocapillus* to attract mates, most likely because of the noise interfering with the male's courtship song (Habib et al. 2007). However, a large proportion of individuals breeding at noisy sites are inexperienced birds breeding for the very first time (Habib et al. 2007). It is possible that well adapted birds have survived earlier breeding attempts and learnt to avoid the marginal habitats, or that young individuals are forced to breed in these habitats because of intense competition for superior sites. In either case, a difference in site occupancy is expected to strengthen selection for older, and potentially better adapted, individuals, thus accelerating the rate of adaptation.

A reduction in effective population size (N_e), because of costs associated with sexually selected traits, can limit genetic variation and, hence, constrain an evolutionary response to environmental change. However, effective population size and genetic variation could increase if environmental change increases randomness in mating success (e.g. through increased cost of mate searching and mate assessment). This could potentially expose new alleles to selection and, in turn, accelerate adaptive change. Thus, an increased cost of sexually selected behaviours may not necessarily endanger population persistence. The ultimate outcome depends on changes to the number and fitness of offspring born into the population.

Invasion of new species because of human activities is a leading cause of extinction worldwide (see also Chapter 14). Sexually selected behaviours can make organisms particularly vulnerable to the negative effects of invaders. For instance, invaders may be able to use sexually selected behaviours to locate their prey or host, like the acoustically orienting parasitoid fly *Ormia ochracea* that kills calling male field crickets *Teleogryllus oceanicus* (Zuk et al. 2006), or they may change the environment and disrupt the cost-benefit balance of sexually selected behaviours. Moreover, the eagerness of males to mate—sometimes with little regard to species identity—can open up the potential for conflicts between species. Misdirected courtship can be a waste of time, energy, or gametes, and reduce individual fitness, eventually leading to the displacement of the inferior species, a process known as sexual exclusion (Groening and Hochkirch 2008).

Similarly, disrupted mate choice behaviour that results in hybridization between closely-related species can have serious implications for species persistence. A now classic example is seen in the African cichlid fishes of Lake Victoria, where a breakdown in species isolation mechanisms was attributed to eutrophication, resulting in a marked decline in species diversity (Seehausen et al. 1997). Likewise, pollution with sewage effluent and agricultural runoff has been implicated in hybridization between closely related species of swordtail fish *Xiphophorus* spp. due to interference with odour cues important in species recognition (Fisher et al. 2006; Chapter 2). On the other hand, a breakdown of reproductive isolation can also lead to novel phenotypes that may be better adapted to altered conditions than the original parental species (Seehausen et al. 2008).

15.4 How can animals respond?

In the previous section, we discussed how environmental changes influence the costs and benefits of sexually selected behaviours at the individual and population levels. We examined the importance of the resource pool and interactions among multiple traits in mediating environmental impacts, and considered potential effects on individual fitness, population dynamics, and evolutionary processes. We now take a closer look at the responses to environmental change, both phenotypic and genetic, and discuss their longer term consequences for population persistence.

If the net benefit of a sexually selected behaviour decreases, animals have two behavioural options at their disposal to avoid a reduction in fitness: move and search for more favourable habitats, or stay and adjust behaviour through phenotypic plasticity. Over longer time periods, a third option exists at the population level: adaptation through genetic changes across generations. The possibility of movement and dispersal will not be considered further, as this is dealt with in Chapter 5. We concentrate, instead, on phenotypic adjustment of behaviour and genetic changes (see also Chapters 11 and 16).

15.4.1 Phenotypic adjustment of behaviour

When an animal encounters changed conditions, the first response is usually behavioural, as behaviours can be highly plastic (Tuomainen and Candolin 2011). How animals respond depends on their genetically determined behavioural reaction norm, that is, the phenotypic expression of a single genotype across a range of environments. The norm can change over the lifetime of an animal through learning and through environmental effects, and it can evolve across generations (West-Eberhard 2003; Pigliucci 2001). This implies that behavioural responses are determined by the evolutionary history of the population and that the probability of an adaptive response depends on the similarity between past and current conditions. Human disturbances often cause drastic changes to the environment, creating habitats that differ profoundly from earlier ones. Consequently, few species are likely to possess

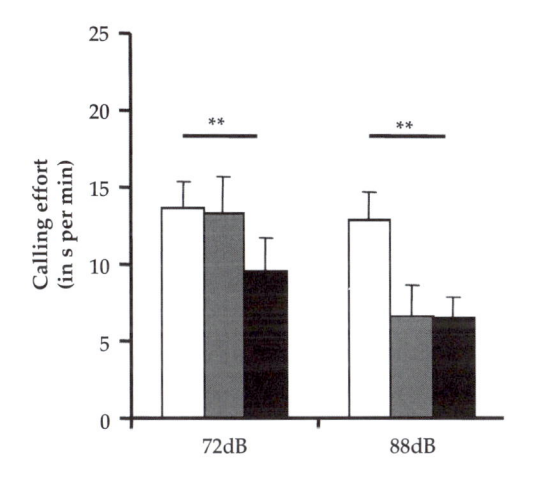

Figure 15.4 Male tree frogs *Hyla arborea* are not able to adjust the structure of their calls in response to traffic noise. Instead they reduce their calling activity. This was demonstrated by Thierry Lengagne (2008), who measured the calling activity of the frogs prior, during, and after exposure to traffic noise (white bar, grey bar and black bar respectively, mean ± SE) at two noise amplitudes, 72 dB and 88 dB (**$P < 0.01$). Adapted with permission from Elsevier.

sexually selected behaviours that are adaptive under human-disturbed environments.

Adaptive plastic alterations of behaviour can improve survival and reproductive success of individuals, either directly or, alternatively, by regulating the cost of less flexible traits, such as morphological traits. For instance, male guppies *Poecilia reticulata* shift displays of colour patterns to the times of the day when visual predation is low, and use the visually less conspicuous sneak copulation strategy when predation risk is high (Endler 1987). However, some sexually selected behaviours are relatively rigid in their expression, such as stereotypical sequences of complex behavioural components (e.g. the visual displays of the male Jacky dragon *Amphibolurus muricatus*; Peters and Evans 2003). These inflexible displays can potentially impose large fitness costs under environmental change.

When human activities result in environmental changes that the species has not encountered during its evolutionary history, adaptive behavioural reaction norms may not exist and individuals may respond maladaptively. For example, male buprestid beetles *Julodimorpha bakewelli* searching for mates are attracted to discarded beer bottles, as the brown glass surface resembles both the reflectance spectrum and texture of real females (Gwynne and Rentz 1983). Another possibility is that individuals alter their mating behaviour in a maladaptive way. Male green frogs *Rana clamitans melanota*, for instance, produce fewer advertisements calls and move more frequently when exposed to artificial light sources (Baker and Richardson 2006), while in Northern cardinals *Cardinalis cardinalis*, a preference for nesting in an exotic shrub heightens nest predation and reduces reproductive performance (Rodewald et al. 2011). In other instances, adaptive adjustments of behaviour may not even be possible because of neural or morphological constraints. Tree frogs *Hyla arborea*, for example, cannot adjust the structure of their mating calls to match increased background noise from human traffic; they reduce their calling activity instead (Fig. 15.4; Lengagne 2008).

Behavioural patterns often change over the lifetime of an individual, particularly through learning and the social transmission of new behavioural patterns (Chapter 4). The propensity to learn is genetically determined and the result of historical selection, but it can also be modified by environmental conditions. Species that have evolved a high aptitude for learning may quickly adjust to changed conditions, while less adept species may have to rely on existing behavioural patterns (Roth et al. 2010). In support of the importance of learning in facilitating behavioural adjustment, birds with relatively larger brains are more successful at establishing themselves in novel environments (Sol et al. 2005). Social living, in this regard, may facilitate behavioural adjustments through the process of cultural transmission. Such a possibility has been demonstrated in a wide range of contexts and species, from the cultural transmission of tool use in dolphins *Tursiops* sp. (Krutzen et al. 2005) to vocal culture in Caledonian crows *Corvus moneduloides* (Bluff et al. 2010).

Behavioural responses that prevent drastic declines in the size of a population can provide time for evolutionary changes to take place. Plasticity can also expose new phenotypes to selection and direct the course of subsequent adaptive evolution, transferring non-heritable environmentally induced variation into adaptive heritable variation (the Baldwin effect and Waddington's theory of genetic assimilation; Pigliucci 2001). In the next section, we discuss the possibility and significance of genetic adaptation of sexually selected behaviours to changing environments (see also Chapter 16 on evolutionary rescue).

15.4.2 Genetic changes

Evolutionary alteration of behaviour in response to environmental change (i.e. a genetic change over generations) depends on the strength of selection and the nature and adequacy of genetic variation. More precisely, the response depends on both the magnitude and direction of standing additive genetic variation in behaviour and genetic constraints. Genetic correlations among multiple traits, in particular, can strongly constrain evolution. The probability that pre-existing genetic variation will be favourable and allow an evolutionary response depends on past evolutionary processes (Barrett and Schluter 2008). Populations that have experienced similar environmental conditions in the past are more likely to encompass adaptive genetic variation than populations encountering new environments for the very first time. However, strong selection or bottlenecks (reductions in population size) can exhaust appropriate genetic variation (Frankham 2005).

When appropriate genetic variation does not exist, new genetic variation can arise through mutations and gene flow. Favourable mutations are rare, and there is only a small likelihood that they will spread and increase in numbers across a large geographic area (Eldredge et al. 2005; Futuyma 2010). Nevertheless, an example of its occurrence can be seen in the rapid spread of a mutation responsible for the loss of singing ability in field crickets, which prevent males from being located by a deadly parasitoid fly (Tinghitella and Zuk 2009). Gene flow, on the other hand, is an important source of genetic variation, particularly when neighbouring populations have experienced similar changes. The degree to which genetic variation limits responses to selection is currently in dispute. Additive genetic variation in sexually selected behaviours is expected to be depleted under directional selection (the lek paradox; Kirkpatrick and Ryan 1991). Yet, ample genetic variation is found in most sexually selected traits, including behavioural traits. It is likely that responses to selection are mostly restricted by pleiotropic genetic associations among traits and correlational selection (Walsh and Blows 2009). In guppies, for instance, female mate preferences do not respond to experimental selection, most likely because of antagonistic correlations between male attractiveness and other fitness components, such as female fecundity and the survival of sons (Hall et al. 2004). Adaptive evolutionary change of integrated traits requires simultaneous changes in many genes, which necessitates genetic variation in the right direction in all of the genes concerned, as well as the lack of antagonistic correlations.

Currently, little information exists on the degree to which sexually selected behaviours can be genetically adjusted to human-induced environmental change. In cases where behavioural changes have been recorded, as in the song duration of the little greenbul *Andropadus virens* in human-altered secondary forests (Smith et al. 2008), these have been consistent with both plasticity and genetic divergence (Fig. 15.5). Complicating the estimation of evolutionary responses, genetic changes can be hidden by phenotypic changes. For instance, Soay sheep *Ovis aries* from the islands of St Kilda are getting smaller even though there is selection for higher body weight. This is because of ecological factors that retard the growth of sheep and impose countergradient variation, which overwhelms evolutionary change (Ozgul et al. 2009).

15.4.3 Population responses

In earlier sections, we discussed how phenotypic adjustment of sexually selected behaviours can help ensure population persistence during the critical early stages of environmental change. If the plastic response of individuals moves the phenotypic value of the population all the way across to a new peak on the adaptive landscape, then stabilizing selection on the trait can follow and no genetic differentiation is needed (Ghalambor et al. 2007). For

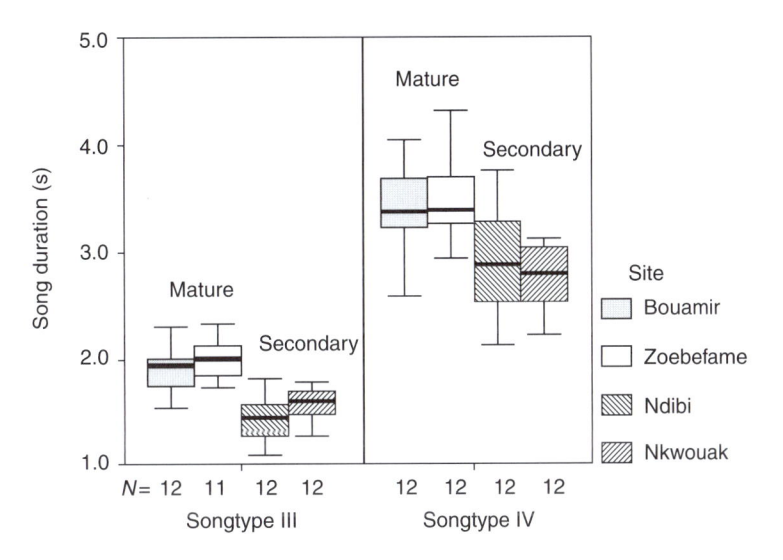

Figure 15.5 Song duration of an African rainforest bird species, the little greenbul *Andropadus virens*, differs between mature and human-altered secondary forest. The song duration is significantly longer in primary forest sites than in secondary forest sites. This is the case for both the relatively short and stereotypic song type III and the relatively long, complex, and variable song type IV. The observed shifts were consistent with divergent selection on heritable variation, but a role for plasticity cannot be ruled out. Adapted from Smith et al. (2008) with permission from Wiley-Blackwell.

example, altered courtship activity in response to changed predation risk can maximize fitness under changing conditions. However, if the behavioural response results in a change in the right direction, but the population is still displaced from the optimum peak, then directional selection towards the optimum will follow. The behavioural response can then facilitate genetic responses by giving the population additional time to adapt genetically, and by exposing new phenotypes to selection (Pigliucci 2001). Thus, using the example of courtship under increased predation risk, a change in courtship behaviour can reduce the cost of conspicuous colours, which can prevent population decline and provide time for evolution of less bright colours. On the other hand, if the behavioural response moves the population further from the optimum (i.e. a maladaptive response), then population viability will decrease and, in turn, endanger population persistence (Badyaev 2005). For instance, a reduction in courtship activity under environmental change can result in a large proportion of females failing to mate if courtship activity needs to exceed a threshold level to entice choosy females.

The degree to which evolution can track ecological changes depends on the rate and scale of the ecological change itself, as well as the generation time and genetic make-up of the population, including any underlying genetic constraints. As discussed above, genetic adaptation of sexually selected traits to rapid environmental change is challenging, as genetic variation can be low and antagonistic correlations can constrain the process. Moreover, anthropogenic disturbances, such as habitat fragmentation, often reduce genetic variation. Here, any mechanisms that can help to facilitate the evolutionary process, such as plastic behavioural responses, could have a major impact on success. The rate of adjustment in relation to the rate of environmental change will likely be important in determining ultimate population consequences. Rapid adjustments (whether phenotypic or genetic) can result in a new equilibrium being reached between the costs and benefits of sexually selected behaviours and, in doing so, allow the population to persist, while slow responses can be fatal under rising costs, causing the population to decline. In the worse case scenario, slow responses may even lead to extinction.

Adjustment of behaviour to changed ecological conditions can contribute to the divergence of a population into new ecotypes and, eventually, new species. Sexually selected behaviours are believed to have played an important role in ecologically-driven divergence by reinforcing isolation and preventing hybridization. Three-spined sticklebacks in Canadian lakes, for instance, have diverged in mate preferences according to microhabitat, which has resulted in the evolution of limnetic and benthic species pairs (Boughman et al. 2005). However, changes in sexually selected behaviours can also remove pre-mating isolation mechanisms and cause the merging of ecotypes or species. The species-pair of sticklebacks, for instance, are currently hybridizing because of environmentally-induced changes in female mate choice behaviour, analogous to what is happening with cichlids in the African rift lakes (Seehausen et al. 2008). Moreover, genotype-by-environment interactions and gene flow can restrict female–male coevolution and constrain ecological divergence. For example, experimental work with *Drosophila mojavensis* shows that genotype-by-environment interactions that disrupt signal reliability impede the divergence of the species' courtship song between two diverging populations (Etges et al. 2007). Thus, the consequences of behavioural responses for populations can be difficult to predict. It is clear, however, that behavioural responses can play a fundamental role in determining the fate of populations and species in a rapidly changing world.

15.5 What next?

In this chapter, we have explained how environmental change can influence sexually selected behaviours, and the consequences these effects can have for individuals, populations, and species. It is obvious that sexually selected behaviours can hinder (and possibly even worsen) the effect of environmental change on populations, and that the ultimate impact depends on a number of factors, such as the rate and scale of the change, the degree of plasticity in behaviour, and the potential for genetic changes. In the final section of our chapter, we turn to questions that warrant closer empirical attention.

15.5.1 Taking account of the complexity of environmental change

Studies of sexual selection tend to focus on only one environmental variable at a time, such as changed predation risk, food availability, or population density. Of course, the reality of environmental changes (whether human-induced or not) is far more complex. Take climate change as an example, which not only affects weather patterns, but other variables as well—from the availability of resources to the distribution of species (prey, predators, parasites). Such an example highlights the need to consider the role of sexually selected behaviours within an appropriate context by taking into account complex interactions across a multitude of variables. Sexual selection is a potent evolutionary force, and sexually selected behaviours can profoundly impact population dynamics. Thus, sexually selected behaviours need to be included in studies of ecological-evolutionary processes under environmental change.

15.5.2 Multiple signals and multiple sensory modalities

Sexual communication typically involves multiple signals operating across different sensory modalities (see Chapter 2 on communication and Candolin 2003). Yet, there is still an overwhelming tendency for studies of sexual selection to focus on single traits, and on visual signals in particular. In the context of environmental change, such a narrow focus can severely underestimate the impact of human activities on sexually selected traits. To fully appreciate the consequences of changing environments on sexual selection, we need studies that are better informed by sensory ecology and less reliant on our own perceptual biases (Lim et al. 2008). We also have to consider the possibility that environmental changes influence different traits in different ways. Such changes can result in traits varying in reliability or even conveying contradictory information.

More work is needed in understanding the occurrence and consequences of genotype-by-environment interactions and their influence on signal reliability, and, in particular, how animals cope with these changes.

15.5.3 Is population rescue possible?

In this chapter, we highlighted the possibility of behavioural and genetic responses potentially rescuing populations from the impacts of environmental changes. To what extent can animals adjust their sexually selected behaviours to environmental change and, more importantly, will it be enough? What is the relative role of these plastic responses in relation to evolutionary changes? How fast can sexually selected behaviours be genetically adapted to environmental change, considering the plausibility of antagonistic genetic correlations? And when will sexually selected behaviours have the opposite effect, potentially undermining population persistence? Several studies suggest that sexually selected behaviours can become less beneficial, and even maladaptive, under environmental change (see also Chapter 11), and that the strength of sexual selection could be relaxed. Will this increase genetic variation and promote adaptation of populations to changing conditions, or will it hinder adaptation (Candolin and Heuschele 2008)? Clearly, much more information is needed before we can answer the crucial question of whether behavioural responses will facilitate or hamper plastic adjustment and genetic adaptation to environmental change, both in the short and longer terms, and how the pattern could vary among species and environments.

Acknowledgements

We thank Hanna Kokko and Gil Rosenthal for comments on an earlier draft of our manuscript. BBMW also wishes to acknowledge funding support from the Australian Research Council.

References

Andersson, M. (1994). *Sexual Selection*. Princeton, Princeton University Press.

Badyaev, A. V. (2005). Stress-induced variation in evolution: from behavioural plasticity to genetic assimilation. *Proceedings of the Royal Society of London B, Biological Sciences*, 272, 877–86.

Baker, B. J. and Richardson, J. M. L. (2006). The effect of artificial light on male breeding-season behaviour in green frogs, Rana clamitans melanota. *Canadian Journal of Zoology*, 84, 1528–32.

Barrett, R. D. H. and Schluter, D. (2008). Adaptation from standing genetic variation. *Trends in Ecology and Evolution*, 23, 38–44.

Bluff, L. A., Kacelnik, A., and Rutz, C. (2010). Vocal culture in New Caledonian crows Corvus moneduloides. *Biological Journal of the Linnean Society*, 101, 767–76.

Booksmythe, I., Detto, T., and Backwell, P. R. Y. (2008). Female fiddler crabs settle for less: the travel costs of mate choice. *Animal Behaviour*, 76, 1775–81.

Boughman, J. W., Rundle, H. D., and Schluter, D. (2005). Parallel evolution of sexual isolation in sticklebacks. *Evolution*, 59, 361–73.

Candolin, U. (2003). The use of multiple cues in mate choice. *Biological Reviews of the Cambridge Philosophical Society*, 78, 575–95.

Candolin, U. and Heuschele, J. (2008). Is sexual selection beneficial during adaptation to environmental change? *Trends in Ecology and Evolution*, 23, 446–52.

Charlesworth, D. and Charlesworth, B. (1987). Inbreeding depression and its evolutionary consequences. *Annual Review of Ecology and Systematics*, 18, 237–68.

Clutton-Brock, T. H., Coulson, T. N., Milner-Gulland, E. J., Thomson, D., and Armstrong, H. M. (2002). Sex differences in emigration and mortality affect optimal management of deer populations. *Nature*, 415, 633–7.

Cothran, R. D. and Jeyasingh, P. D. (2010). Condition dependence of a sexually selected trait in a crustacean species complex: importance of the ecological context. *Evolution*, 64, 2535–46.

Cox, R. M. and Calsbeek, R. (2009). Sexually antagonistic selection, sexual dimorphism, and the resolution of intralocus sexual conflict. *American Naturalist*, 173, 176–87.

Darwin, C. (1871). *The Descent of Man, and Selection in Relation to Sex*. London, Murray.

Denoel, M. and Doellen, J. (2010). Displaying in the dark: light-dependent alternative mating tactics in the Alpine newt. *Behavioral Ecology and Sociobiology*, 64, 1171–7.

Eldredge, N., Thompson, J. N., Brakefield, P. M., Gavrilets, S., Jablonski, D., Jackson, J. B. C., Lenski, R. E., Lieberman, B. S., Mcpeek, M. A., and Miller, W. (2005). The dynamics of evolutionary stasis. *Paleobiology*, 31, 133–45.

Endler, J. A. (1987). Predation, light intensity and courtship behavior in Poecilia reticulata (Pisces, Poeciliidae). *Animal Behaviour*, 35, 1376–85.

Etges, W. J., De Oliveira, C. C., Gragg, E., Ortiz-Barrientos, D., Noor, M. A. F., and Ritchie, M. G. (2007). Genetics of incipient speciation in Drosophila mojavensis. I. Male courtship song, mating success, and genotype x environment interactions. *Evolution*, 61, 1106–19.

Fisher, H. S., Wong, B. B. M., and Rosenthal, G. G. (2006). Alteration of the chemical environment disrupts communication in a freshwater fish. *Proceedings of the Royal Society of London B, Biological Sciences*, 273, 1187–93.

Fisher, R. A. (1930). *The Genetical Theory of Natural Selection*. Oxford, Clarendon Press.

Frankham, R. (2005). Genetics and extinction. *Biological Conservation*, 126, 131–40.

Futuyma, D. J. (2010). Evolutionary constraint and ecological consequences. *Evolution*, 64, 1865–84.

Ghalambor, C. K., Mckay, J. K., Carroll, S. P., and Reznick, D. N. (2007). Adaptive versus non-adaptive phenotypic plasticity and the potential for contemporary adaptation in new environments. *Functional Ecology*, 21, 394–407.

Groening, J. and Hochkirch, A. (2008). Reproductive interference between animal species. *Quarterly Review of Biology*, 83, 257–82.

Gwynne, D. T. and Rentz, D. C. F. (1983). Beetles on the bottle—male Buprestids mistake stubbies for females (Coleoptera). *Journal of the Australian Entomological Society*, 22, 79–80.

Habib, L., Bayne, E. M., and Boutin, S. (2007). Chronic industrial noise affects pairing success and age structure of ovenbirds Seiurus aurocapilla. *Journal of Applied Ecology*, 44, 176–84.

Hall, M., Lindholm, A. K., and Brooks, R. (2004). Direct selection on male attractiveness and female preference fails to produce a response. *BMC Evolutionary Biology*, 4, 1–10.

Heuschele, J., Mannerla, M., Gienapp, P., and Candolin, U. (2009). Environment-dependent use of mate choice cues in sticklebacks. *Behavioral Ecology*, 20, 1223–7.

Houle, D. and Kondrashov, A. S. (2002). Coevolution of costly mate choice and condition-dependent display of good genes. *Proceedings of the Royal Society of London B, Biological Sciences*, 269, 97–104.

Hunt, J., Brooks, R., Jennions, M. D., Smith, M. J., Bentsen, C. L., and Bussiere, L. F. (2004). High-quality male field crickets invest heavily in sexual display but die young. *Nature*, 432, 1024–7.

Ingleby, F. C., Hunt, J., and Hosken, D. J. (2010). The role of genotype-by-environment interactions in sexual selection. *Journal of Evolutionary Biology*, 23, 2031–45.

Jarzebowska, M. and Radwan, J. (2010). Sexual selection counteracts extinction of small populations of the bulb mites. *Evolution*, 64, 1283–9.

Jennions, M. D. and Petrie, M. (2000). Why do females mate multiply? A review of the genetic benefits. *Proceedings of the Royal Society of London B, Biological Sciences*, 75, 21–64.

Kirkpatrick, M. and Ryan, M. J. (1991). The evolution of mating preferences and the paradox of the lek. *Nature*, 350, 33–8.

Kokko, H. and Rankin, D. J. (2006). Lonely hearts or sex in the city? Density-dependent effects in mating systems. *Philosophical Transactions of the Royal Society B: Biological Sciences*, 361, 319–34.

Krutzen, M., Mann, J., Heithaus, M. R., Connor, R. C., Bejder, L., and Sherwin, W. B. (2005). Cultural transmission of tool use in bottlenose dolphins. *Proceedings of the National Academy of Sciences of the USA*, 102, 8939–43.

Lengagne, T. (2008). Traffic noise affects communication behaviour in a breeding anuran, Hyla arborea. *Biological Conservation*, 141, 2023–31.

Lim, M. L. M., Sodhi, N. S., and Endler, J. A. (2008). Conservation with sense. *Science*, 319, 281–1.

Lohrey, A. K., Clark, D. L., Gordon, S. D., and Uetz, G. W. (2009). Antipredator responses of wolf spiders (Araneae: Lycosidae) to sensory cues representing an avian predator. *Animal Behaviour*, 77, 813–21.

Mills, S. C., Alatalo, R. V., Koskela, E., Mappes, J., Mappes, T., and Oksanen, T. A. (2007). Signal reliability compromised by genotype-by-environment interaction and potential mechanisms for its preservation. *Evolution*, 61, 1748–57.

Møller, A. P. and Alatalo, R. V. (1999). Good-genes effects in sexual selection. *Proceedings of the Royal Society of London Series B-Biological Sciences*, 266, 85–91.

Ozgul, A., Tuljapurkar, S., Benton, T. G., Pemberton, J. M., Clutton-Brock, T. H., and Coulson, T. (2009). The dynamics of phenotypic change and the shrinking sheep of St. Kilda. *Science*, 325, 464–7.

Peters, R. A. and Evans, C. S. (2003). Design of the Jacky dragon visual display: signal and noise characteristics in a complex moving environment. *Journal of Comparative Physiology A*, 189, 447–59.

Pigliucci, M. (2001). *Phenotypic Plasticity: Beyond Nature and Nurture*. Baltimore, John Hopkins University Press.

Rankin, D. J. and Arnqvist, G. (2008). Sexual dimorphism is associated with population fitness in the seed beetle Callosobruchus maculatus. *Evolution*, 62, 622–30.

Rankin, D. J., Dieckmann, U., and Kokko, H. (2011). Sexual conflict and the tragedy of the commons. *American Naturalist*, 177, 780–91.

Real, L. (1990). Search theory and mate choice: I. Models of single-sex discrimination. *American Naturalist*, 136, 376–405.

Ribeiro, P. D., Daleo, P., and Iribarne, O. O. (2010). Density affects mating mode and large male mating advantage in a fiddler crab. *Oecologia*, 164, 931–41.

Rodewald, A. D., Shustack, D. P., and Jones, T. M. (2011). Dynamic selective environments and evolutionary traps in human-dominated landscapes. *Ecology*, 92, 1781–8.

Roth, T. C., Ladage, L. D., and Pravosudov, V. V. (2010). Learning capabilities enhanced in harsh environments: a common garden approach. *Proceedings of the Royal Society of London B, Biological Sciences*, 277, 3187–93.

Searcy, W. A. and Nowicki, S. (2005). *The Evolution of Animal Communication*. Princeton, New Jersey, Princeton University Press.

Seehausen, O., Alphen, J. J. M., and Witte, F. (1997). Cichlid fish diversity threatened by eutrophication that curbs sexual selection. *Science*, 277, 1808–11.

Seehausen, O., Takimoto, G., Roy, D., and Jokela, J. (2008). Speciation reversal and biodiversity dynamics with hybridization in changing environments. *Molecular Ecology*, 17, 30–44.

Slabbekoorn, H. and Ripmeester, E. A. P. (2008). Birdsong and anthropogenic noise: implications and applications for conservation. *Molecular Ecology*, 17, 72–83.

Smith, T. B., Mila, B., Grether, G. F., Slabbekoorn, H., Sepil, I., Buermann, W., Saatchi, S., and Pollinger, J. P. (2008). Evolutionary consequences of human disturbance in a rainforest bird species from Central Africa. *Molecular Ecology*, 17, 58–71.

Sol, D., Duncan, R. P., Blackburn, T. M., Cassey, P., and Lefebvre, L. (2005). Big brains, enhanced cognition, and response of birds to novel environments. *Proceedings of the National Academy of Sciences of the USA*, 102, 5460–5.

Tinghitella, R. M. and Zuk, M. (2009). Asymmetric mating preferences accommodated the rapid evolutionary loss of a sexual signal. *Evolution*, 63, 2087–98.

Tomkins, J. L., Radwan, J., Kotiaho, J. S., and Tregenza, T. (2004). Genic capture and resolving the lek paradox. *Trends in Ecology and Evolution*, 19, 323–8.

Tuomainen, U. and Candolin, U. (2011). Behavioural responses to human-induced environmental change. *Biological Reviews of the Cambridge Philosophical Society*, 86, 640–57.

Vitousek, M. N. (2009). Investment in mate choice depends on resource availability in female Galapagos marine iguanas (Amblyrhynchus cristatus). *Behavioral Ecology and Sociobiology*, 64, 105–13.

Walsh, B. and Blows, M. W. (2009). Abundant genetic variation plus strong selection = multivariate genetic constraints: a geometric view of adaptation. *Annual Review of Ecology Evolution and Systematics*, 40, 41–59.

West-Eberhard, M. J. (2003). *Developmental Plasticity and Evolution*. New York, Oxford University Press.

Wong, B. B. M., Candolin, U., and Lindström, K. (2007). Environmental deterioration compromises socially-enforced signals of male quality in three-spined sticklebacks *American Naturalist*, 170, 184–9.

Woods, W. A., Hendrickson, H., Mason, J., and Lewis, S. M. (2007). Energy and predation costs of firefly courtship signals. *American Naturalist*, 170, 702–8.

Zuk, M. and Kolluru, G. R. (1998). Exploitation of sexual signals by predators and parasitoids. *Quarterly Review of Biology*, 73, 415–38.

Zuk, M., Rotenberry, J. T., and Tinghitella, R. M. (2006). Silent night: adaptive disappearance of a sexual signal in a parasitized population of field crickets. *Biology Letters*, 2, 521–4.

CHAPTER 16

Evolutionary rescue under environmental change?

Rowan D.H. Barrett and Andrew P. Hendry

> ## ⊃ Overview
>
> When environmental conditions change, the persistence of populations will depend on phenotypic responses that better suit individuals for the new conditions. Such responses can occur through individual-level behavioural or plastic changes, or population-level evolutionary changes (including population-level changes in behaviour and plasticity). Many studies have now documented adaptive phenotypic responses to environmental change, but very few have investigated their potential role in making the difference between population persistence versus extirpation (i.e. evolutionary rescue). We explore these topics by focusing on key questions about evolutionary rescue and the limitations of this process. In doing so, we outline pressing research questions and potential empirical approaches to their resolution.

16.1 Introduction

For populations to persist, they need to be at least reasonably well adapted for the local environmental conditions. That is, the set of phenotypic traits present in the population must yield high enough mean population fitness to maintain a stable population size. This is presumably the situation in most natural populations found in relatively constant environments. When environmental conditions change, however, existing phenotypes are expected to be less well adapted, potentially causing population declines to the point of local extirpation or even range-wide extinction. This mismatch between current phenotypes and those that would yield high fitness in the new environment imposes a pressure on individuals and populations to reduce the mismatch, which can then increase fitness and population size. In short, phenotypic changes can make the difference between persistence and extirpation in the face of environmental change.

Phenotypes can become better matched to altered environmental conditions in several different ways. Most immediately, individuals can alter their behaviour to reduce exposure to the new conditions—or otherwise ameliorate its potentially detrimental effects. Many instances of such immediate behavioural responses are discussed in this book, and so we here mention only a few exemplars: individuals can (1) move to more appropriate locations (e.g. Bowler and Benton 2005) (see Chapter 5), (2) alter their behaviour to reduce susceptibility to new predators (e.g. Poethke et al. 2010) (see Chapters 7, 10, and 14), or select a more appropriate nesting site (e.g. Eggers et al. 2006; Rushbrook et al. 2010) (see Chapter 8). But such immediate behavioural responses will not always be sufficient to prevent population declines. For instance, behaviours that evolved under previous conditions may be maladaptive under new conditions (e.g. evolutionary 'traps': Schlaepfer et al. 2002; Visser 2008). Or behaviours might simply be ineffective, such as when appropriate locations are no longer available. Our goal in the present chapter is to ask what happens in these situations where immediate behavioural responses will not do the trick.

Behavioural Responses to a Changing World. First Edition. Edited by Ulrika Candolin and Bob B.M. Wong.
© 2012 Oxford University Press. Published 2012 by Oxford University Press.

When immediate behavioural responses are insufficient, two other options are possible for reducing the mismatch between existing phenotypes and those favoured under new conditions. First, phenotypes can be altered developmentally, such as through phenotypic plasticity, maternal effects, or various other non-genetic phenotypic alterations (Bonduriansky and Day 2009; Pigliucci 2001, see also Chapter 11). Although these developmental changes (henceforth just 'plasticity') can be a particularly rapid way to recover fitness following environmental change (Parmesan 2006), they are not all powerful. The reason is that plasticity is sometimes maladaptive (Grether 2005) or, when adaptive, can be subject to costs (DeWitt et al. 1998) and limits (Visser 2008). Thus, although plasticity will be an important response to environmental change, it will not always be sufficient to maintain or recover high fitness.

The ultimate solution to environmental change is for populations to evolve such that their phenotypes are better suited for the new conditions (Fig. 16.1). In contrast to the individual-level phenotypic changes achievable through behaviour or plasticity, evolutionary changes in phenotypes require cross-generational shifts in allele frequency. It is obvious that evolutionary changes have been a critical part of the evolution of biological diversity through Earth's history, but the extent to which they will be helpful in the face of future environmental change remains uncertain. Historically, evolution was considered too slow to be relevant on short time scales, such as years or decades. Recently, however, a number of examples have come to light of adaptive evolution occurring on precisely these time frames (reviewed in Hendry and Kinnison 1999; Reznick and Ghalambor 2001; Hendry et al. 2008). Uncertainty remains, however, as to just how common such changes are and how important they are for population persistence: that is, so-called 'evolutionary rescue' (Gomulkiewicz and Holt 1995; Bell and Gonzalez 2009, 2011).

In the present chapter, we investigate how evolutionary (genetic) change can alter phenotypes (adaptively or maladaptively) in ways that can then influence population persistence. We do so by examining several key questions and then a set of potential constraints on this process. We do not focus on immediate behavioural responses or on individual phenotypic plasticity, because these are covered elsewhere in the book (e.g. Chapter 11). We do, however, consider the *evolution* of behaviour and plasticity in response to environmental change. The difficulty in determining the genetic basis and evolution of behaviour (Skinner 1966; Clutton-Brock and Harvey 1985) means that such examples are relatively rare. Thus, although we must usually refer to non-behaviour examples, the basic principles should apply to any sort of trait, including behaviour.

16.2 Key questions

To discuss issues surrounding evolutionary responses to environmental change, we focus on five key questions: how important is genetic (as opposed to plastic) change, to what extent will plasticity evolve, is evolution fast enough to prevent substantial maladaptation, does evolution stem from standing genetic variation or new mutations, and how many genes are likely to be involved?

16.2.1 How important is genetic (as opposed to plastic) change?

The answer to this question will depend on properties of the environment (e.g. the rate and magnitude of change and availability of alternative habitats), the organism (e.g. natural history, population size, and dispersal ability), and the relevant traits (e.g. genetic variance and the potential for plasticity) (Holt 1990). No simple generalizations are possible given that empirical evidence for the importance of genetic change versus plasticity is limited and opinions are divided (Bradshaw and McNeilly 1991; Gienapp et al. 2008; Visser 2008). Part of the uncertainty stems from a scarcity of research simultaneously quantifying both phenotypic and genetic responses to a given environmental change. Instead, the vast majority of studies simply quantify phenotypic changes, the genetic basis for which remains uncertain (reviewed in Darimont et al. 2009; Hendry et al. 2008; Hendry and Kinnison 1999; Root et al. 2003; Westley 2011). To illustrate issues surrounding

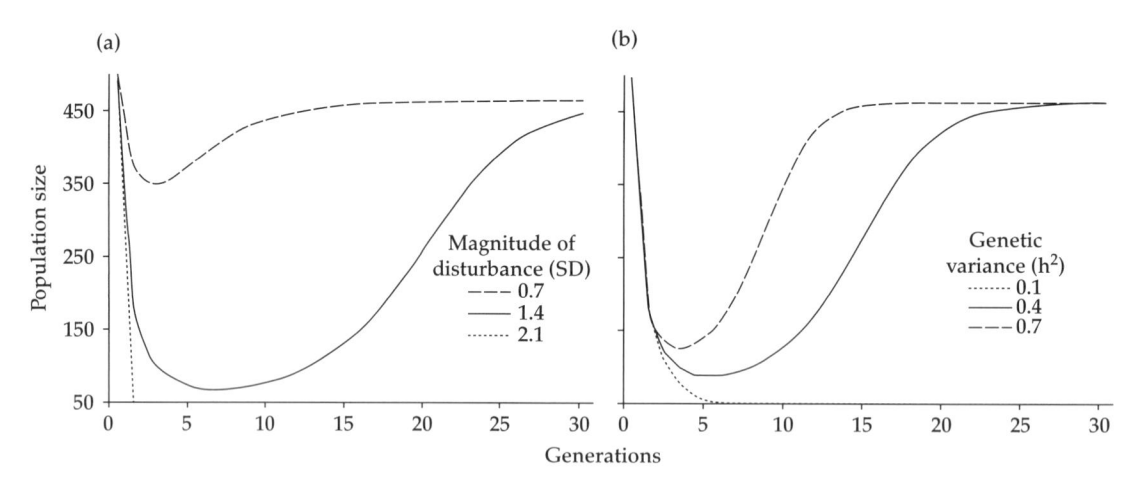

Figure 16.1 Illustrations of evolutionary rescue derived from a deterministic quantitative-genetic model. A well-adapted population experiences an environmental perturbation that causes a decline in average fitness and a reduction in population size. As in Gomulkiewicz and Holt (1995), extinction is assumed to occur stochastically when the population size drops to below a certain size, here 50 individuals. Without evolution (not shown), the maladapted population declines quickly toward extirpation. With evolution, the population can adapt and eventually recover population size. The potential for this evolutionary rescue depends on a number of factors. Those illustrated here are the magnitude of the environmental disturbance (a) and the amount of genetic variance in the population (b). The magnitude of disturbance is indexed as the difference between the current and optimal population mean value in units of phenotypic standard deviation (SD). The genetic variance is measured as heritability ($h^2 = V_A/V_P$). In (a), $h^2 = 0.3$. In (b) the magnitude of environmental change is 0.6 SD. These results are obtained by perturbing the optimum trait value in the model of Hendry (2004).

this endeavour, we here concentrate on studies of responses to climate warming.

One reason why so few studies have disentangled the genetic versus plastic responses to climate change is that the standard inferential approaches are difficult to implement in a temporal context. For instance, common-garden experiments are typically used to confirm a genetic foundation for phenotypic differences, but such experiments are difficult for groups sampled in different years, because they cannot then be raised or grown together. One potential solution is to provide identical common-garden conditions in the different years, such that the different groups experience the same conditions despite not being contemporaneous. This approach is rarely applied but, in one example, Bradshaw and Holzapfel (2001) reared pitcher plant mosquitoes *Wyeomyia smithii* from the same populations under identical laboratory conditions in 1972 and 1996. They found that the photoperiodic response (developmental timing in relation to photoperiod) had evolved in the direction expected to be adaptive under climate warming. A related strategy is to use dormant seeds (Franks et al. 2007) or resting eggs (Cousyn et al. 2001) to resurrect individuals from the past for direct comparison to individuals in the present. More such studies are needed.

An alternative to common-garden studies is to use long-term data from pedigreed natural populations in statistical models that can separate maternal, plastic, and genetic contributions to phenotypic change. Specifically, by comparing patterns seen across years within individuals to those seen across generations, it becomes possible to disentangle the contributions of plasticity and evolution (Kruuk 2004; Nussey et al. 2005). In an early example of this approach, Przybylo et al. (2000) analysed individual breeding performance in collared flycatchers *Ficedula albicollis* and demonstrated that population level responses to climate warming could be entirely attributed to phenotypic plasticity. Similar work in other species has confirmed that birds do often possess the ability to respond plastically to changing environmental conditions (Charmantier et al. 2008; Nussey et al. 2005; Reed et al. 2006; see also Chapter 11). At the same time, these methods do sometimes document evolutionary change. For example, Réale and colleagues (2003) showed that changes in parturition dates in Canadian red squirrels *Tamiasciurus hudsonicus* experiencing warmer spring temperatures were 87% plastic and 13% evolutionary (genetic). Some evolution has clearly occurred in this and other situations but the relative amount remains an open question in the majority of studies (Hadfield et al. 2010).

Although phenotypic plasticity can accomplish a substantial amount of phenotypic change, it will not always be enough. For example, Phillimore et al. (2010) showed that the plastic responses of common frog *Rana temporaria* breeding times to temperature are much lower than those required to maintain locally-adaptive phenotypes under expected climate warming. Evolution will thus be necessary. In addition to such arguments that 'plasticity will not be enough', several studies have confirmed changes in allele frequencies in response to climate change. These include temporal shifts in chromosomal arrangements and candidate gene allele frequencies for *Drosophila* populations experiencing climate warming (Balanya et al. 2006; Umina et al. 2005). Similarly, studies of the tree species *Fagus sylvatica* have shown predictable allele frequency changes with temperature (Jump et al. 2006). But the inferences possible in such studies remain limited—because they typically lack information relating specific genetic changes to phenotypes relevant for fitness under changing conditions. Overall, however, these and the above studies demonstrate that a number of natural populations have evolved in response to altered climate.

In summary, many studies have documented phenotypic responses to climate change, but few have been able to separate genetic from plastic effects. Indeed, our presentation focused on climate change to highlight a situation where this ambiguity is particularly acute. By contrast, studies documenting phenotypic responses to other types of environmental change have generally had an easier time confirming genetic effects (reviewed by Hendry and Kinnison 1999; Kinnison and Hendry 2001; Hendry et al. 2008). And a number of these other contexts involve examples of the evolution of behaviour, such as antipredator behaviour in Trinidadian guppies *Poecilia reticulata* (Magurran et al. 1992) and *Daphnia* (Cousyn et al. 2001), migratory behaviour in birds (Berthold et al. 1992) and toads (Phillips et al. 2010), and host plant choice in insects (Singer et al. 1993). Overall, then, evolution is an important part of responses to environmental change, including in some climate change situations (Bradshaw and Holzapfel 2006; Gienapp et al. 2008; Skelly et al. 2007). However, much more work needs to be done on interactions between evolution and plasticity, particularly because various authors have suggested that plasticity (and behaviour) can either dampen or enhance selection, and thereby modify evolutionary trajectories (e.g. Huey et al. 2003; Ghalambor et al. 2007).

16.2.2 Will plasticity evolve?

In the previous section, we set plasticity and evolution against each other as alternative ways in which phenotypes might change in response to altered environments. We also noted that both processes can jointly contribute to phenotypic change. Here we wish to make the additional point that plasticity can itself evolve—and this process could make a key contribution to evolutionary rescue.

Lande (2009) modelled an abrupt shift in the optimum phenotype for a population showing plasticity in that phenotype. As expected, mean fitness of the population at first declined dramatically because the original phenotypes were poorly suited to the new environment. Fitness then increased to some extent owing to the adaptive plastic responses of individuals—but not greatly so because the plastic response was limited. Over the generations that followed, the most dramatic evolutionary response was in the trait's plasticity: individuals with the highest plasticity were favoured because they were the ones that could produce phenotypes closest to the new optimum. This evolution of plasticity led to a rapid recovery of fitness. Then, once most individuals could plastically achieve the new phenotype, plasticity decreased through time (because plasticity was not very efficient) to be replaced with genetic changes in the non-plastic component of the trait (a process called genetic assimilation). This model, and others that followed (Chevin and Lande

2010; Chevin et al. 2010), thus predict that the evolution of plasticity should be particularly important for populations facing environmental change.

Empirical studies have certainly shown that plasticity often evolves in natural populations facing environmental change (Crispo et al. 2010; Van Doorslaer et al. 2009)—but this is not universal. Recent long-term studies of two pedigreed populations of great tits *Parus major* provide an opportunity to illustrate the alternatives. In a Dutch population (Nussey et al. 2005), (1) individuals differed dramatically in their plasticity for reproductive date, (2) selection driven by climate warming favoured increased plasticity, and (3) current levels of plasticity were insufficient for attaining optimal reproductive timing. In this case, we would expect the evolution of plasticity to be an important part of future responses to warming conditions. In a UK population (Charmantier et al. 2008), (1) individuals did not differ strongly in plasticity, (2) plasticity was not under selection, and (3) the existing plastic response was sufficient for attaining optimal reproductive timing. In this case, we would not expect the evolution of plasticity, perhaps because selection is acting to maintain plasticity at a close to optimal value. In short, the importance of the evolution of plasticity in responding to environmental change could vary widely—even among populations of a single species.

16.2.3 Is evolution fast enough?

Adaptive evolution might prevent population declines and extirpation if it increases mean absolute fitness enough to counter the fitness decline caused by environmental change (Burger and Lynch 1995; Gomulkiewicz and Holt 1995; Bell and Gonzalez 2009, 2011). A first point to consider in this process of evolutionary rescue is the extent to which environmental change initially decreases mean absolute fitness. In particular, a greater shift in the optimum and stronger stabilizing selection around that optimum both increase 'selection load' (i.e. the reduction in mean fitness incurred as a result of selection), which can then decrease population size. It is difficult to directly assess this process in nature. Indirectly, however, most extirpations must ultimately be the consequence of maladaptation of one form or another—and environmental changes have certainly caused many extirpations (Hughes et al. 1997). As an illustrative example, in freshwater lakes that became acidified owing to industrial pollution, only a minority of species in the original community were able to persist by evolving higher tolerance to the lower pH (Bradshaw and McNeilly 1991). And several other studies have related declines in population size to increasing maladaptation of key traits (Both et al. 2006; Portner and Knust 2007).

A second point to consider in the process of evolutionary rescue is the potential for adaptive evolutionary responses, which will generally depend on the availability of relevant additive genetic variance (Fisher 1930). Standing genetic variance is expected to increase with increasing population size and gene flow, and to decrease with increasing directional or stabilizing selection (Bell 1997; Futuyma 2010; Hartl and Clark 1997). Although standing genetic variation in fitness related traits is common in natural populations, it still might be insufficient to accomplish the necessary evolution—as will be discussed later. In such cases, the probability of adaptation will depend on the supply rate of new beneficial mutations, and population size will again be a crucial parameter (Gomulkiewicz and Holt 1995; Lynch and Lande 1993; Orr and Unckless 2008). Regardless of its source (pre-existing standing variation or *de novo* mutations—more about this below), higher levels of relevant genetic variance will typically allow population persistence in the face of greater environmental change—by making it possible for the population to more closely and quickly match the phenotypic optimum (Fig. 16.1; Gomulkiewicz and Holt 1995; Holt and Gomulkiewicz 2004).

Environmental change can be either abrupt or sustained. In the abrupt case, adaptive evolution needs to arrest the population decline quickly enough to forestall extirpation. If evolutionary rescue occurs, phenotypes in the population are expected to eventually match the new optimum—as long as that optimum remains stable (Gomulkiewicz and Holt 1995; Lande 2009; Fig. 16.1). In the sustained case, adaptive evolution needs to prevent the mean phenotype from lagging so far behind the changing optimum that severe negative population growth occurs. If evolutionary rescue occurs, the expectation is a steady state in which the mean phenotype lags a reasonably consistent distance behind the continually changing optimum. A model by Bürger and Lynch (1995) predicted that the 'critical rate' of environmental change (change in the phenotypic optimum) required to prevent extirpation is on the order of 10% of a phenotypic standard deviation per generation, although under some conditions this rate drops closer to 1%.

The specification of a 'critical rate', makes it tempting to compare with observed rates of phenotypic change in natural populations. For example, strong natural selection on Darwin's finches *Geospiza fortis* during a drought resulted in evolutionary responses of 0.66 'haldanes' (phenotypic standard deviations per generation) for beak depth and 0.71 haldanes for body size (Grant and Grant 2006). These rates clearly exceed the suggested critical rate of 0.10 and, indeed, 85% of the population died in a single year. By contrast, rates of phenotypic change in most other studies are typically less than 0.10 haldanes (Hendry and Kinnison 1999; Hendry et al. 2008) and are often not accompanied by population declines (Reznick and Ghalambor 2001). Perhaps in these cases, phenotypic change has been sufficiently fast to counter the demographic costs of initial maladaptation. Despite this illustrative argument, we caution against straight-up comparisons of theory-derived critical rates to observed rates of change in natural populations. The reason is that the specification of a critical rate requires many unrealistic assumptions, such as perpetual persistence under constant environmental change. Critical rates for natural populations over time frames of conservation interest could be very different.

Adaptation to a new environment will almost always involve many phenotypic traits. Given the impossibility of measuring all such traits, most studies tend to focus on one or a few traits thought to be of critical importance. As one example, Pulido and Berthold (2010) monitored migratory behaviour in a population of blackcaps *Sylvia atricapilla* to test for evolutionary responses to recent climate change. Using a common garden experiment, they demonstrated a genetic change towards residency (as opposed to migration) and the evolution of phenotypic plasticity in migration timing. They then used an artificial selection experiment to show that residency evolves in initially migratory populations under selection for a shorter migration. As a second example, Barrett and colleagues (2011) estimated the rate of evolution of temperature tolerance in three-spined sticklebacks *Gasterosteus aculeatus*. They first documented heritable differences in cold tolerance between natural marine and freshwater populations: the latter could tolerate the colder conditions typical of that environment. They then showed that marine fish introduced into freshwater ponds converged on the cold tolerance typical of freshwater populations in only three generations (Fig. 16.2). The observed rate of change of 0.63 haldanes was among the fastest rates recorded in natural populations (by comparison to the data in Hendry and Kinnison 1999). Even so, all of the experimental stickleback populations went extinct during a particularly cold winter, and so evolution, even if very rapid, might not save populations facing altered environmental conditions.

The study of key traits is certainly valuable, but it is also useful to consider overall adaptation by examining changes in major fitness components. The reason is that evolutionary rescue will depend most directly on fitness itself, to which many phenotypic traits will generally contribute. So a more direct route to inferring the potential for evolutionary rescue might be to measure how evolution improves mean population fitness. This is very difficult to accomplish for natural populations but some studies at least have been able to measure how evolution alters major fitness components of individuals. For instance, Gordon et al. (2009) showed that evolution in introduced guppy populations improved survival rates by up to 50% over 13–26 generations, although plasticity might also have made a contribution. Even more dramatically and directly, Kinnison et al. (2008) showed that the evolution of introduced Chinook salmon *Oncorhynchus tshawytscha* populations improved reproductive output in local environments by up to 150% over 26 generations. In both cases, changes in

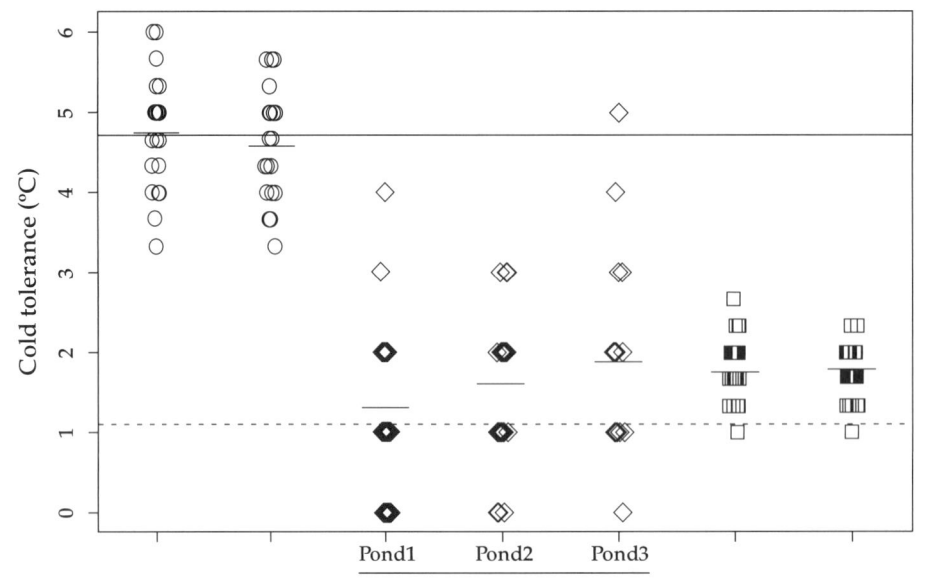

Figure 16.2 Experimental evolution of cold tolerance in three-spined stickleback. Circles and squares indicate individual fish from natural marine and freshwater populations, respectively. Diamonds indicate individual fish from three experimental freshwater ponds that were colonized with wild marine fish three generations previously and have evolved to have similar cold tolerance as natural freshwater populations. (Although evolution is likely, common-garden experiments were not used to confirm it.) Dashed lines show minimum temperatures in nature from 11 marine sites (solid) and 14 freshwater lakes (dashed). Bars indicate population mean values. Modified from Barrett et al. 2011.

major fitness components were markedly greater than changes in individual phenotypic traits thought to be under divergent selection.

In summary, factors likely to influence whether or not adaptive evolution will save populations from extirpation are now reasonably well known from theoretical models. Evidence also now exists from natural populations of noteworthy adaptive changes in specific traits, many of which could contribute to population growth. However, few of these studies involved populations in any apparent danger of extirpation—and so the direct implications for evolutionary rescue are uncertain. We therefore need studies of natural populations that specifically relate the factors thought to be important for evolutionary rescue to whether or not evolutionary rescue actually occurs. This is not a trivial endeavour, of course, because it requires replicated experimental studies, such as have been implemented in the laboratory (e.g. Bell and Gonzalez 2009; Bell and Gonzalez 2011). At present, the closest analogues we have are *ad hoc* 'experiments' that consider which species survive when environments change and whether those species showed the evolution of traits thought to be important in the new conditions (Bradshaw and McNeilly 1991).

16.2.4 Standing genetic variation versus new mutations?

As noted in the previous section, evolutionary responses to environmental change could occur through two different genetic routes: selection on pre-existing ('standing') genetic variation or on new mutations (Barrett and Schluter 2008). Between these two options, standing variation is likely to allow much faster evolution because it is immediately available when selective conditions change and because the higher initial frequency of pre-existing alleles reduces average fixation time

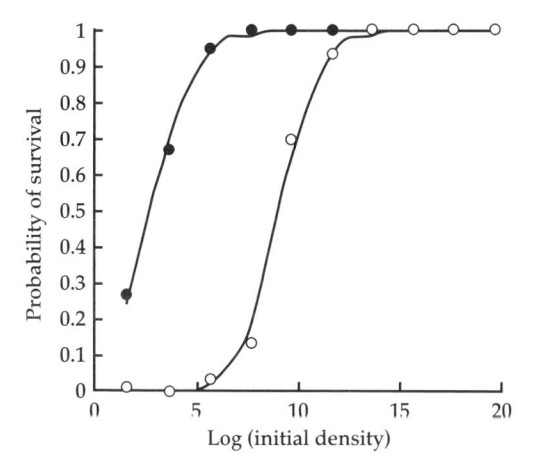

Figure 16.3 Probability of survival over 25 generations of experimental yeast populations abruptly transferred to salt (a stressful environment), in relation to the log of initial population size. Filled circles represent the control (low salt) treatment and open circles represent the high salt treatment. Each point is based on 60 replicate populations. Lines are fitted logistic regressions. Modified from Bell and Gonzalez (2009).

(Hermisson and Pennings 2005; Przeworski et al. 2005). In addition, beneficial alleles present in the standing variation will be older, and so might have accumulated multiple advantageous mutations (McGregor et al. 2007). These alleles might also have been pre-tested by selection in relevant environments (Michel et al. 2007), which increases the chance they will be helpful in the future. All of these arguments suggest that immediate adaptation to environmental change will be predominantly fuelled by standing genetic variation, an assertion supported by some case studies of ecologically relevant genes (Colosimo et al. 2005; Feder et al. 2003; Steiner et al. 2009; Tishkoff et al. 2007).

However, standing genetic variation will not always be sufficient, particularly when populations are very small, inbreeding is high, or genetic variance has been reduced by past selection (Bell and Collins 2008; Lynch and Lande 1993). In these cases, new mutations become crucial and their supply rate will depend heavily on population size—another knock against small populations. Bell and Gonzalez (2009) tested this expectation by conducting experimental evolution studies with yeast to determine how the frequency of mutations capable of rescuing the population under environmental stress (high concentrations of salt) varied as a function of population size. They first confirmed the expected (Bell and Collins 2008; Bell and Gonzalez 2011; Gomulkiewicz and Holt 1995; Orr and Unckless 2008) U-shaped trajectory of population size: a rapid decline owing to initial maladaptation followed by an increase as new better-adapted genotypes increase in frequency. They then confirmed that evolutionary rescue was more likely to occur in larger populations (Fig. 16.3)—because of the reduced stochastic loss of new beneficial mutations. These results confirm that sufficiently large populations can generate the beneficial mutations necessary for evolutionary rescue under environmental stress. It remains to be seen whether these findings regarding new mutations will hold under more natural conditions in the field.

In summary, adaptation to changing environments will probably often involve standing genetic variation—simply because it is common and immediately available (Hansen and Houle 2008; Roff 2007—but see below). By contrast, new beneficial mutations take more time to arise and then increase in frequency (Hermisson and Pennings 2005; Przeworski et al. 2005), making their contribution most likely when appropriate standing genetic variation is absent, generation times short (e.g. bacteria, viruses, unicellar plants, some arthropods), and the environmental change is not too rapid and abrupt.

16.2.5 How many genes and of what effect?

Based on his geometric model of the adaptive process, Fisher (1930) argued that mutations of small effect have a nearly 50% chance of moving a population toward the local optimum, whereas mutations of large effect have a much lower chance of doing so. This reasoning underlies the common assumption that adaptation consists of fine-tuning the phenotype with very large numbers of genes carrying mutations of relatively small effect (reviewed in Orr 2005). Whether this is typically the case, or whether few genes of large effect are often important, has significant implications for the expected rate of evolutionary response to environmental change.

Reasonable arguments have been advanced both for and against the above two possibilities. On the one hand, theory suggests that advantageous alleles of small effect will take nearly as long as neutral alleles to spread across a subdivided population (Cherry and Wakeley 2003). Waiting for new minor mutations to fix at large numbers of genes might therefore require more time than is available to populations confronted with rapid environmental change (Lynch and Lande 1993). On the other hand, because mutations of small effect are so slow to fix, the genes that carry these mutations are more likely to be polymorphic to start with (Hansen and Houle 2004). Adaptation from standing genetic variation might therefore involve mutations of small effect. In addition, greater numbers of genes contributing to adaptation will reduce the consequences of losing some mutations to stochastic processes. By contrast, evolutionary rescue through a large effect mutation at a single locus requires population sizes large enough and environmental changes gradual enough to prevent stochastic loss of the mutation while it is still rare (Gomulkiewicz and Holt 1995). However, this limitation might be less critical if adaptation can be accomplished by major-effect mutations at any one of multiple interchangeable loci (Orr and Unckless 2008). Similarly, the danger of losing a beneficial allele at a single major effect locus can be reduced if the allele is maintained at relatively high frequency because it is recessive (Orr and Betancourt 2001; Przeworski et al. 2005), mutations producing the allele occur recurrently at the same locus (Pennings and Hermisson 2006a; Pennings and Hermisson 2006b), or selection is balancing (Charlesworth et al. 1997; Hermisson and Pennings 2005).

Overall, theory thus suggests a fairly narrow range of parameter space that permits evolutionary rescue through large effect mutations at single loci. Empirical evidence does suggest, however, that genes of large effect can contribute to adaptation. Examples include genes functionally associated with flowering time in *Arabidopsis thaliana* (FLC; Ehrenreich and Purugganan 2006; Scarcelli et al. 2007), body armour and colouration in three-spined stickleback (*Pitx1*, *Eda*, *Kitlg*; Colosimo et al. 2005), swimming ability and development rate in killifish (*Fundulus heteroclitus*; LDH-B; reviewed by Powers and Schulte 1998), a variety of traits in sulphur butterflies (*Colias eurytheme*; PGI; reviewed by Wheat et al. 2006), crypsis in oldfield mice (*Peromyscus polionotus*; Mc1r; Hoekstra et al. 2006), albinism in cavefish (*Astyanax mexicanus*; Oca2; Protas et al. 2006), and beak dimensions in Darwin's finches (*Geospiza* sp.; BMP4; Abzhanov et al. 2004).

With respect to major effect loci affecting behaviour, a good example is the *forager* gene in *Drosophila melanogaster*, which has two alternate alleles that affect movement patterns of larvae between feeding patches (Sokolowski et al. 1997). These alleles are thought to be maintained in natural populations by frequency dependent selection, with the rarer of the two behavioural strategies at advantage in nutrient-poor conditions (Fitzpatrick et al. 2007). Another gene contributing to important behavioural variation is the *v1ar* locus in prairie voles *Microtus ochrogaster*, which has been shown to influence the transition from a polygamous to a more monogamous behaviour through differential expression in the brain (Donaldson and Young 2008). The same gene has been associated with partner fidelity in humans—in a large Swedish population, men homozygous for the *v1ar* variant allele were twice as likely to experience marital discord (Walum et al. 2008). Despite these few compelling examples, studies demonstrating the effects of specific genes on behaviour, whether large or small, have been limited, largely due to the challenges associated with dissecting behaviour genetically.

Studies of single large-effect genes, such as those listed above, have attracted much of the limelight, and in doing so have biased current perceptions about the typical genetics of adaptation. First, in most cases, it isn't clear how many mutations have contributed to the effects of a particular gene. Second, most studies identifying genes through quantitative trait locus (QTL) mapping have only pursued and functionally verified the single gene of largest effect, while ignoring most of the other, perhaps more typical, genes of modest to small effect. Third, candidate gene approaches are, by definition, limited to testing the role of single genes controlling known phenotypes (Haag and True

2001). Even lateral plate number in stickleback, which is often cited as a prime example of a 'single locus' trait because of the major effects of *Eda*, is determined by a more complicated inheritance pattern (Cresko et al. 2004). In addition, *Eda*, or a closely linked gene, has effects on other fitness components (Barrett et al. 2008; Marchinko 2009). Moreover, genome scans have made it increasingly clear that many traits that differ adaptively among populations are controlled by numerous genes of very small effect, such as for human height (Yang et al. 2010) and a number of plant traits (Moose et al. 2004, Thumma et al. 2010). Finally, and perhaps most important, all of the existing studies examine the genetics of single traits, whereas overall adaptation to a given environment will often depend on many traits. Thus, even if a single major-effect gene controls a particular adaptive trait, overall adaptation (i.e. fitness) in a given environment will almost certainly be influenced by many genes of small to modest effect.

In summary, the genetics of adaptation to changing environments will vary depending on the circumstances, sometimes involving only a few genes of large effect but other times involving many genes of small effect. At present, we have examples of both situations but their frequency and importance is not yet certain. In the future, increased marker density provided by new genomic tools will greatly facilitate the discovery of minor effect loci and the epistatic interactions between them (Hohenlohe et al. 2010). Also sorely needed are more studies of the genes that determine changes in *fitness*.

16.3 Constraints on evolutionary responses to environmental change

Given that evolution will often be necessary if populations are to maintain high fitness in the face of environmental change, the potential limits to this process will be crucial to understand. These limits must be common given the frequent instances in which populations have been extirpated when environments have changed. We here consider three basic possibilities: limited genetic variation, trait correlations, and ultimate constraints.

16.3.1 Limited genetic variation

Insufficient genetic variation can hamper evolutionary responses to environmental change because the mean phenotype is unable to closely match the optimum, thus increasing the extent to which mean population fitness is compromised (Bürger and Lynch 1995; Orr and Unckless 2008). To make matters worse, the resulting population decline then initiates a detrimental feedback loop: small population sizes increase genetic drift, which reduces genetic variation, which further reduces evolutionary potential (Lynch and Lande 1993).

So just how much relevant genetic variation is present in natural populations? Meta-analyses have repeatedly shown that most traits in most populations do show substantial additive genetic variation and heritability (Hansen and Houle 2008; Roff 2007). However, many of the genetic variants most relevant to adaptation might have been fixed by selection, leaving mostly standing variation that has only small effects on fitness. In addition, levels of variation, and therefore evolvability, depend on environmental conditions and might therefore be higher or lower in particular situations (Hoffmann and Merilä 1999). For all of these reasons, levels of genetic polymorphism (e.g. heritability) of individual traits do not necessarily indicate the potential for evolutionary responses to environmental change. We need more estimates of genetic variance in fitness-related traits—and fitness itself—under relevant ecological conditions.

More direct inferences about the sufficiency of genetic variation can be obtained by testing whether or not populations evolve in response to environmental change. One route to such inference is by comparing species that do or do not persist through an environmental change. In one example, only five plant species were able to survive contamination from a copper refinery, and these species showed large evolved changes in copper tolerance (Bradshaw 1984). A number of other plant species that previously coexisted in the same area went extinct, suggesting that a lack of variation for copper tolerance prevented their adaptation to the novel stress. Another route to the above inference is by imposing artificial selection in the direction expected to be favoured under environmental

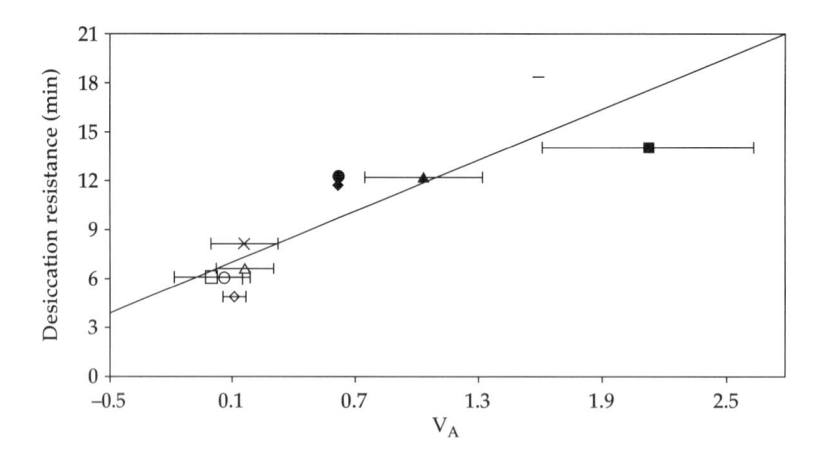

Figure 16.4 Desiccation resistance and additive genetic variance in different *Drosophila* species –, *D. melanogaster*; ●, *D. simulans*; ▲, *D. repleta*; ■, *D. hydei*; ◆, *D. serrata*; ○, *D. birchii*; *D. bunnanda*; □, *D. pseudoananassae*; ×, *D. sulfurigaster*; and ◊, *D. bipectinata*. The key different species have dramatically different genetic variances, with some being essentially zero. Error bars represent 1 SE. Modified from Kellerman et al. (2009). Reprinted with permission from AAAS.

change. In such experiments, adaptive evolution is often observed (Van Doorslaer et al. 2009; Pulido and Berthold 2010) but this is not always so. For example, Hoffman and colleagues (Hoffmann et al. 2003; Kellermann et al. 2009) showed that genetic variance for desiccation resistance, a trait expected to be important under climate change, varies dramatically among *Drosophila* species (Fig. 16.4). At the extreme, the rainforest specialist *Drosophila birchii* failed to evolve increased desiccation resistance even after intense selection for over 30 generations. Existing work thus suggests no general rule can be advanced regarding the likelihood of evolutionary constraints imposed by limited genetic variation. Such constraints will need to be considered on a case-by-case basis.

16.3.2 Trait correlations

Even if relevant genetic variation is ample for individual phenotypic traits, evolution can be constrained by genetic correlations among traits. These correlations could arise through (1) epistatic interactions between genes, (2) genes with pleiotropic effects, or (3) linkage disequilibrium between alleles at loci affecting different traits (Lande 1980; Otto 2004; Weinreich et al. 2005). In the realm of behaviour, genetic correlations likely underlie so-called 'behavioural syndromes' (Sih et al. 2004), where a basic behavioural type (e.g. shy versus bold) dictates that individuals cannot simultaneously optimize behaviours for multiple contexts, such as bold during mating but shy in the presence of predators. It is also clear from behavioural screens for mutants that most genes that affect behaviour also have pleiotropic effects on other traits (Pflugelder 1998), often expressed as developmental abnormalities (Sokolowski 2002). These results suggest that the evolution of behaviour could be constrained by deleterious pleiotropic consequences.

When selection is not in the same direction as the correlation between traits ('antagonistic selection'), the evolutionary response will be constrained (Hansen and Houle 2008; Lande 1979). Such constraints might be relevant in the case of environmental change. Hellmann and Pineda-Krch (2007) simulated the lag between observed phenotypes and optimal phenotypes in a changing environment when fitness was determined by two pleiotropically linked traits. Their analysis confirmed that genetic correlations running counter to the direction of selection increase the lag, and thereby decrease the amount of environmental change a population can tolerate before extirpation. This result echoes the long-standing assumption that greater interdependence of traits (or 'complexity') leads to greater levels of constraint (Fisher 1930). But this assumption might not always hold: for instance, pleiotropy could increase the mutational target size and therefore the evolvability of the trait (Hansen and Houle 2004).

So just how important are these potential constraints for natural populations experiencing environmental change? As one suggestive example, Etterson and Shaw (2001) found that *Chamaecrista fasciculate* legumes subject to drought treatments in a reciprocal transplant experiment were under selection for more and thicker leaves. However, these two traits were negatively genetically correlated, and so most of the variation was orthogonal to the direction of selection. The predicted adaptive evolution was therefore slower than would be expected in the absence of negative genetic correlations. On the flip side, genetic covariances that are high in the same direction as selection could accelerate evolutionary responses—because selection acting on each trait will not only directly influence selection on that trait but also indirectly enhance selection on the other trait (Agrawal and Stinchcombe 2009).

In addition to changing the rate of evolution, genetic correlations could also influence the *direction* of evolution. In particular, evolution could be biased toward the most genetically variable aspects of phenotype (e.g. the 'genetic lines of least resistance', Futuyma et al. 1995; Schluter 1996). If so, correlations among traits could cause populations to evolve in directions that are 'easiest' but not necessarily the best for improved adaptation. However, empirical studies are strongly divided as to whether or not the direction of evolution is routinely biased by genetic correlations (see citations in Berner et al. 2010). Overall then, it is not yet clear if genetic correlations that limit evolutionary responses to environmental change represent a widespread phenomenon.

16.3.3 Ultimate constraints

Evolutionary constraints can also arise through unbreakable functional or performance trade-offs or strict physiological limits. In the case of trade-offs, adaptation can be limited owing to competing performance requirements, such as speed versus endurance (e.g. Wilson et al. 2002), speed versus force (e.g. Herrel et al. 2009), or stability versus manoeuvrability (e.g. Weihs 2002). As a particularly ubiquitous trade-off, parents must always compromise between the number of offspring they produce and the size (or quality) of those offspring (Charnov 1995). Similarly, few organisms grow at their physiological maximum (Calow 1982) because they must also allocate resources to other functions, including starvation resistance (Gotthard et al. 1994), metabolic efficiency (Stevens et al. 1998), and skeletal development (Arendt and Wilson 2001).

In the case of strict physiological limits, temperature tolerance has been suggested as a possibility in the case of climate change. For example, although climate warming is expected to be less extreme in the tropics than at higher latitudes, the greater physiological sensitivity of tropical ectotherms means they could experience greater fitness declines (Fig. 16.5; Deutsch et al. 2008). For aquatic organisms, higher temperatures increase oxygen demand but also decrease oxygen supply, until aerobic metabolism becomes insufficient. Pörtner and Knust (2007) have argued that this constraint explains inter-annual variation in the population size of eelpout fish *Zoarces viviparus* in the North and Baltic Seas. Of course, a remaining question is the extent to which current thermal limits cannot be surpassed by evolution. For instance, fishes certainly can adapt to higher temperatures than those currently causing problems for eelpout. At the extreme, Lake Migadi Tilapia *Oreochromis alcalicus grahami* live in 42 °C water, in part because they can breathe air (Franklin et al. 1995). On the other hand, even the dramatic evolution of temperature tolerance in the aforementioned stickleback experiment in subsection 16.2.3 was not enough to prevent extinction under extreme temperatures (Barrett et al. 2011). So temperature tolerance can clearly evolve—but perhaps not always or not quickly enough or far enough.

In summary, shifts in phenotypic optima as a result of environmental change sometimes might not be reachable owing to trade-offs or strict limits imposed by the functional or physiological architecture of whole organisms. Whether or not these limits actually hamper evolutionary rescue in nature is unknown—although upper temperature tolerance is a strong candidate in the case of climate change. In addition, bioenergetic models suggest that geographic distributions may be limited by hard physi-

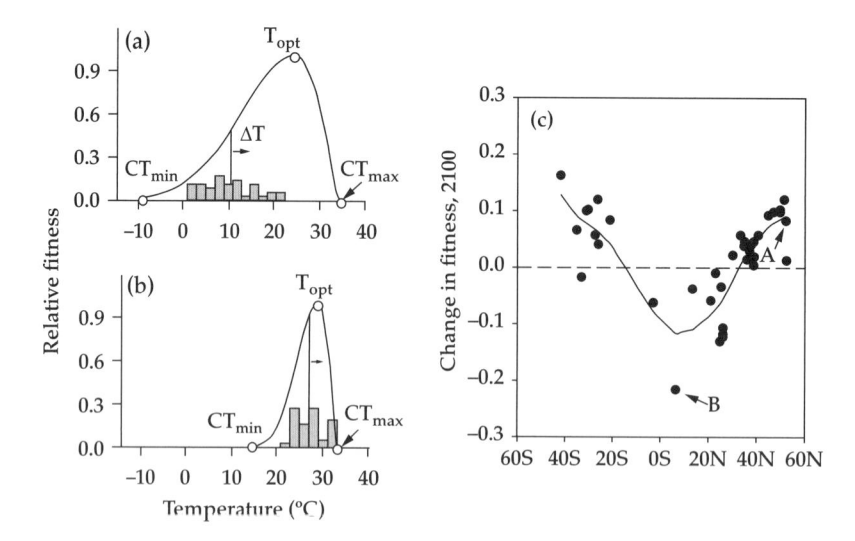

Figure 16.5 Fitness curves for representative insect taxa from temperate (a) and tropical (b) locations, and (c) the estimated change in fitness because of climate warming for all insect species studied, as a function of latitude. Fitness curves were derived from measured intrinsic population growth rates versus temperature for 38 species, including *Acyrthosiphon pisum* (Hemiptera) from 52°N (England) (a), and *Clavigralla shadabi* (Hemiptera) from 6°N (Benin) (b). CTmin, ΔT, Topt, and CTmax are indicated on each curve. Climatological mean annual temperature from 1950–1990 (ΔT, drop lines from each curve), its seasonal and diurnal variation (grey histogram), and its projected increase because of warming in the next century (ΔT, arrows) are shown for the collection location of each species. In (c), negative values indicate decreased rates of population growth in 2100 AD and are found mainly in the tropics. Positive values are found in mid- and high-latitudes. Modified from Deutsch et al. (2008); (c) (2008) National Academy of Sciences, USA.

ological limits imposed by seasonal energetic bottlenecks (Humphries et al. 2002).

16.4 Conclusions

Environmental change is occurring at an unprecedented rate across the globe. These changes are expected to cause a mismatch between the current phenotypes of populations and the phenotypes best suited for the new conditions. This mismatch can cause decreased fitness and declines in population size that might lead to extirpation and extinction. If populations are to reverse these declines, they will need to reduce the mismatch through immediate behavioural responses, phenotypic (developmental) plasticity, or evolution. Empirical studies of populations in altered environments have documented each of these types of phenotypic change, although it has been difficult to confirm the contribution of evolution in certain scenarios, such as climate change. Moreover, the consequences of adaptive trait evolution for population size and persistence are unknown in natural populations.

Theoretical conditions do exist under which adaptive evolution makes the difference between population persistence versus extirpation, but no studies of natural populations have tested the resulting predictions—and only a few have done so in the laboratory. In general, it is expected that adaptation to new conditions will often, although not always, be fuelled by standing genetic variation rather than new mutations. In either case, this adaptation can involve a few genes of large effect all the way through to many genes of minor effect—or different combinations thereof. Further studies utilizing new genomic tools in natural populations should help to establish the frequency and relevance of different types of genetic variation in adaptive responses to environmental change. Although evolution will almost certainly be important in mitigating the negative effects of climate change—it is not all powerful. Evolution can be constrained by insufficient genetic variation, correlations between traits under antagonistic selection, or functional/physiological trade-offs or limits.

Overall, our understanding of evolutionary responses to environmental change, and of evolutionary rescue, is rudimentary. In particular, we require an improved understanding of the role of population sizes, gene flow, mutation, the genetic architecture of relevant traits, trade-offs, and physiological limits. And it seems likely that the greatest advances will be made through the integration of different research approaches that target several functional and biological levels (e.g. mutations, genes, phenotypes, individuals, fitness, populations, communities, ecosystems). As this knowledge grows, it should improve our ability to predict evolutionary responses to climate change—and the resulting consequences of biological diversity.

References

Abzhanov, A., Protas, M., Grant, B. R., Grant, P. R., and Tabin, C. J. (2004) *Bmp4* and morphological variation of beaks in Darwin's finches. *Science*, 305, 1462–5.

Agrawal, A. F. and Stinchcombe, J. R. (2009). How much do genetic covariances alter the rate of adaptation? *Proceedings of the Royal Society of London B, Biological Sciences*, 276, 1183–91.

Arendt, J. and Wilson, D. S. (2001). Scale strength as a cost of rapid growth in sunfish. *Oikos*, 72, 152–77.

Balanya, J., Oller, J. M., Huey, R. B., Gilchrist, G. W., and Serra, L. (2006). Global genetic change tracks global climate warming in *Drosophila subobscura*. *Science*, 313, 1773–5.

Barrett, R. D. H., Paccard, A., Healy, T., Bergek, S., Schulte, P., Schluter, D., and Rogers, S. (2011). Rapid evolution of cold tolerance in stickleback. *Proceedings of the Royal Society of London B, Biological Sciences*, 278, 233–8.

Barrett, R. D. H., Rogers, S. M., and Schluter, D. (2008) Natural selection on a major armor gene in threespine stickleback. *Science*, 322, 255–7.

Barrett, R. D. H. and Schluter, D. (2008) Adaptation from standing genetic variation. *Trends in Ecology & Evolution*, 23, 38–44.

Bell, G. (1997). *Selection: The Mechanism of Evolution*. New York, Chapman and Hall.

Bell, G. and Collins, S. (2008) Adaptation, extinction and global change. *Evolutionary Applications*, 1, 3–16.

Bell, G. and Gonzalez, A. (2009) Evolutionary rescue can prevent extinction following environmental change. *Ecology Letters*, 12, 942–8.

Bell, G. and Gonzalez, A. (2011) Adaptation and evolutionary rescue in metapopulations experiencing environmental deterioration. *Science*, 332, 1327–30.

Berner, D., Stutz, W., and Bolnick, D. (2010). Foraging trait (co)variances in stickleback evolve deterministically and do not predict trajectories of adaptive diversification. *Evolution*, 64, 2265–77.

Berthold, P., Helbig, A. J., Mohr, G., and Querner, U. (1992). Rapid microevolution of migratory behavior in a wild bird species. *Nature*, 360, 668–70.

Bonduriansky, R. and Day, T. (2009) Nongenetic inheritance and its evolutionary implications. *Annual Review of Ecology, Evolution, and Systematics*, 40, 103–25.

Both, C., Bouwhuis, S., Lessells, C., and Visser, M. (2006) Climate change and population declines in a long-distance migratory bird. *Nature*, 441, 81–3.

Bowler, D. and Benton, T. (2005). Causes and consequences of animal dispersal strategies: relating individual behaviour to spatial dynamics. *Biological Reviews of the Cambridge Philosophical Society*, 80, 205–25.

Bradshaw, A. (1984) Adaptation of plants to soils containing toxic metals—a test for conceipt. *Ciba Foundation Symposium*, 102, 4–14.

Bradshaw, A. and McNeilly, T. (1991) Evolutionary response to global climatic change. *Annals of Botany*, 67(Suppl.), 5–14.

Bradshaw, W. E. and Holzapfel, C. M. (2001) Genetic shift in photoperiodic response correlated with global warming. *Proceedings of the National Academy of Sciences of the USA*, 98, 14509.

Bradshaw, W. E. and Holzapfel, C. M. (2006) Evolutionary response to rapid climate change. *Science*, 312, 1477–8.

Bürger, R. and Lynch, M. (1995) Evolution and extinction in a changing environment: a quantitative-genetic analysis. *Evolution*, 49, 151–63.

Calow, P. (1982) Homeostasis and fitness. *American Naturalist*, 120, 416–19.

Charlesworth, B., Nordborg, M., and Charlesworth, D. (1997) The effects of local selection, balanced polymorphism and background selection on equilibrium patterns of genetic diversity in subdivided populations. *Genetical Research*, 70, 155–74.

Charmantier, A., McCleery, R. H., Cole, L. R., Perrins, C., Kruuk, L. E. B., and Sheldon, B. C. (2008). Adaptive phenotypic plasticity in response to climate change in a wild bird population. *Science*, 320, 800–3.

Charnov, E. L. (1995). A trade-off-invariant life-history rule for optimal offspring size. *Nature*, 376, 418–19.

Cherry, J. and Wakeley, J. (2003). A diffusion approximation for selection and drift in a subdivided population. *Genetics*, 163, 421.

Chevin, L. and Lande, R. (2010). When do adaptive plasticity and genetic evolution prevent extinction of a density-regulated population? *Evolution*, 64, 1143–50.

Chevin, L., Lande, R., and Mace, G. (2010) Adaptation, plasticity, and extinction in a changing environment: towards a predictive theory. *Public Library of Science Biology*, 8, e1000357.

Clutton-Brock, T. H. and Harvey, P. H. (1985) Comparative approaches to investigating adaptation. In Krebs J.R. and Davies, N.B. *Behavioural Ecology: An Evolutionary Approach*. Sunderland, MA, Sinauer.

Colosimo, P. F., Hosemann, K. E., Balabhadra, S., Villarreal, G., Dickson, M., Grimwood, J., Schmutz, J., Myers, R. M., Schluter, D., and Kingsley, D. M. (2005) Widespread parallel evolution in sticklebacks by repeated fixation of ectodysplasin alleles. *Science*, 307, 1928–33.

Cousyn, C., De Meester, L., Colbourne, J. K., Brendonck, L., Verschuren, D., and Volckaert, F. (2001) Rapid, local adaptation of zooplankton behavior to changes in predation pressure in the absence of neutral genetic changes. *Proceedings of the National Academy of Sciences of the USA*, 98, 6256–60.

Cresko, W. A., Amores, A., Wilson, C., Murphy, J., Currey, M., Phillips, P., Bell, M. A., Kimmel, C. B., and Postlethwait, J. H. (2004) Parallel genetic basis for repeated evolution of armor loss in Alaskan threespine stickleback populations. *Proceedings of the National Academy of Sciences of the USA*, 101, 6050–5.

Crispo, E., DiBattista, J., Correa, C., Thibert-Plante, X., McKellar, A., Schwartz, A., Berner, D., DeLeon, L., and Hendry, A. (2010). The evolution of phenotypic plasticity in response to anthropogenic disturbance. *Evolutionary Ecology Research*, 12, 47–66.

Darimont, C., Carlson, S., Kinnison, M., Paquet, P., Reimchen, T., and Wilmers, C. (2009). Human predators outpace other agents of trait change in the wild. *Proceedings of the National Academy of Sciences of the USA*, 106, 952–954.

Deutsch, C., Tewksbury, J., Huey, R., Sheldon, K., Ghalambor, C., Haak, D., and Martin, P. (2008). Impacts of climate warming on terrestrial ectotherms across latitude. *Proceedings of the National Academy of Sciences USA*, 105, 6668–72.

DeWitt, T., Sih, A., and Wilson, D. (1998). Costs and limits to phenotypic plasticity. *Trends in Ecology & Evolution*, 13, 77–81.

Donaldson, Z. R. and Young, L. J. (2008). Oxytocin, vasopressin and the neuro-genetics of sociality. *Science*, 322, 900–4.

Eggers, S., Griesser, M., Nystrand, M., and Ekman, J. (2006) Predation risk induces changes in nest-site selection and clutch size in the Siberian jay. *Proceedings of the Royal Society of London B, Biological Sciences*, 273, 701–6.

Ehrenreich, I. M. and Purugganan, M. D. (2006) The molecular genetic basis of plant adaptation. *American Journal of Botany*, 93, 953–62.

Etterson, J. and Shaw, R.G. (2001). Constraint to adaptive evolution in response to global warming. *Science*, 294, 151–4.

Feder, J. L., Berlocher, S. H., Roethele, J. B., Dambroski, H., Smith, J. J., Perry, W. L., Gavrilovic, V., Filchak, K. E., Rull, J., and Aluja, M. (2003). Allopatric genetic origins for sympatric host-plant shifts and race formation in Rhagoletis. *Proceedings of the National Academy of Sciences of the USA*, 100, 10314–19.

Fisher, R. (1930). *The Genetical Theory of Natural Selection*. Claredon Press, Oxford, U.K.

Fitzpatrick, M. J., Feder, E., Rowe, L., and Sokolowski, M. B. (2007). Maintaining a behaviour polymorphism by frequency-dependent selection on a single gene. *Nature* 447, 210–15.

Franklin, C., Johnston, I., Crockford, T., and Kamunde, C. (1995). Scaling of oxygen consumption of Lake Magadi tilapia, a fish living at 37°C. *Journal of Fish Biology*, 46, 892–34.

Franks, S. J., Sim, S., and Weis, A. E. (2007) Rapid evolution of flowering time by an annual plant in response to a climate fluctuation. *Proceedings of the National Academy of Sciences of the USA*, 104, 1278–82.

Futuyma, D., Keese, M., and Funk, D. (1995) Genetic constraints on macroevolution: the evolution of host affiliation in the leaf beetle genus *Ophraella*. *Evolution*, 49, 797–809.

Futuyma, D. J. (2010) Evolutionary constraint and consequences. *Evolution*, 64, 1865–84.

Ghalambor, C.K., McKay, J.K., Carroll, S.P., and Reznick, D.N. (2007). Adaptive versus non-adaptive phenotypic plasticity and the potential for contemporary adaptation in new environments. *Functional Ecology*, 21, 394–407.

Gienapp, P., Teplitsky, C., Alho, J. S., Mills, J. A., and Merilä, J. (2008). Climate change and evolution: disentangling environmental and genetic responses. *Molecular Ecology*, 17, 167–78.

Gomulkiewicz, R. and Holt, R. (1995) When does evolution by natural selection prevent extinction? *Evolution*, 49, 201–7.

Gordon, S., Reznick, D., Kinnison, M., Bryant, M., Weese, D., Rasanen, K., Millar, N., and Hendry, A. (2009) Adaptive changes in life history and survival following a new guppy introduction. *The American Naturalist*, 174, 34–45.

Gotthard, K., Sörensen, N., and Christer, W. (1994) Adaptive variation in growth rate: life history costs and consequences in the speckled wood butterfly, *Pararge aegeria*. *Oecologia*, 99, 281–9.

Grant, P. and Grant, B. (2006). Evolution of character displacement in Darwin's finches. *Science*, 313, 224–26.

Grether, G. F. (2005). Environmental change, phenotypic plasticity, and genetic compensation. *The American Naturalist*, 166, E115–23.

Haag, E. S. and True, J. R. (2001). From mutants to mechanisms? Assessing the candidate gene paradigm in evolutionary biology. *Evolution*, 55, 1077–84.

Hadfield, J., Wilson, A., Garant, D., Sheldon, B., and Kruuk, L. (2010). The misuse of BLUP in ecology and evolution. *The American Naturalist*, 175, 116–25.

Hansen, T. and Houle, D. (2004) Evolvability, stabilizing selection, and the problem of stasis. In Pigliucci, M. and Preston, K. (eds) *The Evolutionary Biology of Complex Phenotypes*. Oxford, Oxford University Press.

Hansen, T. and Houle, D. (2008). Measuring and comparing evolvability and constraint in multivariate characters. *Journal of Evolutionary Biology*, 21, 1201–19.

Hartl, D. L. and Clark, A. G. (1997) *Principles of Population Genetics*. Sunderland, M.A, Sinauer Assoc.

Hellmann, J. J. and Pineda-Krch, M. (2007). Constraints and reinforcement on adaptation under climate change: Selection of genetically correlated traits. *Biological Conservation*, 137, 599–609.

Hendry, A. P. (2004) Selection against migrants contributes to the rapid evolution of ecologically-dependent reproductive isolation. *Evolutionary Ecology Research*, 6, 1219–36.

Hendry, A. P., Farrugia, T. J., and Kinnison, M. T. (2008). Human influences on rates of phenotypic change in wild animal populations. *Molecular Ecology*, 17, 20–9.

Hendry, A. P. and Kinnison, M. T. (1999). The pace of modern life: Measuring rates of contemporary microevolution. *Evolution*, 53, 1637–53.

Hermisson, J. and Pennings, P. S. (2005). Soft sweeps: Molecular population genetics of adaptation from standing genetic variation. *Genetics*, 169, 2335–52.

Herrel, A., Podos, J., Vanhooydonck, B., and Hendry, A.P. (2009). Force-velocity trade-off in Darwin's finch jaw function: a biomechanical basis for ecological speciation? *Functional Ecology*, 23, 119–25.

Hoekstra, H. E., Hirschmann, R. J., Bundey, R. A., Insel, P. A., and Crossland, J. P. (2006). A single amino acid mutation contributes to adaptive beach mouse color pattern. *Science*, 313, 101–4.

Hoffmann, A. A., Hallas, R. J., Dean, J. A., and Schiffer, M. (2003). Low potential for climatic stress adaptation in a rainforest *Drosophila* species. *Science*, 301, 100–2.

Hoffmann, A. A. and Merila, J. (1999). Heritable variation and evolution under favourable and unfavourable conditions. *Trends in Ecology & Evolution*, 14, 96–101.

Hohenlohe, P. A., Bassham, S., Etter, P., Stiffler, N., Johnson, E., and Cresko, W. (2010). Population genomics of parallel adaptation in threespine stickleback using sequenced RAD tags. *PLOS* 6, e1000862.

Holt, R. (1990). The microevolutionary consequences of climate change. *Trends in Ecology and Evolution*, 5, 311–15.

Holt, R. D. and Gomulkiewicz, R. (2004). Conservation implications of niche conservation and evolution in heterogeneous environments. In Ferriere, R., Dieckmann, U., and Couvet, D. (eds) *Evolutionary Conservation Biology*. Cambridge, Cambridge University Press.

Huey, R. B., Hertz, P. E., and Sinervo, B. (2003). Behavioural drive versus behavioural intertia in evolution: a null model approach. *The American Naturalist*, 161, 357–66.

Hughes, J. B., Daily, G. C., and Ehrlich, P. R. (1997) Population diversity: Its extent and extinction. *Science*, 278, 689–92.

Humphries, M. M., Thomas, D. W., and Speakman, J. R. (2002) Climate-mediated energetic constraints on the distribution of hibernating mammals. *Nature*, 418, 313–16.

Jump, A., Hunt, J., Martinez-Izquierdo, J., and Pevuelas, J. (2006). Natural selection and climate change: temperature-linked spatial and temporal trends in gene frequency in *Fagus sylvatica*. *Molecular Ecology*, 15, 3469–80.

Kellermann, V., van Heerwaarden, B., Sgro, C. M., and Hoffmann, A. A. (2009). Fundamental evolutionary limits in ecological traits drive *Drosophila* species distributions. *Science*, 325, 1244–6.

Kinnison, M.T. and Hendry, A.P. (2001). The pace of modern life II: from rates of contemporary microevolution to pattern and process. *Genetica*, 112–13, 145–6.

Kinnison, M.T., Unwin, M., and Quinn, T. P. (2008). Eco-evolutionary vs. habitat contributions to invasion in salmon: experimental evaluation in the wild. *Molecular Ecology*, 17, 405–14.

Kruuk, L. E. B. (2004). Estimating genetic parameters in natural populations using the 'animal model'. *Philosophical Transactions of the Royal Society B: Biological Sciences*, 359, 873–90.

Lande, R. (1979). Quantitative genetic analysis of multivariate evolution, applied to brain: body size allometry. *Evolution*, 33, 402–16.

Lande, R. (1980). The genetic covariance between characters maintained by pleiotropic mutations. *Genetics*, 94, 203–15.

Lande, R. (2009) Adaptation to an extraordinary environment by evolution of phenotypic plasticity and genetic assimilation. *Journal of Evolutionary Biology*, 22, 1435–46.

Lynch, M. and Lande, R. (1993) Evolution and extinction in response to environmental change. In Kareiva, P., Kingsolver, J.G., and Huey, R. (eds) *Biotic Interactions and Global Change*, pp. 235–50. Sunderland: Sinauer.

Magurran, A. E., Seghers, B. H., Carvalho, G. R., and Shaw, P. W. (1992). Behavioural consequences of an artificial introduction of guppies (*Poecilia reticulata*) in N.

Trinidad: evidence for the evolution of anti-predator behaviour in the wild. *Proceedings of the Royal Society of London B, Biological Sciences*, 248, 117–22.

Marchinko, K. (2009). Predation's role in repeated phenotypic and genetic divergence of armor in threespine stickleback. *Evolution*, 63, 127–38.

McGregor, A. P., Orgogozo, V., Delon, I., Zanet, J., Srinivasan, D. G., Payre, F., and Stern, D. L. (2007). Morphological evolution through multiple cis-regulatory mutations at a single gene. *Nature*, 448, 587–90.

Michel, A. P., Rull, J., Aluja, M., and Feder, J. L. (2007). The genetic structure of hawthorn-infesting *Rhagoletis pomonella* populations in Mexico: implications for sympatric host race formation. *Molecular Ecology*, 16, 2867–78.

Moose, S. P., Dudley, J. W., and Rocheford, T. R. (2004). Maize selection passes the century mark: a unique resource for 21st century genomics. *Trends In Plant Science*, 9, 358–64.

Nussey, D. H., Postma, E., Gienapp, P., and Visser, M. E. (2005). Selection on heritable phenotypic plasticity in a wild bird population. *Science*, 310, 304–6.

Orr, H. A. (2005). The genetic theory of adaptation: A brief history. *Nature Reviews Genetics*, 6, 119–27.

Orr, H. A. and Betancourt, A. J. (2001). Haldane's sieve and adaptation from the standing genetic variation. *Genetics*, 157, 875–84.

Orr, H. A. and Unckless, R. L. (2008). Population extinction and the genetics of adaptation. *The American Naturalist*, 172, 160–9.

Otto, S. P. (2004). Two steps forward, one step back: the pleiotropic effects of favoured alleles. *Proceedings of the Royal Society of London B, Biological Sciences*, 271, 705–14.

Parmesan, C. (2006). Ecological and evolutionary responses to recent climate change. *Annual Review of Ecology, Evolution, and Systematics*, 37, 637–69.

Pennings, P. S. and Hermisson, J. (2006a). Soft Sweeps II—Molecular population genetics of adaptation from recurrent mutation or migration. *Molecular Biology and Evolution*, 23, 1076–84.

Pennings, P. S. and Hermisson, J. (2006b). Soft Sweeps III: The signature of positive selection from recurrent mutation. *PLOS Genetics*, 2, e186.

Pflugelder, G. O. (1998). Genetic lesions in *Drosophila* behavioural mutants. *Behaviour and Brain Research*, 95, 3–15.

Phillimore, A., Hadfield, J., Jones, O., and Smithers, R. (2010). Differences in spawning date between populations of common frog reveal local adaptation. *Proceedings of the National Academy of Sciences of the USA*, 107, 8292–7.

Phillips, B. L., Brown, G. P., and Shine, R. (2010). Evolutionarily accelerated invasions: the rate of dispersal evolves upwards during the range advance of cane toads. *Journal of Evolutionary Biology*, 23, 2595–601.

Pigluicci, M. (2001) Phenotypic plasticity. In Fox, C.W., Roff, D., and Fairbairn, D.J. (eds) *Evolutionary Ecology: Concepts and Case Studies*. Oxford, Oxford University Press.

Poethke, H. J., Weisser, W. W., and Hovestadt, H. (2010). Predator-Induced dispersal and the evolution of conditional dispersal in correlated environments. *The American Naturalist*, 175, 577–86.

Portner, H. O. and Knust, R. (2007). Climate change affects marine fishes through the oxygen limitation of thermal tolerance. *Science*, 315, 95–7.

Powers, D. A. and Schulte, P. M. (1998). Evolutionary adaptations of gene structure and expression in natural populations in relation to a changing environment: A multidisciplinary approach to address the million-year saga of a small fish. *The Journal of Experimental Zoology*, 282, 71–94.

Protas, M. E., Hersey, C., Kochanek, D., Zhou, Y., Wilkens, H., Jeffery, W. R., Zon, L. I., Borowsky, R., and Tabin, C. J. (2006). Genetic analysis of cavefish reveals molecular convergence in the evolution of albinism. *Nature Genetics*, 38, 107–11.

Przeworski, M., Coop, G., and Wall, J. D. (2005). The signature of positive selection on standing genetic variation. *Evolution*, 59, 2312–23.

Przybylo, R., Sheldon, B., and Merilä, J. (2000). Climatic effects on breeding and morphology: evidence for phenotypic plasticity. *Journal of Animal Ecology*, 69, 395–403.

Pulido, F. and Berthold, P. (2010). Current selection for lower migratory activity will drive the evolution of residency in a migratory bird population. *Proceedings of the National Academy of Sciences of the USA*, 107, 7341–6.

Réale, D., Berteaux, D., McAdam, A. G., and Boutin, S. (2003). Lifetime selection on heritable life-history traits in a natural population of red squirrels. *Evolution*, 57, 2416–23.

Reed, T., Wanless, S., Harris, M., Frederiksen, M., Kruuk, L., and Cunningham, E. (2006). Responding to environmental change: plastic responses vary little in a synchronous breeder. *Proceedings of the Royal Society of London B, Biological Sciences*, 273, 2713–19.

Reznick, D. N. and Ghalambor, C. K. (2001). The population ecology of contemporary adaptations: what empirical studies reveal about the conditions that promote adaptive evolution. *Genetica*, 112–113, 183–98.

Roff, D. (2007). *Evolutionary Quantitative Genetics*. New York, Chapman and Hall.

Root, T. L., Price, J. T., Hall, K. R., Schneider, S. H., Rosenzweig, C., and Pounds, J. A. (2003). Fingerprints of global warming on wild animals and plants. *Nature*, 421, 57.

Rushbrook, B., Head, M., Katsiadaki, I., and Barber, I. (2010). Flow regime affects building behaviour and nest structure in sticklebacks. *Behavioral Ecology and Sociobiology*, 64, 1927–35.

Scarcelli, N., Cheverud, J., Schaal, B., and Kover, P. (2007). Antagonistic pleiotropic effects reduce the potential adaptive value of the FRIGIDA locus. *Proceedings of the National Academy of Sciences of the USA*, 104, 16986–16991.

Schlaepfer, M.A., Runge, M.C., and Sherman, P.W. (2002). Ecological and evolutionary traps. *Trends in Ecology and Evolution*, 17, 474–80.

Schluter, D. (1996). Adaptive radiation along genetic lines of least resistance. *Evolution* 50, 1766–74.

Sih, A., Bell, A., Johnson, J., and Ziemba, R. (2004). Behavioral syndromes: an integrative overview. *Quarterly Review of Biology*, 79, 241–77.

Singer, M.C., Thomas, C.D., and Parmesan, C. (1993). Rapid human induced evolution of insect-host interactions. *Nature*, 366, 681–3.

Skelly, D. K., Joseph, L. N., Possingham, H. P., Freidenburg, L. K., Farrugia, T. J., Kinnison, M. T., and Hendry, A. P. (2007). Evolutionary responses to climate change. *Conservation Biology*, 21, 1353–5.

Skinner, B. F. (1966). The phylogeny and ontogeny of behaviour. *Science*, 153, 1205–13.

Sokolowski, M. B. (2002). Neurobiology—social eating for stress. *Nature*, 419, 893–4.

Sokolowski, M. B., Pereira, H. S., and Hughes, K. (1997). Evolution of foraging behavior in *Drosophila* by density-dependent selection. *Proceedings of the National Academy of Sciences of the USA*, 94, 7373–7.

Steiner, C. C., Rompler, H., Boettger, L. M., Schoneberg, T., and Hoekstra, H. E. (2009). The genetic basis of phenotypic convergence in beach mice: similar pigment patterns but different genes. *Molecular Biology and Evolution*, 26, 35–45.

Stevens, E. D. and Cook, T. (1998). Respiratory metabolism and swimming performance in growth hormone transgenic Atlantic salmon. *Canadian Journal of Fisheries & Aquatic Sciences*, 55, 2028–35.

Thumma, B., Southerton, S., Bell, J., Owen, J., Henery, M., and Moran, G. (2010). Quantitative trait locus (QTL) analysis of wood quality traits in *Eucalyptus nitens*. *Tree Genetics & Genomes*, 6, 305–17.

Tishkoff, S. A., Reed, F. A., Ranciaro, A., Voight, B. F., Babbitt, C. C., Silverman, J. S., Powell, K., Mortensen, H. M., Hirbo, J. B., Osman, M., Ibrahim, M., Omar, S. A., Lema, G., Nyambo, T. B., Ghori, J., Bumpstead, S., Pritchard, J. K., Wray, G. A., and Deloukas, P. 2007 Convergent adaptation of human lactase persistence in Africa and Europe. *Nature Genetics*, 39, 31–40.

Umina, P., Weeks, A., Kearney, M., McKechnie, S., and Hoffmann, A. A. (2005). A rapid shift in a classic clinal pattern in *Drosophila* reflecting climate change. *Science*, 308, 691.

Van Doorslaer, W., Stoks, R., Duvivier, C., Bednarska, A., and De Meester, L. (2009). Population dynamics determine adaptation to temperature in *Daphnia*. *Evolution*, 63, 1867–78.

Visser, M. (2008). Keeping up with a warming world; assessing the rate of adaptation to climate change. *Proceedings of the Royal Society of London B, Biological Sciences*, 275, 649–59.

Walum, H., Westberg, L., Henningsson, S., Neiderhiser, J.M., Reiss, D., Igl, W., Ganiban, J.M., Spotts, E.L., Pedersen, N.L., Friksson, E., and Lichtenstein, P. (2008). Genetic variation in the *vasopressin receptor 1a* gene (*AVPR1A*) associates with pair-bonding behaviour in humans. *Proceedings of the National Academy of Sciences of the USA*, 105, 14153–6.

Weihs, D. (2002). Stability versus maneuverability in aquatic locomotion. *Integrative and Comparative Biology*, 42, 127–34.

Weinreich, D. M., Watson, R. A., and Chao, L. (2005). Sign epistasis and genetic constraint on evolutionary trajectories. *Evolution*, 59, 1165–74.

Westley, P. A. H. (2011). What invasive species reveal about the rate and form of contemporary phenotypic change in nature. *The American Naturalist*, 177, 496–509.

Wheat, C. W., Watt, W. B., Pollock, D. D., and Schulte, P. M. (2006). From DNA to fitness differences: sequences and structures of adaptive variants of *Colias Phosphoglucose Isomerase* (*PGI*). *Molecular Biology and Evolution*, 23, 499–512.

Wilson, R. S., James, R. S., and Van Damme, R. (2002). Trade-offs between speed and endurance in the frog *Xenopus laevis*: a multi-level approach. *Journal of Experimental Biology*, 205, 1145–52.

Yang, J., Benyamin, B., McEvoy, B., Gordon, S., Henders, A., Nyholt, D., Madden, P., Heath, A., Martin, N., Montgomery, G., Goddard, M., and Visscher, P. (2010). Common SNPs explain a large proportion of the heritability for human height. *Nature Genetics*, 42, 565–9.

CHAPTER 17

Ecotourism, wildlife management, and behavioural biologists: changing minds for conservation

Richard Buchholz and Edward M. Hanlon

> ⊃ **Overview**
>
> Disturbance of wildlife by ecotourists is thought to have negative effects on the population viability of some species. This topic is well suited to investigation by the integrative approach of animal behaviourists. We review the recent wildlife disturbance literature to see if it has benefited from the origination of the new field of 'conservation behaviour'. Using previous reviews as a stepping off point, we also address the adequacy of the methodological approaches taken to assess the impact of anthropogenic disturbance on wild animals. Despite some evidence that studies of wildlife disturbance are becoming more integrative, the quality of the science is rather poor and generally not suitable for use by wildlife managers and policy-makers. We hope our review motivates animal behaviourists to offer their unique services to conservation efforts; without better behavioural science, we fear that natural areas may slowly lose the very species that humans hope to see there.

'One of the penalties of an ecological education is that one lives alone in a world of wounds.'
–Aldo Leopold, *The Round River - A Parable.*

17.1 Introduction

It is easy to be pessimistic about the fate of Nature on planet Earth. The human population continues to grow, as does our resource use (McKee et al. 2003). Truly wild areas are rare and the human footprint is pervasive (Sanderson et al. 2002). Many species have been driven to extinction by human actions (e.g. Steadman 1995), and large percentages of well-known taxa are threatened and endangered (Pereira et al. 2010). The extinction of small and less obvious taxa, such as the invertebrates, is greater than the loss of vertebrates, and the passing of these 'little things that run the world' (Wilson 1987, p. 344) is generally unnoticed (Dunn 2005; Regnier et al. 2009). Unfortunately, even when a species' endangerment is recognized and recovery efforts are planned, political posturing and misguided prioritization of finances allow taxa to slip into oblivion (Male and Bean 2005). In light of all this bad news, what should we do?

As citizens of the planet, we should change our individual behaviour so that we live sustainably and insist that our policy-makers implement strategies to protect and restore biodiversity and functional ecosystems. As scientists, however, our role is to police the quality of science-based decision making through peer review, and to provide unbiased data and expert interpretation to policy-makers. Recently, animal behaviourists (a broad umbrella that includes behavioural ecologists, comparative psychologists, ethologists, and a variety of other researchers studying behaviour in particular taxa or contexts) have shown a strong interest in understanding the behavioural consequences of conservation management actions.

Starting nearly 15 years ago in a series of symposia and edited volumes, animal behaviourists began to formalize ways to approach conservation problems that had issues of animal behaviour at their core (Clemmons and Buchholz 1997; Caro 1998; Gosling and Sutherland 2000; Festa-Bianchet and Apollonio 2003; Linklater 2004). This nascent discipline, commonly known as 'conservation behaviour', is widely touted as holding great promise for helping to conserve animal diversity (Blumstein and Fernández-Juricic 2010; Buchholz 2007), but very little behavioural theory seems to be making its way into conservation practice. A decade into the life of conservation behaviour, Caro (2007, p. 395) concluded that: 'Ideas proposed by theory-driven academics working outside the conservation arena are generally not solutions to specific practical conservation problems, but notions about what might matter under a set of hypothetical conditions.' Along the same lines, Angeloni et al. (2008) contrasted the glaring disconnect between the behavioural and conservation literatures with the comparatively strong links that exist between the fields of ecology and conservation biology. They found that articles in the journal *Ecology* were nearly 25 times more likely to mention conservation in their titles or abstracts than were publications in animal behaviour journals. Similarly, *Ecology* authors cited conservation journal publications at a rate 14 times that of authors in animal behaviour journals. Finally, in a comparison of North American membership overlap in professional societies, ecologists were twice as likely as animal behaviourists to also be members of a conservation society. Angeloni et al. (2008, p. 736) suggest that one reason for the lack of overlap between the fields of animal behaviour and conservation biology might be that animal behaviourists feel that, 'they have little to contribute to conservation biology, or that the applied nature of the subject makes it less intellectually challenging and objective'.

Here we explore how animal behaviourists need to change to allow the theoretical promise of conservation behaviour to become an applied reality that benefits biodiversity. We build on a proposal by Angeloni et al. (2008, p. 736) that: 'conservation would be advanced if behavioural ecologists chose to study animals that occur in environments with human disturbance'. Indeed, there already exists an extensive wildlife management literature on how human presence changes the behaviour of wildlife, although little of this research appears to have been done by classically trained animal behaviourists. Older reviews of this wildlife disturbance literature have suggested fundamental problems with the design and analysis of those studies. We propose that research into anthropogenic disturbance of wildlife behaviour is an area with immediate management implications that would benefit greatly from greater involvement by theoretical animal behaviourists.

The specific goals of our chapter are to: (1) describe the problem of behavioural disturbance of wildlife by humans, (2) review how the new field of conservation behaviour might be able to help in a way that has 'practicality in application' (Caro 2007, p. 395), (3) evaluate the completeness of the methodology and statistical approaches taken by recent investigations of wildlife disturbance, and (4) determine if the quality of the science in these published studies is adequate for wildlife management and conservation policy making.

17.2 Anthropogenic behavioural disturbance of wildlife

Humans often disturb wildlife, even when that is not our intention. Natural resource managers have suspected for some time that recreational activities of visitors to national parks and nature refuges disrupt the activities of wild animals. We use the moniker 'ecotourist' loosely in reference to any non-consumptive human visitation to a mostly uninhabited (by humans, that is) natural area. In the United States alone, the U.S. Fish and Wildlife Service (2010) reports that almost 41 million people visit national wildlife refuges each year. We feel it is safe to say that the vast majority of ecotourists have no intention of harming the natural places that they visit, but they do. In a review of avian, mammalian, and herpetofaunal reaction to a variety of forms of recreational use, Boyle and Samson (1985) found that negative impacts on wildlife were 2.5 times more common than harmless and inconclusive outcomes (n = 189 studies). Some combinations of taxon and disturbance type had only negative outcomes (bat disturbance by spelunkers), while others had equivocal results (mammalian responses to wildlife observers). In only one of 16 taxon-recreation type combinations did harmless/inconclusive outcomes outnumber negative findings (avian reactions to rock-climbing). The review authors pointed out that the negative responses of wildlife to the activities of humans did not necessarily indicate a threat to population persistence or a loss of biodiversity. Despite a plethora of studies in the 25 years since Boyle and Samson's review, there appears to be no standard definition of, or approach to, how anthropogenic disturbance should be studied.

We focus on how to study whether the behavioural changes in wildlife caused by humans are harmful to population viability. Of course, non-consumptive human disturbance of wildlife is not just behavioural; recreationists also alter habitat and inadvertently cause direct mortality. For example, off-road vehicles not only cause a noisy disruption to wildlife, they also alter substrates, which can impair plant growth, and crush a variety of small animal life (Grant and Doherty 2009). Nevertheless, obviously disruptive disturbances that may upset human nature-lovers are not necessarily bad for particular wild animal species. For example, loud off-road vehicles may scare away wildlife in the short term, but the longer term effects on population viability could theoretically be positive; the trails cleared by mechanized vehicles may enable some wildlife species to suffer lower travel costs between foraging patches, or provide unencumbered escape routes to prey species when they are attacked by predators. Similarly just because animals change their behaviour in response to human presence does not mean they are harmed by it. Ecologists may document habitat destruction and changes in species richness due to human activities, but the behavioural influences of disturbance will be difficult to quantify and weigh without training in behavioural biology. Sih et al. (2011, p. 367) show

how sensory and cognitive approaches are needed to understand the short- and long-term consequences of the multiple stressors caused by 'human-induced rapid environmental change'. We do not mean to suggest that animal behaviourists have all the answers, rather we share the belief that behavioural ecologists should be a part of an integrative, multi-disciplinary conservation team (Arcese et al. 1997).

17.3 Is behavioural change bad?

Animals are forced to change their behaviour all the time. Animal behaviourists generally operate under the assumption that animals are genetically adapted to optimize the trade-offs between actions that are costly to individual fitness and actions that benefit lifetime reproductive success. That is, individuals should attempt to minimize costs while maximizing benefits. Prey species, for example, sporadically interrupt their foraging actions to lift their heads and scan for predators. This vigilance comes at the opportunity cost of decreased foraging; animals cannot do both at the same time. Thus, a prey animal may consider whether any predator cues are present, how many others in their group are also vigilant, and their internal caloric needs, among other factors, when deciding how often to scan for predators. When ecotourists invade wildlife reserves, their sudden, novel appearance may cause animals to behave maladaptively in what has effectively become a new environment (see also Chapter 11). Behaviour is maladaptive if those actions reduce the lifetime reproductive success of individuals. The salient concern is learning how to identify those anthropogenic changes and wildlife responses that warrant management intervention.

To do this, we must uncover the circumstances in which the intensity of human disturbance is untenably additive to natural disruptors, and tease out the forms of disturbance that act as super-normal stimuli. Chronic disturbance may lead to the accumulation of an allostatic load (McEwen 2000) wherein repeated, stressful deviations from homeostasis result in poor health. We must also learn to predict where existing animal adaptations to disturbance are maladaptive in the face of novel, but ostensibly harmless change. To some degree, we expect animals to adapt readily, both proximately and ultimately, to momentary, minor, or incremental environmental change. They should learn to ignore harmless change (e.g. a gust of wind rustling the leaves of a bush, see also Chapter 4), and evolve to optimize their response to potentially harmful change (e.g. whether to run away when evidence of a predator is perceived, even though the predator is, as of yet, not attacking; see Chapter 16). With respect to prey, McLean (1997, p. 144) states 'Even if a new predator shows up, the individual's general behaviour patterns incorporate at least some potential for coping with the new problem because of prior adaptation to a similar problem'. Indeed, Berger (2007) has demonstrated that in ungulate populations where predators were extirpated generations ago, seemingly absent predator-avoidance strategies rapidly reassert themselves when the historical predators are reintroduced. Therefore, if some forms of anthropogenic behavioural disturbance are similar in type and frequency of occurrence to previous challenges, our visits to natural areas may not impair population viability at all for some wildlife. However, if our disturbance is additive, wild animals may face trade-offs in survival or reproduction.

In Yellowstone National Park (USA), the reintroduced timber wolf *Canis lupus* is a constant threat to the survival of individual elk *Cervus elaphus*. Elk populations survive in the park despite attacks by this keen predator because the wolves' trophic strategy embodies trade-offs that allow elk to escape them often (Fortin et al. 2005). Because they lack fossil fuel technologies, wolf attacks on elk can never be as abundant in frequency, or as persistent in pursuit, as a human photographer on a snowmobile. If animal behaviourists consider wildlife disturbance questions intellectually uninteresting, they should look more closely at this conservation issue. Elk were abundant in Yellowstone despite snowmobile disturbance before wolves were reintroduced. Now that wolves have returned, elk population numbers have declined; but how much of this decline is due to wolf predation alone as opposed to the additive costs of frequent disturbance by humans and wolves? Likewise, although they may come in a few different pelage colors, their physical appearance and hunting strategy does not vary so much that elk ponder the intention of wolves. Anthropogenic disturbance, on the other hand, is often not consistent and predictable. Through rapid cultural and technological innovation, humans introduce an ever-changing set of novel stimuli (i.e. sights, sounds, and odours) to natural environments. These stimuli are accompanied by a focused attention on the animal that might mimic a predator's approach, but is without predatory intent. The unique sensory worlds of non-humans makes it extremely difficult for us to understand how they perceive the onslaught of signals that come from human disturbance. Huang et al. (2011), for example, demonstrated that the sound of a camera shutter induces predator-avoidance-like interruption to the display behaviour of anoles, while other harmless stimuli, including the camera's flash, have no significant effect. If wild animals are unable to acclimate to constant interruption and anthropogenic stimuli, these disturbances may prevent them from engaging in fitness-enhancing tasks, such as foraging, nursing young, and finding mates, presumably leading to population decline. Also if prey habituate to human disturbance, this may possibly make them more susceptible to their natural predators because they become less alert to predation threats.

17.4 What is conservation behaviour and how can it help?

Conservation behaviour uses a scientific understanding of animal behaviour both to determine the reasons for differential susceptibility to extinction among animal species, and to manage threatened animals and their habitats to promote population viability. Description of the natural history of an animal may be a component of conservation behaviour studies, but it is not its objective. Instead, conservation behaviour uses the framework of ethology (Tinbergen 1963), to allow conservation issues to be investigated from four complementary behavioural perspectives: mechanisms, ontogeny, adaptive function, and phylogeny. The application of Tinbergen's framework to conservation has been discussed in

Table 17.1 Tinbergen's 'four questions' provide an integrative framework for investigating conservation issues. Two behavioural studies of disturbance effects on different species of penguins (Walker et al. 2006; Ellenberg et al. 2009) are used to illustrate the complementary nature of proximate (how) and ultimate (why) questions about the cause and origin of disturbance-induced behaviours. In the case of penguin responses to colony visitation by humans, the data required to adequately answer the ultimate questions are lacking, which was the case in nearly all studies reviewed.

	CAUSE	ORIGIN
PROXIMATE	How do tourist visits alter short term penguin physiology?	How does prior experience with tourist disturbance affect subsequent stress responses to it?
	Experimental disturbance of yellow-eyed penguin females caused an average increase in heart rate from 78 beats/min to 184 beats/min. The average time it took females to recover a normal heart rate following a disturbance was almost twelve minutes (Ellenberg et al. 2009).	After an experimental 15 minute tourist-type visitation, magellanic penguins exposed to similar treatment over the previous five days had significantly lower plasma corticosterone levels than penguins visited only once (Walker et al. 2006).
ULTIMATE	Why are certain reactions to disturbance maladaptive now, and what makes them adaptive in the absence of humans?	Why do some penguin species share a negative reaction to tourism, while others do not?
	The answer to this question is currently not well investigated.	The answer to this question is currently not well investigated.
	'…How do effects of human disturbance compare with other challenges experienced during the animal's life' (Walker et al. 2006, p. 152)?	'Habituation appears to require a maximum of predictable low-level disturbance, although what 'low-level disturbance' actually is appears to depend on species, location and even individual' (Ellenberg et al. 2009, p. 295).

detail elsewhere (Buchholz 2007; Buchholz et al. 2008), but a brief introduction is warranted here.

Tinbergen's four questions about behaviour are divided into proximate and ultimate approaches. Proximate questions of behaviour consider how an individual is able to perform an activity. They ask about the mechanisms within an organism that make certain behaviors possible. The proximate causes of behaviour include the sensory and endocrine mechanisms that regulate behaviour (e.g. Chapters 2 and 3). However, we know that those mechanisms may be modified by individual experience, thus we must consider the proximate origins of behaviour as well (i.e. how learning and ontogeny modify behaviour; e.g. Chapter 4). In contrast, ultimate questions of behaviour address the reasons why animal species have evolved particular proximate systems. The ultimate cause of a behaviour is the means by which it helps the individual survive and reproduce (adaptive function). If we ponder the ultimate origin of that same behaviour pattern, we are examining its evolutionary history by comparing how that behaviour differs across a phylogeny of related species. We use the controversial issue of ecotourism at penguin nesting colonies to show (Table 17.1) how the framework of conservation behaviour can be used to understand which research questions necessary to developing an effective management strategy remain unanswered. This comprehensive approach holds promise for ameliorating conservation problems only if it can be translated into behavioural management actions that enable threatened populations to remain genetically and demographically viable. Research that entertains all four of Tinbergen's complementary questions provides a comprehensive understanding of the hows and whys of behavioural variation. The study of non-lethal wildlife disturbances must incorporate the Tinbergen framework if we are to fully understand the impact of human behaviour on biodiversity.

17.5 Recent literature in recreational disturbance of wildlife

We assessed the behavioural disturbance literature with four objectives: (1) to investigate the degree to which the framework of conservation behaviour had become incorporated into this literature, (2) to characterize the methodological approaches being used to study the anthropogenic changes in wildlife behaviour, (3) to assess whether the quality of the scientific research design and analysis is sufficient to provide valuable data to conservation policymakers, and (4) to decide whether the work allows for comparison across taxa and disturbances.

We searched the Biological Abstracts and Science Citation Index reference data bases for the keywords: disturbance, anthropogenic, human, and wildlife. We limited our search to 2005–2010 because this time span should cover literature that would reflect the innovations from the origination of conservation behaviour in the mid-1990s, as well as the critiques in the mid-2000s. Papers were reviewed and accepted for inclusion if they met four criteria. First, the study in question had to be a primary empirical source and not a review of other studies. Second, the study needed to be explicitly behavioural in nature. This behavioural requirement excluded population surveys, species richness studies, and species diversity research, regardless of whether human disturbance influenced these variables. Studies of behavioural mechanistic responses to disturbance, such as stress hormone levels or heart rate changes, were accepted into the study. Third, studies that investigated only mechanistic responses to disturbance were not included if differences in habitat, even if due to human activity, comprised the only method of concluding that study subjects had been subjected to disturbance. We excluded these studies because we thought it likely that habitat alteration would have direct nutritional effects on individual condition that were not the focus of our review of how the behavioural disturbance alone by humans causes changes in animal behaviour. In total, 91 papers were identified by the data base searches. After careful review, only 33 of these papers met our criteria for inclusion in our assessment of the science used to study anthropogenic disturbance of wildlife behaviour.

17.5.1 Conservation behaviour and the wildlife disturbance literature

Is the recent wildlife disturbance literature taking advantage of integrative approaches to these conservation problems, as proposed by conservation behaviour's framework? We found that 5 studies used mechanistic approaches, 9 studies examined ontogenetic questions, and all 33 studies were interested in the functional response of the animal to human disturbance. Eleven studies simultaneously used two approaches to investigate the problem, while only three studies used three of Tinbergen's methods. There was a decrease in the proportion of studies that relied on only one approach over time, a positive sign that conservation behaviour's framework is making inroads with the wildlife disturbance research community. Categorized by the first year of reported fieldwork, 83% of studies conducted from 1996–1999 used only a single approach. That percentage dropped to 66% for years 2000–2003, and was even lower (63%) for 2004–2007. Of course, using multiple measures of disturbance impacts is not always feasible. Tarlow and Blumstein (2007) reviewed some of the methodological and fiscal constraints of common measures of anthropogenic stressors in wild animals.

Although some studies contrasted responses between species and suggested explanations based on differing natural histories (e.g. Geist et al. 2005), no empirical study used a comparative, phylogenetic approach to understand which selective factors predispose species to loss of population viability in response to non-lethal human behavioural disturbance. Nevertheless, the literature demonstrates that some authors have begun to investigate evolutionary patterns of fearfulness. Blumstein (2006) compiled studies of 'flightiness' in birds, and concluded that larger species, and those that are carnivorous or social, may be more sensitive to human disturbance. These results suggest that more investigators are attempting an integrated approach to solving wildlife disturbance problems, but that the literature still lacks the comprehensive use of Tinbergen's framework promoted by the discipline of conservation behaviour.

17.5.2 Methodological problems in the wildlife disturbance literature

Are the proper methodological approaches being used to design studies of wildlife disturbance? Gutzwiller (1991) outlines a set of statistical and biological considerations necessary to design an effective study of human disturbance of wild animals. In our detailed review of the recent disturbance

Table 17.2 Recently published studies of behavioural disturbance of wildlife by humans generally lack complete methodological information, suffer from poor statistical design and do not consider likely covariates. The studies included in the review were: Bisson et al. (2009), Bouton et al. (2005), Broseth and Pedersen (2010), Crook et al. (2009), Dooley et al. (2010), Ellenberg et al. (2006, 2009), Geist et al. (2005), Goudie (2006), Guillemain et al. (2007), Heyman et al. (2010), Holm and Laursen (2009), Hughes et al. (2008), Jayakody et al. (2008), Lafferty et al. (2006), Larsen and Laubek (2005), Madsen et al. (2009), Martinez-Abrain et al. (2008), Moore and Seigel (2006), Morse et al. (2006), Naylor et al. (2009), Neumann et al. (2010), Rumble et al. (2005), Sastre et al. (2009), Tarr et al. (2010), Thiel et al. (2008), van Polanen Petel et al. (2007), Vidya and Thuppil (2010), Walker et al. (2006), Weston and Elgar (2005), Wolf and Croft (2010), Wrege et al. (2010), and Zuberogoitia et al. (2008).

a. Basic Methodological description

	Reported	Not reported
Year & Dates	88%	12%
Study Duration	85%	15%
Observation Period	61%	39%

b. Statistical Design

Scientific approach	Observation only 42%	Experimental only 52%	Both 06%
Data independence	Individually marked 46%	Not marked 54%	

c. Potential covariates

	Reported	Not reported			
Territory size	06%	94%			
Age	06%	94%			
Distance to cover	09%	91%			
	Female	Male	Both/Unclear	Adult with Juvenile	
Sex	03%	06%	82%	09%	
	None	Incomplete	Partial	Mostly complete	Complete
History Disturbance	04%	30%	06%	42%	18%
Habitat type	13%	15%	12%	27%	33%

literature, we were shocked to find the state of the science to be quite poor. Many papers lacked the basic methodological information necessary to interpret the results and replicate the study. For example some studies never mentioned when the work was performed, the duration of the research, or how long animals were observed (Table 17.2a). Methodological parameters are chosen with no apparent justification. For example, of those that did report study length, this parameter differed by several orders of magnitude ($\bar{x}$ = 154 days, range 2–1460) without any justification. We thought the length of study might be related to the generation time of the study organism, but this was not the case (Fig. 17.1). The observation periods allocated to individual animals varied even more extremely ($\bar{x}$ = 4,193 min, 5–32,820) than study length, also with little biological justification.

Many statistical tests assume that individual data points reflect independent observations. This assumption is widely violated in the wildlife disturbance literature. Fewer than half of recent studies collected data from known individuals (artificially or naturally marked; Table 17.2b), suggesting that most of these studies did not meet assumptions of data independence required by the statistical tests. Powerful statistical analyses that utilize repeated measures are not possible without known individuals.

Although Gutzwiller (1991) described how the most efficient and expedient means of detecting negative impacts of human disturbance is through an experimental approach wherein the investiga-

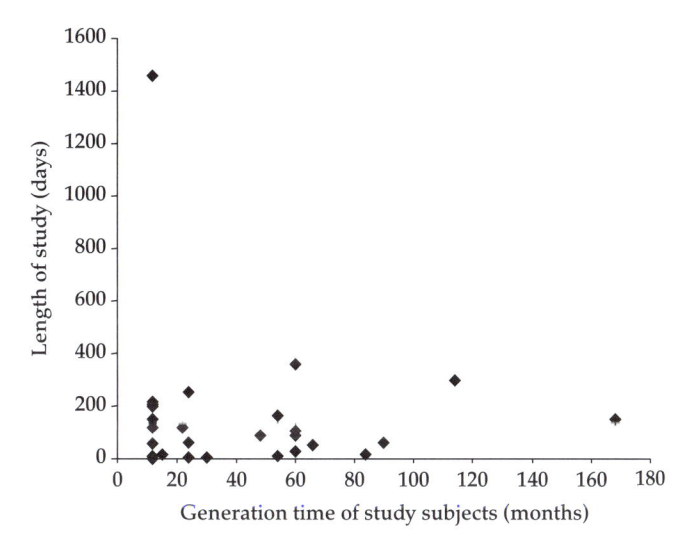

Figure 17.1 The duration of field work is poorly justified in the wildlife disturbance literature. Hypothetically, longer-lived species may not exhibit the full fitness costs of a disturbance until months or years after it occurs, but our review showed no relationship between age of maturity and study lengths.

tors control the type and timing of disturbance, a large proportion of recent studies still rely on purely observational data (Table 17.2b). Nevertheless, experimental manipulation of disturbance variables is more common in the literature than when he first expressed the need for this research to rely less on observational (correlational) data sets.

Similarly, a statistically well designed study should consider how covariates and interaction effects determine our ability to identify whether disturbance is threatening the well being of wildlife populations. We found that few studies considered standard covariates such as age, sex, territory size and habitat type (Table 17.2c). Rarely did investigators consider the options that animals had when responding to disruption. For example, documentation of the distance to cover for disturbed animals was rare (Table 17.2c) as was consideration of group size, but surely these covariates determine the degree of risk that individual animals perceive when encountering a threatening disruption by humans?

Obviously, the history of disturbance must be considered to understand whether individuals have had time to learn to not fear humans, or whether populations have had time to adapt genetically to become less fearful over generations.

In our examination of covariates, we quantified how thorough a study was in its reporting on the history of disturbance to the study populations (or to specific individuals). A complete history was defined as one having an approximate number of seasonal or yearly human visits to the study area, or a record of the number of years since routine visitation, causing sizable disturbance, began. A minority of studies provided a complete descriptive history of disturbance (Table 17.2c). At most, studies with incomplete disturbance histories gave a short list of some disturbances that could possibly be occurring in the area. We found that those that were classified as having incomplete, partial, or non-existent information on the local history of human disturbance, were in sum nearly as numerous as those that had mostly complete information. Studies that incorporate a variety of possible interactions among variables generate valuable insight. For example, Stankowich (2008) conducted a meta-analysis of ungulate flight responses to human disturbance from which he was able to conclude that subjects in open habitats, and those with young offspring, perceived greater risk from humans. The lack of analyses of covariation and interaction in the literature we reviewed does not bode well for the usefulness of these studies for

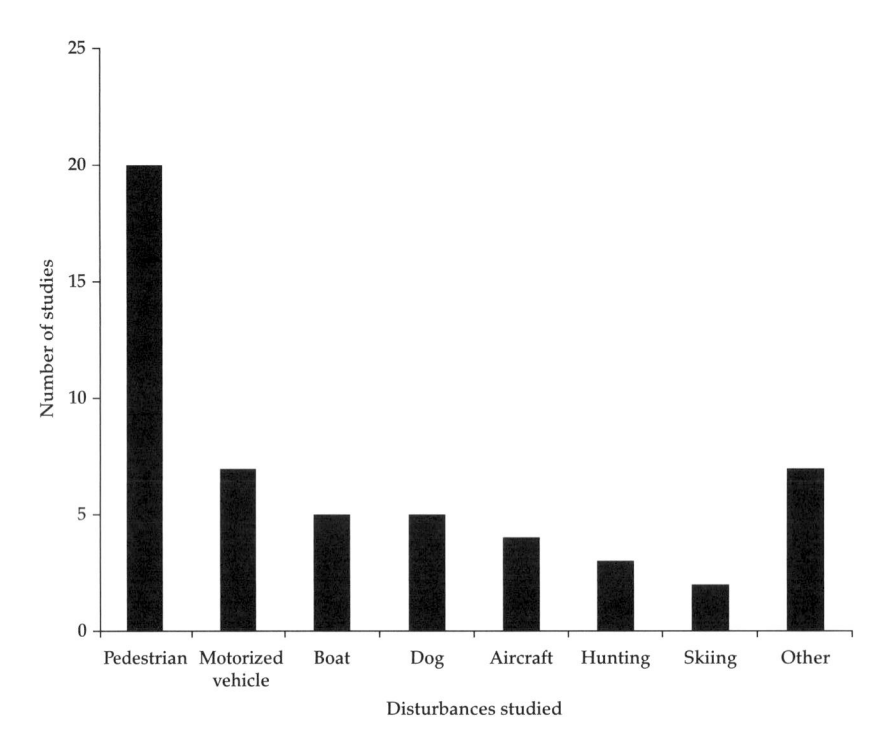

Figure 17.2 Most behavioural disturbance studies assess the impact of pedestrians (the 'pedestrian' category includes humans walking, running, or jogging) on wildlife. The 'other' category includes disturbances that were usually seen in one study only (e.g. predator decoys, scuba-divers, dynamite blasts, etc.).

understanding our non-consumptive effects on wildlife.

Gutzwiller (1991, p. 253) was concerned that 'subtle stimuli' were being ignored by researchers. In our review of the literature, we found this to be true; and even more concerning was that the basic quantification parameters for disturbance stimuli were often not recorded nor analysed for their effects. The specific character of disturbances is important to quantify if valuable data are to be gathered. For example, the passing of two different aircraft over a colony of breeding birds will likely not have the same effects. The variables of height, speed, and sound will interact in different ways, potentially having very different effects on behaviour.

The majority of disturbance agents included humans on foot or skis (Fig. 17.2), but most did not consider the speed, noise, appearance, or orientation of the disturbance relative to the animals (Table 17.3). Human disturbance is multimodal, and yet, less than half of the publications considered multiple disruption agents. Surprisingly, 14 of the 33 publications gave so little information on disturbances or study periods, that the number (count) of daily disturbance events subjects were exposed to could not be inferred. Where disturbance rate was given or calculable, the number per day ranged widely ($\bar{x}$ = 17, range = 0.5–65). Only half the studies considered the temporal patterning of disturbances, and more than three-quarters did not mention the amplitude (e.g. decibels, km/h) of disturbance stimuli or the individual or cumulative duration of the disturbances (Table 17.3). Mean disturbance duration in those studies that did provide the information was 1.54 hours (0.02–4.1). Approximately one-third of the studies considered the spatial range of the disturbance stimulus and thus how pervasive the disturbance effects might be on a geographical scale.

Just as investigators performed poorly at fully describing the multivariate nature of the disturbances themselves, they failed to measure the

Table 17.3 Most studies did not consider the multiple ways that disturbances may be experienced by wildlife, nor the multiple ways that wildlife behavioural responses to disturbances may vary and interact.

	Disturbance		Response	
	Measured	Not measured	Measured	Not measured
Count	58%	42%	70%	30%
Duration	18%	82%	52%	48%
Amplitude	21%	79%	67%	33%
Directionality	25%	75%	09%	91%
Spatial scale	39%	61%	39%	61%
Temporal Pattern	46%	54%	42%	58%
Interaction Effects	27%	73%	27%	73%

breadth of the types of responses of the wildlife to the human disturbance. The most common measurements of response were to simply count how many animals reacted and to measure the amplitude of their response (Table 17.3). Duration of response to disturbance was measured in one-half of the studies, but the spatial scale of response was not indicated in nearly two-thirds of the investigations, and thus we infer that the actual duration of response of individuals was probably not measured accurately in many cases. Just as the latency or duration of the animal's response to disturbance might be crucial to elucidating the fitness consequences of exposure to human activities, the spatial scale of the animal's reaction is biologically important. If researchers note that animals ran, flew, or swam away, the cost of locomotion (and thus the possible trade-offs) cannot be estimated without understanding how far they fled. Likewise, we need to know whether the reactions of individual study subjects to disturbance are immediate, or whether delayed responses occur. Given that more than one-third of studies reporting the dates of fieldwork spanned a year or less (including the off-season), it would appear current research is not assessing delayed fitness consequences. Even when investigating the response duration of marked individuals, it is not clear how long animals should be observed after disturbance to search for lags in response.

The vast majority of studies ignored the directionality of the animal's response to disturbance. It may seem obvious, and thus not worthy of analysis, that animals should exhibit negative taxis to threatening stimuli. This assumption may not hold for many reasons: the habitat type and physical features of the landscape; the territoriality, group demographics, and social dynamics of the species; the modes of locomotion available to the species; and the health and body condition of individuals will no doubt play a role in where they move relative to the disturbance. Also if human disturbance is an annoyance rather than a threat to survival, wildlife do not necessarily need to maintain the greatest distance between themselves and humans, and non-perpendicular movement might be more likely. Even time of day may play an important role in determining if individual animals feel the need to flee in a particular direction. Unfortunately, less than one-half of the studies considered how the responses of wildlife varied with time of day, and none considered both directionality and temporal variation. Obviously a statistical design that considers the multiple, interacting dimensions of wildlife responses would be most effective at revealing how managers must manage human disturbances, and yet only approximately one-quarter of the studies we reviewed considered interaction effects (Table 17.3). Technological methods for automated data collection and statistical approaches to interpreting the data are constantly evolving (e.g. Preisler et al. 2006), but these new methods are of no value if the fundamental measures of disturbance and disturbance response are flawed or absent, as in so many of the studies we examined. Overall we must conclude that despite Gutzwiller's (1991) warnings two decades ago, the wildlife disturbance literature today still fails to aspire to the rigorous and thorough methodological and statistical principles needed to study animal behaviour properly.

17.5.3 Wildlife disturbance science and conserving biodiversity

Our review of the recent wildlife disturbance literature raises great concerns about the scientific quality of this work as the basis for making policy decisions. We also found more general concerns. Despite the greater diversity in tropical ecosystems, 64% of the recent wildlife disturbance research we evaluated was done in non-tropical Europe and North America; 76% of the studies were done in highly developed nations (Europe, North America, New Zealand, Australia), despite the importance of ecotourism to conservation and economic sustainability in developing countries. Likewise, most studies of wildlife disturbance focused not on threatened or endangered species but on species categorized by the IUCN as of 'least concern' (64%). Most studies were of birds, with no studies considering the well being of invertebrate animals (Fig. 17.3). Although conservation biology is described as a 'crisis discipline' in which rapidity is needed to prevent extinction, we found that the results of wildlife disturbance studies were delayed by slow publication. On average, there is over a one-half-decade delay from fieldwork to publication ($\bar{x}$ = 5.68 years, range 2–14). Most of the disturbance studies are published in conservation and wildlife management literature (60%), with intermediate numbers in taxon-oriented journals (15%; mostly the secondary ornithological literature), a few in behavioural journals (6%), and the remaining 19% in a variety of habitat- or ecosystem-oriented publications. The typical venue for publication of wildlife disturbance research raises the question of whether this work is evaluated properly by reviewers with experience and training in behavioural biology.

If we were to assume, despite the evidence reviewed here, that these behavioural studies were conducted correctly, we would still find the literature to be of little value to conservation. We discovered that few investigators have linked behavioural change to the currency of conservation managers: fitness measurements. Most research quantified the behavioural response of wildlife to disturbance in terms of indirect measures of fitness (58%), for example, as changes in locomotion. The simple displacement of individual animals may be inconvenient to them, but it does not mean that their survival and reproduction have been impinged upon. Indeed, even if fitness is reduced for a few individuals, this does not necessarily mean that population viability has been harmed. Nevertheless, despite mostly considering only indirect measures of fitness, 73% of the studies still concluded that there were negative effects of human disturbance on wildlife. One fifth (21%) of the studies claiming negative effects were concerned enough about the degree of effect to claim that these wildlife responses may threaten the long-term viability of populations. Of those studies 80% made explicit management recommendations to reduce the negative

Class representation in disturbance studies

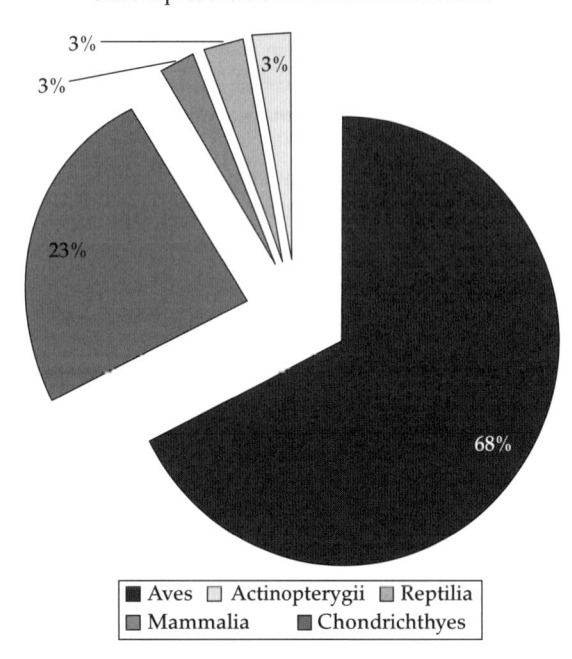

- 68% Aves
- 23% Mammalia
- 3% Actinopterygii
- 3% Reptilia
- 3% Chondrichthyes

Figure 17.3 The overwhelming majority of recent behavioural studies were conducted on birds. Certain taxa may be more readily available or easier to work with, but focusing behavioural research on those undermines our ability to generalize broadly. Effective ecosystem management must base its strategies on multi-taxon studies.

impact of human disturbance. Interestingly the sole study that claimed there might be negative consequences of disturbance on population viability without making management recommendations was the only one of the five that did not use an indirect fitness measure. Clearly, there is indecision by researchers as to which negative responses of wildlife to human disturbance should be acted upon by policy-makers.

17.6 Conclusions

A common political strategy used to justify setting aside protected areas, particularly in under-developed nations, is that they will generate sufficient revenues through visitation to support both park management and the fiscal well being of surrounding communities. Park rules (e.g. do not touch the animals, do not litter, do not leave the trail, limits on visitor group size, etc.) are often put in place to control the obvious negative impacts of throngs of ecotourists descending on natural areas. But because previous studies were not well planned and executed, we have little idea if the mere presence of a single human creates an untenable atmosphere for some animal species, let alone the arrival of brightly coloured buses filled with screaming school children, the constant drone of recreational watercraft, or the besiegement of a lion pride by safari trucks, that characterize the parks that are invaded by nature-hungry humans each day. It will be sadly ironic if ecotourism and other non-consumptive uses of wildlife preserves cause parks to become green deserts, where a natural vegetation structure exists but the intolerant, endemic, and charismatic animals are gone. At issue here is whether the adaptations that animals use for dealing with disturbance may protect them from extinction on a human-dominated world, or predispose them to poorer survival and reproduction.

We reviewed the literature on anthropogenic disturbance of wildlife because we wondered: if animal behaviourists are disinterested by conservation research, what is the scientific quality of the 3–5% of the conservation literature that employs behavioural study (Angeloni et al. 2008)? Also, if animal behaviourists are not participating in conservation-

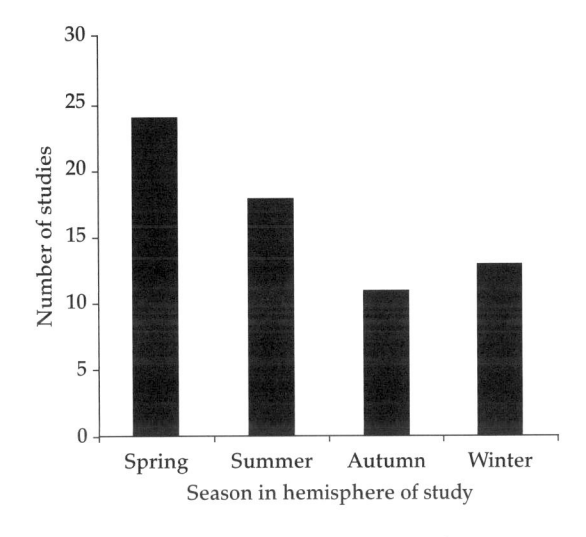

Figure 17.4 Because species have distinct environmental pressures during different times of the year (e.g. food availability in winter, intrasexual competition in spring), it is important to investigate how disturbance affects wildlife differently in different seasons. Multiple studies spanned all four seasons, but the obvious focus of most research was on spring and summer. Without a seasonally balanced reservoir of data, conservation and wildlife management plans may overlook or underestimate problems caused by disturbance.

based professional societies, is the behavioural research published in conservation journals valid and valuable for conservation decision-making? Sadly we must conclude that, generally, this literature is of poor quality and is not suitable for guiding policy decisions. In their literature review of 25 years ago, Boyle and Samson (1985, p. 113) found that the studies were unable to answer questions critical to population viability, namely 'does disturbance cause measurable changes in population fecundity or mortality rates?' and 'is recreation-caused mortality partly or wholly compensatory, permitting the loss of an annual "non-consumptive surplus"?' Unfortunately, not much has improved since then. The scientific literature on anthropogenic behavioural impacts on wildlife is largely based on studies that are anecdotal, involve haphazard or brief data collection, are purely observational and suffer from data independence problems, and/or are not structured to recover data of relevance to population management. Very basic considerations of research design, such as seasonality, are not well balanced in the available literature

(Fig. 17.4). Negative reactions of wildlife to human activities are self-evident, but qualitative assessments are not good science. Without the inclusion of animal behaviourists, who are already trained in both experimental design and proper quantification of stimuli and animal responses to them, it would seem that the wildlife disturbance literature will continue to be of limited value for conservation decision-making.

The question of how animals will respond to human disturbance is multi-faceted. We must consider how the animal perceives the disturbance scenario. Do the animals simply treat humans as curious denizens of the habitat, or do the signals we provide tap into existing adaptive strategies that may result in decreased survival and reproductive success? The answers depend on knowing both the evolutionary history of the species, the antipredatory adaptations of the population under study, the individual experiences, tolerances and learning capacities of animals, and their sensory limitations.

It is our wish that our evaluation of the anthropogenic behavioural disturbance of wildlife issue, will show animal behaviourists with interests in conservation that their assistance is desperately needed, and that the challenges are both intellectual and practical. If those are not good enough reasons alone, the behavioural ecologist should consider that the profession of studying the behaviour of animals in the wild is itself endangered by the pervasive threats to biodiversity (Caro and Sherman 2011). We started our chapter with a quote from pioneering American conservationist Aldo Leopold. Leopold remarks on the painful insight that comes from knowing how destructive human influences can be to Nature. The quoted passage continues 'An ecologist must either harden his shell and make believe that the consequences of science are none of his business, or he must be the doctor who sees the marks of death in a community that believes itself well and does not want to be told otherwise' (Leopold 1978, p.197). In this book on global change, we ask that behaviourists change their ways. Animal behaviourists must reach out to those responsible for monitoring and managing human disturbances of wildlife, and offer their services. Properly applied, behavioural science has the capacity to prevent, ameliorate, and remedy human impacts that harm biodiversity. This topic provides opportunities for powerful collaborations between behavioural ecologists, wildlife biologists, and policy-makers. Our review demonstrates that there is much room for improvement in wildlife disturbance science, and that this improvement can be accomplished through the skillful participation of animal behaviourists.

Acknowledgements

We thank the editors and reviewer Daniel Blumstein for helpful comments that improved this chapter. We dedicate this chapter to animal behaviourist and educator Professor H. Jane Brockmann, on the occasion of her retirement; without her assistance and encouragement to the senior author many years ago, it is unlikely that either us would have been asked to present this chapter on conservation behaviour now.

References

Angeloni, L., Schlaepfer, M.A., Lawler, J.J., and Crooks, K.R. (2008). A reassessment of the interface between conservation and behaviour. *Animal Behaviour*, 75, 731–7.

Arcese, P., Keller, L.F., and Cary, J.R. (1997). Why hire a behaviorist into a conservation or management team? Pages 48–71, in Clemmons, J. R. and Buchholz, R. (eds). *Behavioral Approaches to Conservation in the Wild*. Cambridge, Cambridge University Press.

Berger, J. (2007). Carnivore repatriation and Holarctic prey: narrowing the deficit in ecological effectiveness. *Conservation Biology*, 21, 1105–16.

Bisson, I., Butler, L.K., Hayden, TJ., Romero, L.M., and Wikelski, M.C. (2009). No energetic cost of anthropogenic disturbance in a songbird. *Proceedings of the Royal Society of London B, Biological Sciences*, 276, 961–9.

Blumstein, D.T. (2006) Developing an evolutionary ecology of fear: how life history and natural history traits affect disturbance tolerance in birds. *Animal Behaviour*, 71, 389–99.

Blumstein, D.T. and Fernández-Juricic, E. (2010). *A Primer of Conservation Behavior*., Sunderland, Massachusetts, Sinauer Associates.

Bouton, S., Frederick, P.C., Rocha, C.D., Dos Santos, A.T.B., and Bouton, T.C. (2005). Effects of tourist disturbance on wood stork nesting success and breeding behavior in the Brazilian Pantanal. *Waterbirds*, 28, 487–97.

Boyle, S.A. and Samson, F.B. (1985). Effects of nonconsumptive recreation on wildlife: A review. *Wildlife Society Bulletin*, 13, 110–16.

Broseth, H. and Pedersen, H.C. (2010). Disturbance effects of hunting activity in a willow ptarmigan *Lagopus lagopus* population. *Wildlife Biology*, 16, 241–8.

Buchholz, R. (2007). Behavioural biology: an effective and relevant conservation tool. *Trends in Ecology and Evolution*, 22, 401–7.

Buchholz, R., Yamnik, P., Pulaski, C., and Campbell, C. (2008). Conservation and Behaviour, In: *Encyclopedia of Life Sciences*. John Wiley & Sons, Ltd: Chichester. http://www.els.net/ [DOI: 10.1002/9780470015902.a0021217]

Caro, T. (1998). *Behavioural Ecology and Conservation Biology*, New York: Oxford University Press.

Caro, T. (2007). Behaviour and conservation: a bridge too far? *Trends Ecology and Evolution*, 22, 394–400.

Caro, T. and Sherman, P. (2011). Endangered species and a threatened discipline: behavioural ecology. *Trends Ecology and Evolution*, 26, 111–18.

Clemmons, J. R. and Buchholz, R. (1997). *Behavioral Approaches to Conservation in the Wild*. Cambridge, Cambridge University Press.

Crook, S.L., Conway, W.C., Mason, C. D., and Kraai, K.J. (2009). Winter time-activity budgets of diving ducks on Eastern Texas reservoirs. *Waterbirds*, 32, 548–58.

Dooley, J.L., Sanders, T.A., and Doherty, P.F. (2010). Mallard response to experimental walk-in and shooting disturbance. *Journal of Wildlife Management*, 74, 1815–24.

Dunn, R. R. (2005). Modern insect extinctions, the neglected majority. *Conservation Biology*, 19, 1030–6.

Ellenberg, U., Mattern, T., and Seddon, P.J. (2009). Habituation potential of yellow-eyed penguins depends on sex, character and previous experience with humans. *Animal Behaviour*, 77, 289–96.

Ellenberg, U., Mattern, T., Seddon, P.J., and Jorquera, G.L. (2006). Physiological and reproductive consequences of human disturbance in Humboldt penguins: The need for species-specific visitor management. *Biological Conservation*, 133, 95–106.

Festa-Bianchet, M. and Apollonio, M. (2003). *Animal Behavior and Wildlife Conservation*. Washington, D.C., Island Press.

Fortin, D.L. Beyer, H.L., Boyce, M.S., Smith, D.W., Duchesne, T., and Mao, J.S. (2005). Wolves influence elk movements: behavior shapes a trophic cascade in Yellowstone National Park. *Ecology*, 86, 1320–30.

Geist, C., Liao, J., and Libby, S. (2005). Does intruder group size and orientation affect flight initiation distance in birds? *Animal Biodiversity and Conservation*, 28, 69–73.

Gosling, L. M. and Sutherland, W.J. (2000). *Behaviour and Conservation*. Cambridge University Press, Cambridge.

Goudie, R. Ian. (2006). Multivariate behavioural response of harlequin ducks to aircraft disturbance in Labrador. *Environmental Conservation*, 33, 28–35.

Grant, T. and Doherty, Jr., P.F. (2009). Potential mortality effects of off-highway vehicles on the Flat-tailed Horned Lizard (Phrynosoma mcallii): A manipulative experiment. *Environmental Management*, 43, 508–513.

Guillemain, M., Blanc, R., Lucas, C., and Lepley, M. (2007). Ecotourism disturbance to wildfowl in protected areas: historical, empirical and experimental approaches in the Camargue, Southern France. *Biodiversity and Conservation*, 16, 3633–51.

Gutzwiller, K.J. (1991). Assessing recreational impacts on wildlife: the value and design of experiments. *Transactions of the North American Wildlife and Natural Resources Conference*, 56, 233–7.

Heyman, W.D., Carr, L.M., and Lobel, P.S. (2010). Diver ecotourism and disturbance to reef fish spawning aggregations: It is better to be disturbed than to be dead. *Marine Ecology Progress Series*, 419, 201–10.

Holm, T.E. and Laursen, K. (2009). Experimental disturbance by walkers affects behaviour and territory density of nesting Black-tailed Godwit *Limosa limosa*. *Ibis*, 151, 77–87.

Huang, B., Lubarksy, K., Teng, T., and Blumstein, D.T. (2011). Take only pictures, leave only… fear? The effects of photography on the West Indian anole, *Anolis cristatellus*. *Current Zoology*, 57, 77–82.

Hughes, K.A., Waluda, C. M., Stone, R.E., Ridout, M.S., and Shears, J.R. (2008). Short-term responses of king penguins *Aptenodytes patagonicus* to helicopter disturbance at South Georgia. *Polar Biology*, 31, 1521–30.

Jayakody, S., Sibbald, A.M., Gordon, I.J., and Lambin, X. 2008. Red deer *Cervus elaphus* vigilance behaviour differs with habitat and type of human disturbance. *Wildlife Biology*, 14, 81–91.

Lafferty, K.D., Goodman, D., and Sandoval, C.P. (2006). Restoration of breeding by snowy plovers following protection from disturbance. *Biodiversity and Conservation*, 15, 2217–30.

Larsen, J. K. and Laubek, B. (2005). Disturbance effects of high-speed ferries on wintering sea ducks. *Wildfowl*, 55, 99–116.

Leopold, A. (1970). *A Sand County Almanac. With Essays On Conservation From Round River*. New York, Ballantine Books.

Linklater, W.L. (2004). Wanted for conservation research: behavioural ecologists with a broader perspective. *BioScience*, 54, 352–60.

Madsen, J., Tombre, I.E., and Nina, E. (2009). Effects of disturbance on geese in Svalbard: implications for regulating increasing tourism. *Polar Research*, 28, 376–89.

Male, T.D. and Bean, M.J. (2005). Measuring progress in US endangered species conservation. *Ecology Letters*, 8, 986–92.

Martinez-Abrain, A., Oro, D., Conesa, D., and Jimenez, J. (2008). Compromise between seabird enjoyment and disturbance: the role of observed and observers. *Environmental Conservation*, 35, 104–8.

McEwen, B.S. (2000). Allostasis and allostatic load: Implications for neuropsychopharmacology. *Neuropsychopharmacology*, 22, 108–24.

McKee, J.K., Sciulli, P.W., Fooce, C.D., and Waite, T.A. (2003). Forecasting global biodiversity threats associated with human population growth. *Biological Conservation*, 115, 161–4.

McLean, I.G. (1997). Conservation and the ontogeny of behavior. In Clemmons, J. R. and Buchholz, R. (eds) *Behavioral Approaches to Conservation in the Wild*, pp. 132–56. Cambridge, Cambridge University Press.

Moore, M. J. and Seigel, R. A. (2006). No place to nest or bask: Effects of human disturbance on the nesting and basking habits of yellow-blotched map turtles (*Graptemys flauimaculata*). *Biological Conservation*, 130, 386–93.

Morse, J.A., Powell, A.N., and Tetreau, M.D. (2006). Productivity of Black Oystercatchers: Effects of recreational disturbance in a national park. *Condor*, 108, 623–33.

Naylor, L.M., Wisdom, M.J., and Anthony, R.G. (2009). Behavioral responses of North American elk to recreational activity. *Journal of Wildlife Management*, 73, 328–38.

Neumann, W., Ericsson, G., and Dettki, H. (2010). Does off-trail backcountry skiing disturb moose? *European Journal of Wildlife Research*, 56, 513–18.

Pereira, H.M, Leadley, P.W., Proença, V., Alkemade, R., Scharlemann, J.P.W., Fernandez-Manjarrés, J.F., et al. (2010) Scenarios for global biodiversity in the 21st Century. *Science*, 330, 1496–501.

Preisler, H.K., Ager, A.A., and Wisdom, M.J. (2006). Statistical methods for analyzing responses of wildlife to human disturbance. *Journal of Applied Ecology*, 43, 164–72.

Regnier, C., Fontaine, B., and Bouchet, P. (2009). Not knowing, not recording, not listing: numerous unnoticed mollusk extinctions. *Conservation Biology*, 23, 1214–21.

Rumble, M.A., Benkobi, L., and Gamo, R.S. (2005). Elk responses to humans in a densely roaded area. *Intermountain Journal of Science*, 11, 10–24.

Sanderson, E.W., Jaiteh, M., Levy, M.A., Redford, K.H., Wannebo, A.V., and Woolmer, G. (2002). The human footprint and the last of the wild. *Bioscience*, 52, 891–904.

Sastre, P., Ponce, C., Palacin, C. Martin, C.A., and Alonso, J.C. (2009). Disturbances to great bustards (*Otis tarda*) in central Spain: human activities, bird responses and management implications. *European Journal of Wildlife Research*, 55, 425–32.

Sih, A., Ferrari, M.C.O., and Harris, D.J. (2011). Evolution and behavioural responses to human-induced rapid environmental change. *Evolutionary Applications*, 4, 367–87.

Stankowich, T. (2008). Ungulate flight responses to human disturbance: A review and meta-analysis. *Biological Conservation*, 141, 2159–73.

Steadman, D.W. (1995). Prehistoric extinctions of Pacific Island birds: biodiversity meets zooarchaeology. *Science*, 267, 1123–31.

Tarlow, E.M. and Blumstein, D.T. (2007). Evaluating methods to quantify anthropogenic stressors on wild animals. *Applied Animal Behaviour Science*, 102, 429–51.

Tarr, N.M, Simons, T.R., and Pollock, K.H. (2010). An experimental assessment of vehicle disturbance effects on migratory shorebirds. *Journal of Wildlife Management*, 74, 1776–83.

Thiel, D., Jenni-Eiermann, S., Braunisch,V., Palme, R., and Jenni, L. (2008). Ski tourism affects habitat use and evokes a physiological stress response in capercaillie *Tetrao urogallus*: a new methodological approach. *Journal of Applied Ecology*, 45, 845–53.

Tinbergen, N. (1963). On aims and methods of ethology. *Zeitschrift fur Tierpsychologie*, 20, 410–33.

van Polanen Petel, T.D., Giese, M.A.,Wotherspoon, S., and M.A. Hindell. (2007). The behavioural response of lactating Weddell seals (*Leptonychotes weddellii*) to oversnow vehicles: a case study. *Canadian Journal of Zoology*, 85, 488–96.

Vidya, T.N.C. and Thuppil, V. (2010). Immediate behavioural responses of humans and Asian elephants in the context of road traffic in southern India. *Biological Conservation*, 143, 1891–900.

Walker, B.G., Boersma, P.D., and Wingfield, J.C. (2006). Habituation of adult Magellanic penguins to human visitation as expressed through behavior and corticosterone secretion. *Conservation Biology*, 20, 146–54.

Weston, M.A. and Elgar, M.A. (2005). Disturbance to brood-rearing hooded plover *Thinornis rubricollis*: responses and consequences. *Bird Conservation International*, 15, 193–209.

Wilson, E. O. (1987). The little things that run the world (the importance and conservation of invertebrates). *Conservation Biology*, 1, 344–6.

Wolf, I.D. and Croft, D.B. (2010). Minimizing disturbance to wildlife by tourists approaching on foot or in a car: A study of kangaroos in the Australian rangelands. *Applied Animal Behaviour Science*, 126, 75–84.

Wrege, P.H., Rowland, E.D., Thompson, B.G., and Batruch, N. (2010). Use of acoustic tools to reveal otherwise cryptic responses of forest elephants to oil exploration. *Conservation Biology*, 24, 1578–85.

Zuberogoitia, I., Zabala, J., Martinez, J.A., Martinez, J.E., and Azkona, A. (2008). Effect of human activities on Egyptian vulture breeding success. *Animal Conservation*, 11, 313–20.

Index

Acacia trees 133
acid rain 53
acoustic signals 18–19
　transmission of 21–2
adaptive change 8–10, 145–6
　negative effects of 11–12
　see also evolution
adder, death (*Acanthophis praelongus*) 196
aggressive behaviour 163–5
agricultural changes 93–4
alewife (*Alosa pseudoharengus*) 182–3
Allee effects 163, 194
anadromous life history 80, 182–3
antelope, Saiga (*Saiga tatarica tatarica*) 163
Anthropocene 125–6
anthropogenic disturbance *see* human behavioural disturbance
antipredator behaviour 97–8
　contemporary evolution 180–2
　costs and benefits 99
　ecosystem consequences 179
　foraging behaviour relationships 98–9, 165–6, 177, 179, 180
　trophic cascades and 177
　see also foraging; predation risk
ants
　harvester (*Pogonomyrmex occidental*) 180
　invasive 195–6
aphid, pea (*Acyrthosiphon pisum*) 100, 136
aspartate aminotransferase (AST) 19
atrazine 38

badger (*Meles meles*) 125
Baldwin effect 9, 52
bass (*Micropterus salmoides*) 101
bear
　black (*Ursus americanus*) 54
　brown (*Ursus arctos*) 115, 159, 162–3, 175
beaver (*Castor canadensis*) 101
beech (*Fagus sylvatica*) 219
behaviour
　adaptive change 8–10
　demography and 10–12
　between and within individual variation 7–8
　ecosystem consequences 175–9
　　behavioural plasticity 179–80
　　consumption effects 175, 176–7
　　contemporary evolution 180–3
　　nutrient cycling effects 175–6, 178–9
　　reaction norms and 183–6
　maladaptive behaviour 4, 9–10, 236
　population dynamics relationship 160, 161–2, 171
　　consequences of individual variation 159–61
　　feedback 168–71
　　male versus female effects 162–3
　　studies 163–6
　unidirectional changes 7
　see also behavioural phenotype; behavioural plasticity
behavioural phenotype 4
　causes of phenotypic change 4–8
behavioural plasticity 48, 145–6
　beneficial plasticity 150–2
　　population-level consequences 151–2
　dispersal 65–7
　during invasion 193–4
　　impacts on natives 195–6
　ecosystem consequences 179–80
　environmental change responses 208–10
　evolution of 10, 52
　fitness consequences 147–54
　foraging behaviour 95–7, 100, 101–2
　limits to 155
　maladaptive plasticity 148, 149, 152–4
　migration time 85–7, 88, 151–2
　non-adaptive plasticity 147
　optimal plasticity 149–50
　outlook 154–5
　reversible plasticity 8, 146
　special role of 146–7
　temperature-induced 148
behavioural syndromes 152, 226
behavioural traits
　fitness relationship 6–7
　heritability 9
bioclimatic modelling 125
bisphenol A (BPA) 38
blackbird, European (*Turdus merula*) 37, 53, 110
blackcap (*Sylvia atricapilla*) 85, 170–1, 221
bluebird
　mountain (*Sialia currucoides*) 74, 164–5
　western (*Sialia mexicana*) 66, 74, 164–5
blue eye fish, Pacific (*Pseudomugil signifer*) 24–5
boar, wild (*Sus scrofa*) 94, 109
bowerbird, satin (*Ptilonorhynchus violaceus*) 20, 26
brain size 50–1, 193
breeding behaviour *see* reproductive behaviour
buffalo, cape (*Syncerus caffer*) 121
bullfrog, American (*Rana catesbeiana*) 134
bunting, Indigo (*Passerina cyanea*) 74
butterfly
　bog fritillary (*Boloria/Proclossiana eunomia*) 70, 73

251

butterfly (*cont.*)
  Monarch (*Danaus plexippus*) 80
  orange tip (*Anthocharis cardamines*) 132
  Rocky Mountain parnassian (*Parnassius smintheus*) 70
buzzard, rough-legged (*Buteo lagopus*) 96

capital breeders 81
capture-mark-recapture (CMR) studies 72–3
cardinal, Northern (*Cardinalis cardinalis*) 209
caribou (*Rangifer tarandus*) 135
catadromous life history 80
chemical signals 19–20
  pH effects 24
  salinity effects 24–5
  signal transmission 24–5, 24
chickadee, black-capped (*Poecile atricapillus*) 50
chipmunk (*Tamias striatus*) 54
climate change
  infanticidal behaviour and 114–15
  learned responses 55–7
  migration and 81–2, 84–9
    population consequences 170
  parasite/pathogen–host interactions and 134
  possible consequences 125–6
  sexual selection and 113–14
  singing behaviour and 112–13
  timing of breeding and 35–6
  see also temperature
clown fish (*Amphiprion percula*) 53
cockle, New Zealand (*Austrovenus stutchburyi*) 180
cod, Atlantic (*Gadus morhua*) 170
colour change 108–9
communication 16, 17
  anthropogenic disturbance 16–18, 26
  multiple signals 212–13
  signal detection 25–6
  signal production 18–21
    acoustic signals 18–19
    chemical signals 19–20
    matching signals to altered habitats 20–1
    signals acquired from the human environment 20
    visual signals 19
  signal transmission 21–5
    acoustic signals 21–2
    chemical signals 24–5, 24
    visual signals 22–4

urban environments 55
competition
  conspecific 11, 109, 123
    territorial competition 12
  global environmental change impacts 133–4
conservation behaviour 235, 237–8, 239
consumption, ecosystem effects 175, 176–7
contemporary evolution 180–3
cormorant, great (*Phalacrocorax carbo*) 83
corridors 73–4
costs
  of antipredator behaviour 99
  of being naive 50
  of dispersal 64–5
  of learning 50–1
  of sexual selection 201–2
    population-level consequences 202
cougar (*Puma concolor*) 165
crayfish
  Orconectes limosus 180
  signal (*Pacifastacus leniusculus*) 193
cricket, field
  Teleogryllus commodus 26
  Teleogryllus oceanicus 208
crow
  Caledonian (*Corvus moneduloides*) 209
  carrion (*Corvus corone*) 65
  Torresian (*Corvus orru*) 4
cultural drift 26

danger management *see* antipredator behaviour
Datura wrightii 132
damsel bug (*Nabis* sp.) 100
DDT 37, 39, 97
deer
  red (*Cervus elephus*) 94
  roe (*Capreolus capreolus*) 135
  white-tailed (*Osocoileus virginianus*) 99, 165
demography
  behavioural adaptation and 10–12
  social behaviour adaptation consequences 122–4
diet choice 180
dispersal 63–4, 75–7, 145–6
  changing landscape relationships 69–75
    dispersal as an invasion mechanism 76
    ecological traps 74–5
    habitat fragmentation 71–4

habitat quality 69–71
  multiple impacts 76
  costs versus benefits 64–5
  density dependent dispersal 70
  during invasion 192
  ideal free dispersal 70–1
  information acquisition 67–8
  as a plastic behaviour 65–7
dodo (*Raphus cucullatus*) 97
dolphin (*Tursiops* sp.) 101–2, 209
domestication 107, 115
  reproductive behaviour and 107–9
    mate choice 109
    parental care 109
Drosophila
  D. melanogaster 224
  D. mojavensis 212
duck
  domestic (*Anas platyrhynchos*) 108
  harlequin (*Histrionicus histrionicus*) 102
dugong (*Dugong dugong*) 101–2

eagle
  golden (*Aquila chrysaetos*) 136
  white-tailed sea (*Haliaeetus albicilla*) 103
ecological traps 10–11, 74–5, 153–4
ecosystem consequences of behaviour 175–9
  consumption effects 175, 176–7
  nutrient cycling effects 175–6, 178–9
  rapid behavioural trait change 179–83
    behavioural plasticity 179–80
    contemporary evolution 180–3
  reaction norms and 183–6
eelpout fish (*Zoarces viviparus*) 227
elephant (*Loxodonta africana*) 56
elk
  Cervus canadensis 135, 177
  Cervus elaphus 101
endocrine system 32–3
  chemical pollution effects 37–41
    endocrine disrupting chemicals (EDCs) 18, 39–41
  domestication effects 108
  environmental disruption 33–5, 41–2
  photoperiodism 35–6
  urbanization effects 36–7
environmental noise 25–6, 47
estrogenic chemicals 18, 39–41
eutrophication 153

evolution 217
    constraints on responses to environmental change 225–8
        limited genetic variation 225–6
        trait correlations 226–7
        ultimate constraints 227–8
    contemporary evolution 180–3
    evolved responses during invasion 194–5
    genetic change 210
        number of genes and size of effect 223–5
        relative importance of 217–19
        standing genetic variation versus new mutations 222–3
    learning and 51–2
    of plasticity 10, 52, 219–20
    speed of 220–2
    see also adaptive change
evolutionary rescue 10–11, 217, 218, 228–9
    critical rate of change 221–2
    see also evolution
extinction avoidance 10

fear response, urbanization and 110–11
    see also human proximity responses
fertilizer 94
fiddler crab (*Uca uruguayensis*) 202
finch
    Darwin's (*Geospiza fortis*) 221
    house (*Carpodacus mexicanus*) 55
fishing pressure 53
fitness
    migration time and 82–4
        birds 82–3
        salmon 83–4
    phenotype relationship 6–7
flight distances 107–8
flycatcher
    collared (*Ficedula albicollis*) 71, 114, 218
    pied (*Ficedula hypoleuca*) 82, 87, 131–2, 150
flying fox (*Pteropus poliocephalus*) 55–6
foraging 94
    behavioural flexibility 95–7, 100, 101–2
    behaviour as a diagnosis tool 102–3
    consequences of change 99–102
        for communities and biodiversity 101–2
        ecosystem consequences 179
        for populations 99–101
    food change effects 94–7, 99

human disturbance effects 165–6
    predation risk effects 97–9, 100–1, 165–6, 179, 180
    group foraging 120–1
    urban environments 54–5
    see also antipredator behaviour; predation risk
fox
    Alaskan red (*Vulpes vulpes*) 56
    Arctic (*Vulpes lagopus*) 108
    island (*Urocyon littoralis*) 136
frog
    American bullfrog (*Rana catesbeiana*) 134
    common (*Rana temporaria*) 150, 219
    green (*Rana clamitans melanota*) 209
    leopard frog (*Rana pipiens*) 134
    (*Rana clamitans melanota*) 19
    tree (*Hyla arborea*) 209

gecko
    Hemidactylus frenatus 196, 197
    Hoplodactylus duvaucelii 195
    Lepidodactylus lugubris 54, 196, 197
generalists, versus specialists 53
genetic change 210
    limited genetic variation 225–6
    number of genes and size of effect 223–5
    relative importance of 217–19
    standing genetic variation versus new mutations 222–3
    trait correlations 226–7
Girardinichthys metallicus 19
gizzard shad (*Dorosoma cepedianum*) 178–9
global environmental change (GEC) 129–30, 145
    behavioural interactions and 132–6
        competition 133–4
        consumer–resource interactions 134–6
        interactive effects of multiple drivers 138–9
        mutualisms 132–3
        network architecture consequences 136–8
        parasitism/pathogens 134
    mechanisms of impact on species interactions 130–1
        altered behaviour 132
        ontogenetic changes 132
        range shifts 131
        temporal shifts 131–2
    responses to 145
    see also climate change

glucocorticoid hormones 36–7
goldfish (*Carassius auratus*) 25
goose
    barnacle (*Branta leucopsis*) 94–5, 103
    pink-footed (*Anser brachyrhynchus*) 85
    snow (*Chen caerulescens*) 178
greenbul, little (*Andropadus virens*) 210, 211
group behaviour
    human impacts 121–2
    predation risk and 120–1
    see also social behaviour
grunt
    Haemulon flavolineatum 179
    Haemulon plumieri 179
guinea pig (*Cavia aperea*) 108
gull, yellow-legged (*Larus michahellis*) 19
guppy (*Poecilia reticulata*) 4, 49, 180–2, 209, 210, 219, 221

habitat choice, ecological traps 10–11
habitat fragmentation 71–4
    causes and consequences of fragmentation 72
    connectivity 71–2
        corridors 73–4
    conservation and 73–4
    dispersal and 72–3
habitat quality
    degradation effects 23, 26, 132
    dispersal and 69–71
    see also habitat fragmentation
habituation to signals 26
hawkmoth (*Manduca sexta*) 132
heat island effect 36
herbivore feeding behaviour 134–5
homeostasis 33
honeybee (*Apis mellifera*) 148
hormones 32
    domestication effects 108
    glucocorticoid hormones 36–7
    signal transduction 34
    see also endocrine system
human behavioural disturbance 235
    consequences of behavioural change 236–7
    population consequences 165–6
    recreational disturbance literature 238–45
        methodological problems 239–44
        policy implications 244–6
human harvesting impact 115

human proximity responses
  domestication and 107–8
    flight distances 107–8
    tonic immobility 108
  urbanization and 110–11
  see also human behavioural disturbance
humic acids (HA) 25, 27
hyena, spotted (*Crocata crocata*) 121
hypothalamic–gonadal (HPG) axis 33
hypothalamic–pituitary (HPA) axis 33

iguana, Galapagos marine (*Amblyrhynchus cristatus*) 204
impala (*Aepyceros melampus*) 56
infanticide 109
  climate change effects 114–15
  human harvesting effects 115
innate behaviour 46–8
  interaction with learnt behaviour 49–50
innovation frequency 53
inter-generational change 5–6, *6*
intra-generational change 5, *6*
intraguild predation (IGP) 102
invasions 190–1, 196–8
  behavioural variation effect on spread 193–5
    evolved responses 194–5
    plastic responses 193–4
  dispersal behaviour during spread 192
  impacts on natives 195–6, 208
  mechanics of spread 191–2
  population growth during spread 192–3
invasive species 27, 190, 192
  dispersal as a mechanism for invasion 74
  parasitism and 134

jackal, golden (*Canis aureus*) 121
Jacky dragon (*Amphibolurus muricatus*) 209
Julodimorpa bakewelli 209

kangaroo rat (*Dipodomys spectabilis*) 98
keystone species 102, 176–7, 182–3
killifish, banded (*Fundulus diaphanus*) 20
kite
  black (*Milvus migrans*) 20
  black-shouldered (*Elanus caeruleus*) 121
K reproductive strategy 168
krill 100–1

lamprey (*Petromyzon marinus*) 18
learning 46, 209
  associated costs 50–1
  behavioural development role 46–8
  evolution and 51–2
  innate behaviour interaction 49–50
  learned responses to environmental variation 52–7
    climate change 55–7
    importance of 209
    urbanization 54–5
  social learning 48–9
light environment 23
light pollution
  communication disturbance 18–19
  endocrine effects 37
  migration and 81–2
lion (*Panthera lca*) 121
litter size 109
lizards 152
  common (*Lacerta vivipara*) 68, 75
  side-blotched (*Uta stansburiana*) 168
loosestrife, purple (*Lythrum salicaria*) 133
Lorenz, Konrad 3
lynx (*Lynx lynx*) 135

maladaptive behaviour 4, 9–10, 236
maladaptive plasticity 148, 149, 152–4
marmot, yellow-bellied (*Marmota flaviventris*) 169–70
mate choice
  domestication effects 109
  see also reproductive behaviour
maternal effects 8
melatonin-1-receptor (MC1R) gene 108–9
melatonin 19, 37
methylparathion 19
migration 63, 80–1
  environmental change and 81–2
  population consequences 170–1
  nutrient cycling and 178–9
  strategies, birds 11–12, 170–1
  timing 82
    birds 5, 82–3, 84–6, 151–2
    climate change effects 84–9
    consequences of change 87–9, 151–2
    fitness relationships 82–4
    salmon 83–4
  vertical migration 100–1
minnow, fathead (*Pimephales promelas*) 25–6
monkeyflower (*Mimulus ringens*) 133

moose (*Alces alces*) 179
mosquito, pitcher plant (*Wyeomyia smithii*) 218
mouse (*Peromyscus leucopus*) 180
mussel (*Mytilus californianus*) 176
mutualistic interactions 132–3
  mutualistic networks 136–7

natural selection
  learning and 51–2
  of self-regulatory behaviour 160
neuroendocrine secretion 33
newt
  alpine (*Mesotriton alpestris*) 203
  palmate (*Lissotriton helveticus*) 20
nitrogen cycling 178
noise pollution 21, 22
novel environments 154–5
nutrient cycling, ecosystem effects 175–6, 178–9

ocean acidification 53
ocean currents 82
olfaction 25
Olympic village effect 191
ontogenetic changes 132
otter, sea (*Enhydra lutis*) 177
ovenbird (*Seiurus aurocapillus*) 207
overfishing 53, 99

parasites
  distribution 121
  global environmental change impacts 134
  host switching 134
  links with social hosts 125
parental care, domestication and 109
pathogens
  distribution 121
  global environmental change impacts 134
pea, partridge (*Chamaecrista fasciculata*) 227
penguin, chinstrap (*Pygoscelis antarctica*) 97
perch, Nile (*Lates niloticus*) 195
peregrine (*Falco peregrinus*) 98
permeability 72
personality 166
petrel
  Antarctic (*Thalassoica antarctica*) 55
  giant (*Macronectes giganteus*) 55
pH, chemical communication and 24
phenology
  breeding 35–6, 111, 150, 151
  climate change effects 35–6, 56

INDEX 255

population consequences 170
  species interactions and 131–2
migration 5, 82–9
  birds 5, 82–3, 84–6, 151–2
  climate change effects 84–9
  consequences of change 87–9, 151–2
  salmon 83–4
phenotypic plasticity 217
  evolution of 219–22
  relative importance of 217–19
  see also behavioural plasticity
pheromonal communication 24
phosphorus cycling 178
photoperiodism 35–6
phthalates 38
pike (*Esox lucius*) 70
pikeminnow, Colorado (*Ptychocheilus lucius*) 25
plasticity *see* behavioural plasticity; phenotypic plasticity
pollination 133
pollutants
  communication disturbance 18, 19–20, 208
  endocrine disruption 37–41
polyandry 124–5
population dynamics 161–2
  behaviour relationships 160, 161–2, 171
    consequences of individual variation 159–61
    feedback 168–71
    male versus female effects 162–3
    studies 163–6
  during invasion 192–3
  individual contribution 166–8
  population rescue 213
  sexual selection consequences 202
    environmental change impacts 207–8, 210–12
possum, brush-tailed (*Trichosurus vulpecula*) 125
prairie dog (*Cynomys ludovicianus*) 180
predation risk
  density-mediated effects 100, 101
  foraging behaviour relationships 97–9, 165–6, 177, 179, 180
    ecosystem consequences 179
    trophic cascades and 177
  group behaviour and 120–1
  trait-mediated effects 100, 101
  see also foraging
predator avoidance behaviour *see* antipredator behaviour

predator–prey interactions 135–6
Price equation 4–8
Prunus africana 133

quoll (*Dasyurus hallucatus*) 4

raccoon (*Procyon lotor*) 54
rainbow fish (*Melanotaenia eachamensis*) 121
range shifts 74
  species interactions and 131
rat
  Pacific (*Rattus exulans*) 195
  Rattus norvegicus 49
rattlesnake (*Crotalus horridus*) 73
reaction norms 183–6
recreational disturbance literature 238–45
  methodological problems 239–44
  policy implications 244–6
redstart, American (*Setophaga ruticilla*) 82
reproductive behaviour
  costs of 201–2
  domestication effects 107–9
    mate choice 109
    parental care 109
  global change impacts 106–7, 112–16
    climate change 35–6, 57, 112–15
    sexual selection 113–14
    singing behaviour 112–13
  polyandry 124–5
  resource distribution relationship 122, 123
  timing of breeding 35, 111, 150, 151
  urbanization effects 109–12
  see also sexual selection
reproductive character displacement 27
reproductive suppression 123
resource availability 179–80
resource distribution 121–2, 123
robin, Seychelles magpie (*Copsycus sechellarum*) 4, 122
reproductive strategy 168

salinity, chemical communication and 24–5
salmon
  Atlantic (*Salmo salar*) 50, 86, 168
  chinook (*Oncorhynchus tshawytscha*) 81, 86, 221
  migration time 83–4
  Pacific, (*Oncorhynchus* spp.) 175, 178

sockeye (*Oncorhynchus nerka*) 82, 86, 87
sandpiper, western (*Calidris mauri*) 103
scrub jay, Florida (*Apheloconia coerulescens*) 73
seal
  elephant (*Mirounga leonina*) 56
  fur (*Arctocephalus gazella*) 97
  grey (*Halichoerus grypus*) 113
sea star, purple (*Pisaster ochraceus*) 176–7
sea urchin (*Strongylocentrotus*) 177
secondary sexual characters 19, 20–1, 113–14
selection 6–7
  antagonistic selection 226
  see also natural selection; sexual selection
selective breeding 108
  see also domestication
Selye, Hans 33
sensitization to signals 26
Setaria tundra 134
sexual conflict 11
sexual selection
  climate and 113–14
  costs of 201–2
    population consequences 202
  environmental change consequences 204–8, 212–13
    complexity of environmental change 212
    honesty of behavioural displays 206–7
    interactions among sexually selected traits 205–6
    population consequences 207–8
    resource allocation and trade-offs 204
  environment relationships 202–4
  human harvesting effects 115
  importance of 201–2
  multiple signals 212–13
  responses to environmental change 208–12
    genetic changes 210
    phenotypic adjustment of behaviour 208–10
    population responses 210–12
shark, tiger (*Galeocerdo cuvier*) 102
sheep
  bighorn (*Ovis canadensis*) 165, 167
  Soay (*Ovis aries*) 166, 210
singing behaviour, climate change effects 112–13

skunk (*Spilogale gracilis amphiala*) 136
social behaviour 119–20
  demographic consequences
    individual based models 124–5
    of social behaviour
      adaptation 122–4
  influencing factors 120–2
    human impacts 121–2
    parasites and pathogens 121
    predation risk 120–1
    resource distribution 121–2, 123
  kinship and 123–4
  reproductive suppression and 123
social learning 48–9
sparrow
  song (*Melospiza melodia*) 18
  white-crowned (*Zonotrichia leucophrys*) 35
sparrowhawk (*Accipiter nisus*) 110, 113
spatial heterogeneity 47
spatial sorting 191
specialists, versus generalists 53
species interactions 129–30
  global environmental change
    effects 132–6
    competition 133–4
    consumer–resource
      interactions 134–6
    mutualisms 132–3
    network architecture
      consequences 136–8
    parasitism/pathogens 134
  interactive effects of multiple
    drivers 138–9
  mechanisms of impact of global
    change 130–2
    altered behaviour 132
    ontogenetic changes 132
    range shifts 131
    temporal shifts 131–2
squirrel
  Canadian red (*Tamiasciurus hudsonicus*) 219
  grey (*Sciurus carolinensis*) 54
starling, European (*Sturnus vulgaris*) 18, 35, 40, 41, 97–8
stickleback, three-spined (*Gasterosteus aculeatus*) 20, 153, 206, 207, 212, 221
stress 33, 37
  domestication and 108

sunfish (*Lepomis macrochirus*) 101
swallow
  barn (*Hirundo rustica*) 82, 113
  tree (*Tachycineta bicolor*) 39
swan, Bewick's (*Cygnus bewickii*) 85
swift (*Apus apus*) 96
swordtail fish 208
  Xiphophorus birchmanni 25
  Xiphophorus malinche 27

temperature
  behavioural plasticity and 148
  breeding time and 35
  heat island effect 36
  migration time and 85, 86–7
  sex determination and 57
  see also climate change
temporal heterogeneity 47–8
tern, arctic (*Sterna paradisea*) 80, 113
Tetrahymena ciliates 66
thrush, song (*Turdus philomelos*) 85
tiger (*Panthera tigris*) 107
tilapia, Lake Migadi (*Oreochromis alcalicus grahami*) 227
Tinbergen, Niko 3
  'four questions' framework 237–8
tit, great (*Parus major*) 18, 19, 35, 50, 149, 220
toad, cane (*Rhinella marina*) 4, 97, 191, 192, 196, 197
toadfish (*Halobatrachus didactylus*) 22
tonic immobility 108
trophic cascades 175–6
  behaviourally-mediated 177, 185
  density-mediated 177
  diet choice and 180
trout
  brown (*Salmo trutta*) 86, 108
  coral (*Plectropomus leopardus*) 121
  rainbow/steelhead (*Oncorhynchus mykiss*) 25, 102, 168
tuatara (*Sphenodon guntheri*) 57
tuberculosis, bovine (*Mycobacterium bovis*) 125
tuna, bluefin (*Thunnus thynnus*) 56
turtle, painted (*Chrysemys picta*) 57

urban canyons 22
urbanization 36, 115
  communication responses 55
  endocrine effects 36–7

  fear response changes 110–11
  foraging innovations 54–5
  learned responses 54–5
  life history strategy changes 111–12
  reproductive behaviour and 109–12
  timing of breeding 111

vertical migration 100–1
visual signals 19
  signal transmission 22–4
    habitat degradation impacts 23
    light environment and 23–4
vole
  bank (*Myodes glareolus*) 206
  meadow (*Microtus pennsylvanicus*) 50
  prairie (*Microtus ochrogaster*) 224

wagtail, pied (*Motacilla alba*) 122
warbler
  blue-winged (*Vermivora cyanoptera*) 160
  golden-winged (*Vermivora chrysoptera*) 160
  Seychelles (*Acrocephalus sechellensis*) 123
waste products from animals 178
water dragon, Australian (*Physignathus lesuerii*) 57
water flea, spiny (*Bythotrephes longimanus*) 100
water turbidity 20–1, 153
whales 100–1
  killer (*Orcinus orca*) 21, 177, 180
wildebeest (*Connochaetes taurinus*) 80
wolf (*Canis lupus*) 101, 121, 177, 237
wolf spider
  Hygrolycosa rubrofasciata 26
  Pardosa purbeckensis 65
  Schizocosa ocreata 204
woodpecker, lesser-spotted (*Picoides minor*) 124
wrasse, bluehead (*Thalossoma bifasciatum*) 123

xenobiotics 38

Yellowstone National Park, wolf reintroduction 101, 121, 237

zebrafish (*Danio rerio*) 25